THE NEW
INDEPENDENT HOME

Revised and Expanded Edition

DEAN ROSÉ

THE REAL GOOD SOLAR LIVING BOOKS

Real Goods Trading Company in Ukiah, California, was founded in 1978 to make available new tools to help people live self-sufficiently and sustainably. Through seasonal catalogs, a periodical (*The Real Goods News*), a bi-annual *Solar Living Sourcebook*, as well as retail outlets, Real Goods provides a broad range of tools for independent living.

"Knowledge is our most important product" is the Real Goods motto. To further its mission, Real Goods has joined with Chelsea Green Publishing Company to co-create and co-publish the Real Goods Solar Living Book series. The titles in this series are written by pioneering individuals who have firsthand experience in using innovative technology to live lightly on the planet. Chelsea Green books are both practical and inspirational, and they enlarge our view of what is possible as we enter the next millenium.

Stephen Morris
President, Chelsea Green

John Schaeffer
President, Real Goods

THE NEW INDEPENDENT HOME

People and Houses that Harvest the Sun

REVISED AND EXPANDED EDITION

Michael Potts

CHELSEA GREEN PUBLISHING COMPANY
WHITE RIVER JUNCTION, VERMONT
TOTNES, ENGLAND

On the cover: The Vermont home of Margo and Ian Baldwin, co-founders of Chelsea Green Publishing Company. Originally built by woodstove- and boomerang-designer Eric Darnell then thoroughly renovated, the house is earth-sheltered on the north, with an attached greenhouse on the south. Heat is provided by sunlight, with a masonry heater in the kitchen and a woodstove that also preheats water, in tandem with a solar water heater.

Text and drawings, unless otherwise noted, copyright 1993, 1999 by Michael Potts.
Design by Ann Aspell.
Ornament by Rene Schall.

Printed in Canada

5 4 3 2 1 99 00 01 02 03

Printed on acid-free, recycled paper.

Library of Congress Cataloging-in-Publication Data

Potts, Michael, 1944–
The new independent home : people and houses that harvest the sun / Michael Potts. – Rev. ed.
      Rev. ed. of: The independent home / Michael Potts. 1993.
      Includes bibliographical references and index.
      ISBN 10890132-14-4 (alk. Paper)
      1. Dwellings—Energy conservation. 2. Renewable energy sources.
I. Potts, Michael, 1944–   Independent home. II. Title.
TJ163.5.D86P68   1999   99-39385
696—dc21   CIP

Chelsea Green Publishing Company
P.O. Box 428
White River Junction, Vermont 05001
(800) 639-4099
www.chelseagreen.com

To Sienna and Damiana, my daughters,
and Rochelle,
who remind me of my purpose

CONTENTS

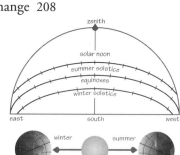

GRAPHS, DIAGRAMS, AND TABLES

FOREWORD

I EXPERIENCED A PERSONAL EPIPHANY ONE MARCH DAY IN TINY Caspar, California. This sleepy hamlet, just off Highway 1 a few miles north of mega-trendy Mendocino, is on a craggy promontory that plunges down into the Pacific. I was using the pay phone on the porch of the Caspar Inn, a venerable roadhouse where the night before a rockabilly legend named Sleepy LaBeef had entertained the locals. One of Sleepy's claims to fame is that he was a contemporary of Elvis Presley and Carl Perkins, musical luminaries who had their moments of fame and glory. Sleepy's other claim to fame is that he is still working hard, plying his trade in backwaters like Caspar. Ultimately, this has proved a more sustainable strategy than the teenage-idol paths of Carl and Elvis.

I called back to my office in Vermont. There, it was a typical March day. A warm front had moved in from the Gulf during the night. Snow flurries turned to drizzle, then froze upon contact with the dirt roads, turning them into skating rinks. For most of those in the office, the morning commute had been a hair-raising encounter with the elements. The ride home, alas, might be worse.

Meanwhile in Caspar, it was similarly a typical March day. From my perch at the pay phone on the deck of the inn, I looked out on a sparkling morning, cottonball clouds against a sky of Pacific blue. Less typically, there was a crisp horizon, meaning that you could clearly see the puffs from the grey whales migrating from Baja back to Puget Sound. I heard waves crashing while seagulls squalled, and the resident colony of sea lions barked and clamored.

"What's it like out there?" my associate in Vermont asked.

"Raining," I replied innate Yankee guilt not permitting me the luxury of a truthful response. In truth, I felt incredibly smug. I had the best of both worlds, a home and job in Vermont, and the freedom to travel to great places like Caspar at the times when Vermont is at her most gruesome.

Michael Potts' book *The Independent Home* and now his newly revised *The New Independent Home* are about people who want the best of both worlds. They want the conveniences and pleasures that technology has made possible, but free from the destructiveness and guilt associated with the plunderers of our planet.

Michael uses Caspar as the stage for his personal drama, much as does David Budbill in another Chelsea Green book, *Judevine.* In this way a small, local story becomes the embodiment of wider and deeper discoveries. This kind of small-stage-in-a-big-world story has oft been told. It is called, variously, "Walden, or Life In the Woods" (Thoreau), "The Good Life" (Scott and Helen Nearing), "The Contrary Farmer" (Gene Logsdon), "Four-Season Farm" (Eliot Coleman), or "Solar Living" (Real Goods Trading Company). The plot is essentially the same. A Man (or a Woman, or frequently a Man and a Woman) sets out to find an ideal life. The quest involves a Place, a Home, a World in Opposition, and a Guiding Philosophy. And the result is—gentle triumph.

In a world that seems to be evolving with ever-increasing speed, the time elapsed between the original *Independent Home* and this new edition has added character lines and nuances that give the drama added depth and dimension. There are still pioneers in the world. These are their stories.

Whenever I visit Michael's modest abode in Caspar, I am struck by the juxtaposition of nature and technology that lends balance to his life. This is a consistent theme with the new generation of homesteaders portrayed in his book. Not many people want to do without, and why would they? If you can live a gracious, responsible, sustainable home without squandering natural resources, then isn't that truly the best of both worlds?

The New Independent Home is, therefore, a book about paradise and how certain individuals are finding it, or creating it, in North America today. For a brief moment on the deck of the Caspar Inn, I realized that I'd caught a glimpse of the future.

Stephen Morris,
Publisher of Chelsea Green

INTRODUCTION

IN 1993, SEEKING A GOOD PLACE TO START THIS BOOK, I SET OUT
along the road inland from my place in Caspar, where a spring of legend-
ary sweetness and stability issues from the hillside above the road. In
nearby clearings there is evidence that for hundreds of years the native
Yuki Indians stopped here in summer and again on their way back, on
their annual trek from a protected inland valley to the coast. Local folks
come and park their pickups beside the spring and fill jugs from the
crystalline flow. As recently as 1984, they also brought along their garbage, and while their jugs filled with sweet water, they heaved the trash—
batteries, paint thinner, tires, disposable diapers, garden clippings, rotten
meat, and innumerable varieties of plastic—over the downhill side into
the swale where the spring's overflow has trickled for millennia down to
the river and the sea.

In the first edition of the book, I offered this story as proof that
amazing leaps of consciousness can take place in a decade, the time it took
our entire community to stop dumping and then clean up that toxic
midden. It may be decades before animals will once again drink from the
spring's overflow below the road, but we somehow changed our collective
human minds by finally understanding: We must not dump where we
drink!

For me, this relatively small, local realization glows as an emblem of
a greater awakening: the urge to return to the home place. At the heart of
such epiphanies, our need and ability to create Community shines.

In the six years since *The Independent Home* was first published, this
community of awareness has grown. During the time between this book's
writing and this year's revision, a growing number of conscious builders
and enlightened homesteaders have joined what in 1993 seemed to be a
romantic minority. There were few signs that the juggernaut of America's
globalized housing industry—proud manufacturer of dwellings perfectly
disconnected from their environment, the needs of their inhabitants, and
any true connection with the home place—would change. All through

the second half the twentieth century, and even now as we enter the new millennium, precious land has been polluted with cookie-cutter houses filled with resource-gobbling conveniences. These excrescences yield profits to subdividers and builders, but make ill use of sun, wind, weather, and whatever natural magnificence might be found. But in the past half-decade, sick building syndrome, environmental illnesses, and widespread social malaise have become a national concern, and more home-buyers are beginning to insist on quality shelter in a healthy, neighborly environment.

This book combines stories told by energy heroes and pioneers with explanations of the ideas that make their stories important. Examples of earth-friendly houses and the systems that power them are scattered along a trail of concepts, beginning with the rudimentary elements of our beloved modern comforts—flowing electrons, pure water and air, light, food, and shelter. Some of the stories describe people who have simplified their needs in order to "get down to the essentials." Later stories describe techniques for designing and building more complicated independent homes, and relate insights concerning the mysteries of learning, work, and community, highlighting ways we can reduce energy dependency in any residence. Finally, the latter chapters' stories and the book's end explore the ways we can reshape our broader communal and ecological structures in the coming Solar Age.

Pioneers might remember their hardships but sometimes neglect to tell us the full extent of unknown territory they traversed, often with great difficulty, before finally settling themselves in a home. Here, the stories help guide us over rough terrain. Between stories I review concepts that may have been forgotten

and bring you up to date on renewable-energy technologies and the marketplace of energy independence. My method is to peel back complexity until the simplest elements are visible, and so in coming chapters we pursue electrons, attempt to think like water molecules, and deconstruct and re-examine the components supposedly required to make us feel at home. This book's subject, the Independent Home, is more globular than linear, so you may find ideas in the book's early chapters that will not make complete sense until you learn more about them in later chapters. For example, you may emerge from the Introduction wondering what, precisely, does Dennis Weaver mean when he calls his home an "earthship"? The truth will be revealed when you reach chapters 8 and 13.

This book bristles with unfamiliar terms, because its topic encompasses many interdisciplinary boundaries where experts typically mark off their territories with jargon. The editors and I had a good time trying to include every unfamiliar term in the extensive glossary and index to be found among the resources at the back of the book.

Out of Darkness

In the United States, we officially started our national scramble to live well without despoiling our land's richness in 1970, with the birth of the Environmental Protection Agency. In thirty years we have reformed much of the industrial dark side of delivering pure water, electricity, fuel for heating and cooking, and the stuff of graceful living. New England's lakes and rivers, once sacrifice zones for Ohio's and Indiana's coal burners and local industry, are reviving. Chesapeake Bay, Cancer Alley along the lower Mississippi, and indeed most near-shore oceanic and riverine zones around the planet, are still poisoned with heavy metals, endocrine disrupters, and bio-hazards, but we know enough not to build our homes on toxic landfills, or downstream from hazards, and we are striving to clean up our messes. Many of us, including the financiers, designers, and operators of polluting facilities, went to school at a time when we were taught that our planet was infinitely giving and forgiving, as if its rivers and oceans, all its crucial bounty, were boundless and magically regenerative. Now coming of age, we are surprised to find, again and again, that efficient design and careful processing can eliminate waste and pollution while improving profitability. As one millennium gives way to another, we are no longer willing to accept "barely legal" standards, but insist on coming home to purity, authenticity, comfort, and efficiency.

A century from now, as our great grandchildren change their calendars to 2100, American energy consumption in the late 1900s may be regarded as unbelievably wasteful, our habits as ridiculous and mannered as the Victorians, as profligate as the Romans, as extreme as the Inquisition—regrettable and destructive episode cited as a warning to builders and planners in training. Historians will explain that humankind during this brief epoch thought itself to be the absolute ruler and shaper of the planetary domain. The explosive development of global consumerism, the replacement of nature by technology, and the headlong quest for wealth through disposability and obsolescence transfixed whole generations, deadening them to the rhythms, powers, and glories of our planetary partners, all other life forms, the grand community we call Nature.

And yet in 2100, the history of the Petroleum Era will include tales of heroic guides, pioneers, and settlers who showed humanity how to get back to the land. Some of their stories are told for the first time in these pages. As one of the early adopters of the Solar Resurgence, the idea of writing a book began for me with a successful search for a like-minded community. Half a dozen years later, I am happy to report that our "green minority" is growing, and more at home than ever. Having recognized our disconnectedness from nature, and yearning for comfortable, self-sustaining living space, right livelihood, and a cordial and cooperative relationship with our biological community, we rebelled and invested our whole lives in reinventing what should be the most natural thing in the world: Home.

Come along on my personal voyage of discovery and reinvention, looking for others who are moving toward a home place where we can all enjoy the best of modern life without taking our comfort at the price of another's misery,

The author's home in Caspar, California.

where we can rejoin the cycles that rule all life on our planet, and where we can extend our vision to include all other life on this wondrous globe.

The name we've given to the study of ecosystems, Ecology, comes from the Greek *oikos*, meaning "house," and the Latin *oeco-*, "household." This book explores homely science, examining the systems we expect to find in places we can rightfully call home. Look deeply, and we discover that in the home as in the greater living community, all terrestrial systems are closed: What we consume today will not be available tomorrow *(depletion)*, and what we discard today will come back to plague us tomorrow *(persistence)*. By the geologic clock, human civilization has risen so explosively that its spatters and blight in only a century have taken down most of the planet's ancient forests and in only a generation have covered half of its agricultural lands. By applying closed-system global logic to the locale of the home, we approach an old frontier with a new toolbox and a comprehensive awareness of our impact. Rather than settling in haste, like locusts (or tract developers), whose slash-and-burn tactics lay the land to waste, solar pioneers settle gently, with regard for the land's secrets and time-honored wisdom. Rather than polluting and encumbering the land, we enrich it. Where long-time inhabitants—trees, plants, animals great and small—already thrive, we seek to conserve diversity. As long as all systems interact smoothly—including those we introduce—Nature's equilibrium remains intact. Out of a heavy-handed method of hasty domination achieved by ruthless elimination of competitive diversity, we are evolving an ethos of stewardship and cooperation.

Living with a house: Dennis Weaver's story

Dennis Weaver, known to most of us as a cowboy actor and advertising spokesman, may seem an unlikely member of a revolutionary vanguard, but his worldly experience and earthy common sense led him to look for a better way to live. His story comes first in this book because many know him, and his search went deeper than the "right and wrong" choices all homesteaders face, through compromises and blind alleys, to the rewards of a better life. Like the pioneering heroes of many of the stories you will read in this book, Dennis sought to regain a home on the land, crossing a frontier to the kind of life that was commonplace in North America no more than a century ago, and that remains so in much of the "underdeveloped" world.

Dennis and his family divide their time between an earthship—a sun-harvesting home built with earthen walls, discarded tires and beverage cans, and local materials into a south-facing slope in the foothills near Ridgway in southwest-

ern Colorado—and a conventional house in California. Here is his account of the houses he lives in.

Our main house is our earthship in Colorado. We've been there for almost a decade, and we spend more and more time there now. The thing that attracted me about the earthship is that it is environmentally friendly, it is self-sufficient, it is energy-efficient, and it uses hardly anything from the conventional systems that make a conventional house functional. We aren't tied to the grid. We aren't plugged in, you might say, like in a hospital, where you are plugged in to life supports. The conventional houses that we live in now are on life support.

There's a very natural feeling in an earthship. The warmth comes from stored energy, stored temperature in the mass of the house, and it just naturally releases at night to keep the air temperature warm or cool. There are no heating ducts or air conditioning, so you don't have artificial temperature control, and you don't have a lot of sound going on that is

Dennis Weaver and his earthship.

man-made and unnatural. We have room to grow a lot of plants in the house, and they grow so beautifully because it is a natural environment for them. It feels good, the air is real good. Of course if it warms up in the house, we've got the capacity to control that by opening skylights, windows, and you get a nice moving-air feeling. And it's very natural moving air; it's not like it's pushed out with a fan.

It's arid in Ridgway, and there's a lot of sage brush over here, but the soil is great. We have a meadow down below us where the llamas graze. There's a lot of rainfall and a lot of snow. We have a little forest behind our house and to the side of it, with all kinds of trees, pine trees and piñon trees, a couple of cedars and ponderosas. We have a garden that is very, very productive.

You know, your house is your partner, it's your companion, it's your shelter, and you have to live with it in that sense. You can't just go over to the wall and set a thermostat, then go off and leave it and get all this artificial stuff going. You have to know that when it gets warm, you open windows and let the heat out, and you have to know that when night comes and it's wintertime, you need to close the windows as the sun is going down so that you hold the warmth in the house. You just have to live with it. I don't feel that's a drawback. I think it's really nice.

Moving back and forth between an on-the-grid and an off-the-grid house, you have to make an adjustment. It's not at all hard.

When we're in southern California, and the electricity goes out, our plug is pulled, and we're helpless, totally helpless: we're at the mercy of the central power system. That's not the case out in Colorado, where the power can go out in the neighborhood around us and we don't even know it. I'd certainly like to install solar power in the California house. That would be terrific, because then we wouldn't be a victim to the system, and it also would be a clean source of power, an inexhaustible source of power, and a cheap source of power. So I'd really like to do that.

The Ridgway experience has just pointed out the fact that we are held captive by the systems, and that we use a lot of energy in our house that comes from dirty sources, coal-fired powered plants or nuclear power, and we are very aware of that. We're also very aware that our power is coming from a limited source. It's something that is not going to last. I may not outlive the fossil fuels, but certainly my grandchildren will outlive them. And I think we should be thinking about our grandchildren. Some scientists are saying we only have something like a ten-year supply of petroleum left. I don't know if that's true or not, but let's say it's forty or fifty: It's going to be gone. Whatever we're going to be forced to do at that time, it just seems sensible to me to start doing it now. Don't wait until the crisis is on us. Let's manage by objective rather than crisis.

There is no question: my major concern is for the place where I live, and the environment where we all have to live. I mean, every living thing needs a proper environment in which to exist, and if it doesn't have it, it dies. Human beings are no different. And the sad part, the absolutely desperate part of what we're doing is, we're destroying our environment, we're destroying our life support. And that is something that I'm extremely concerned about. I talk about it all the time. I go out on the lecture circuit and talk about it, and I try to incorporate solutions in my own lifestyle.

I also have an electric car. They're at a primitive stage right now, but that industry needs to be encouraged. I use it for short trips. You know, we should support that industry because it's a thing of the future, or one of the possibilities anyway, and electric vehicles reduce the amount of energy we use. It's the energy that we consume that really creates the environmental problems we're dealing with, whether it's acid rain or depletion of the ozone layer. We're using a dirty source, we're using an exhaustible source, we're using an expensive source. When you consider that we are held hostage, and have been in the past, by some crank across the ocean, because our energy source, petroleum, is threatened—we've got to get away from that. We've got to be free of that energy source for our own security, and for our own healthy economy; we've got to do it. We have no choice there.

That's not going to be a tragic thing. I believe that there's a huge environmental industry waiting to be tapped, which will create a strong economic base for us, which will produce a tremendous number of jobs. You know, as human beings, we go through realignment all the time; we have to adjust. As we evolve, we continually adjust. Just look what's happened through the industrial revolution: It's been nothing but one adjustment after another. The idea that adjustment is something we have to be fearful of is really silly. Of course industries will change and people will have to retrain, but when the automobile was invented, the farriers had the same reaction. What happened to the horse-shoers? They went into repairing cars. Listen, if we do not create and back an environmental industry in this country, we're going to be swept under the rug, just like we were with automobiles and computers. All that technology that we created is in the hands of foreign countries, and we're buying from them instead

of producing. If we don't get in step with what is needed for the future, and go into an industry supported by another kind of energy source, we're going to be out there in the back of the pack someplace.

I can't give any special advice; there's so much good advice out there, so many good books. Get a simple book, and watch your energy use. And be concerned. Be aware that we're going to have an immense problem if we don't address it now. I mean, it's already an immense problem, but if we continue to stick our heads in the sand and ignore it, live in denial, and say we don't have a problem, it's only going to get worse.

A Short Note on Methodology

Whatever truth and usefulness is found in this book is a gift to me from hundreds of conversations with friends, builders, and stewards, more than 150 questionnaires, 100 formal interviews with energy-conscious informants, and 50 follow-up interviews. Any group of Americans, and certainly a group as critically aware as those represented here, presents diverse and divergent opinions, and on any particular point my own conclusions will differ from those of many of my informants. I have tried to stay within the spectrum of broadly conceived consensus, but my enthusiasm for my topic leads me, at times, into outrageous and even crackpot opinionatedness. In this book few of the ideas, but all of the errors, are mine alone.

To gather information for the book I traveled thirty thousand miles (by fuel-efficient mass transit whenever possible) to satisfy myself that this movement back toward nature and self-sufficiency is happening everywhere. I spent hundreds more hours traveling electronically by telephone and computer network. Everywhere I found extraordinary people. Many were so intent on reducing their energy use and perfecting their relationship to their home planet that they were surprised and gratified to discover they were not alone. In the six years since this book first appeared, I have traveled more widely, retraced many of my steps, and have determined that 1993's tentative tendrils of solar culture are now weaving themselves into a strong, flexible network. The verging awareness of our shared strength and burgeoning influence has fluoresced. Like Dennis Weaver, many innovators and explorers, stewards and architects, visionaries and rainmakers were generous with their time and patient in their explanations. I found myself most comfortable writing myself out of their stories, recording instead the voices and ideas of the storytellers as authentically as I could. Each storyteller understood this book's mission, and all had the opportunity to read and correct the words attributed to them. Each tries to convey what she or he believes is most important for you to know about independent homesteading.

In addition to a glossary that I hope readers will find illuminating, this book includes a bibliography to help readers locate the liveliest nodes of the growing community network of renewably energized home-makers, contacts that can help us take control of the energy systems in our lives. You may also visit the book's website at

http://www.chelseagreen.com/IndHome

where you will find an up-to-date glossary and a means for nominating terms not already included in the printed glossary. Since energy self-sufficiency is ever-changing and rapidly evolving, the website reflects, as no printed book can, the current state of renewable energy and independent living.

About Independent Living

The term "independent living" is proudly used by adults who overcome dire handicaps to live on their own, and I mean to show them no disrespect by borrowing their term. The substance-dependence problems that characterize so many modern American homes may in time be recognized as denaturalizing handicaps in their own right, and meanwhile I want to acknowledge the debt that we TABs (Temporarily Able Bodied) owe to the differently-abled. As we move hesitantly toward independent living, we find ourselves questioning and redefining every task in accordance with our particular personal limitations and a certain awkwardness in our newfound ability to use non-polluting, renewable energy sources. At first, our efforts may make us feel feeble and clumsy, because we are trying to use unfamiliar means to attain a standard of comfort and convenience taken for granted by the energy-careless majority. We and our families may risk mockery and condescension, yet this barely approaches the exposure our differently abled brothers and sisters experience. If and when you feel tentative and vulnerable, I urge you to take a moment to contemplate the experience, remembering that you yourself have chosen this path, and give some thought to others who do not have the good fortune to choose.

For the ideal of an "independent home" to have meaning, we will need to believe in the ultimate truth that every home is an interdependent home.

CHAPTER 1

DECLARING INDEPENDENCE

WE CAN LIVE IN ENERGY INDEPENDENT HOMES. WE CAN CONTINUE to live well without depending on imported energy. By improving energy use in an existing house or by building anew, there are many ways for us to lighten our load on our planet's strained resources. This book tells how energy pioneers have committed their lives to finding better energy paths, meticulously eliminating waste from their energy budgets and harvesting energy from the sun and its minions, water and wind. These pioneers are eager to report good news: the new energy frontier is open, safe, and a rewarding place to explore and settle.

In rapidly increasing numbers, people are moving beyond the reach of powerlines and seeking more natural power sources. Many with limited budgets find that the only property they can afford is unimproved or spoiled land far from the power grid, city water, sewer, and other civilized amenities. Many more are retiring early to take up a new and challenging vocation: responsibility for their own needs amidst nature at the end of the road. On and off the grid, for reasons of conscience and economics, we are cutting our energy consumption and looking for other ways to improve our stewardship of the planet. At times the trail may seem lonely, but in every state, in almost every community, families are following this trail, and taking their rewards in a better and cleaner lifestyle, freedom from the economic tyranny of unsteady power monopolies, and an abiding hope that others will join and help preserve our planet's fragile balance.

This book contains firsthand accounts by energy pioneers, many of whom gambled decades ago that independence was possible, even though it could be attained only by finding a way across a wilderness of unknowns and loneliness. As pathfinders, their stories can now guide us on a well-marked, increasingly well-traveled way. They bring us good tidings: By following the ever-widening path of energy independence, we may continue to enjoy our present comforts without fear that we do so at our children's expense. The tidings they bear apply everywhere the sun shines, promising that a sustainable livelihood can be made available for all the peoples of the world.

On their own personal voyages of discovery and invention, these renewable energy pioneers found romance, intensity, and fulfillment far beyond their hopes. By tracking their sources of energy back behind the wall outlet and supermarket shelf where most people stop, they have discovered an intimate and restorative connection with the sun, the earth, the seasons, and the panoply of nature. They report that their efforts at integrating the best and most modern technology with ageless wisdom about the world is, together with parenting, the most rewarding work they have ever done. As they invent, rediscover, and refine their energy strategies, a shared energy awareness emerges to unite their families and restore the sense of family purpose so often lacking in the plugged-in world. Having traveled the path myself, I enjoy the stories these heroic pioneers tell, appreciate their advice, harken to their mistakes, and gain insight and inspiration from ways of thinking born out of the high and low points of their journeys.

The Dependent Home

The history of homes extends clear back to the beginnings of history itself. Both ideas, home and history, probably occurred to our foremothers while they huddled in a cave, waiting out bad weather. All forces radiated from the cave where these troglodytes conducted their lives, birthing and dying, preparing food, and sheltering in stormy times. Originally these were wanderers, and caves were the center to which they returned year after year, around which they developed the arts that gave birth to culture.

For millennia, we have elaborated that culture, but until quite recently, the home re-

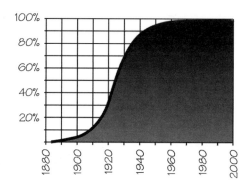

Starting in 1884, electrification was undertaken gradually in the United States, but by 1920 became a rush to connect. The most expensive and far-ranging expansion was accomplished by the Rural Electrification Administration during the Depression, when connecting to the grid was deemed part of the national economic recovery plan. The war effort slowed connections, but by 1945, nine of every ten American homes were connected. In the mid-1990s, the trend to remote homes resulted in a slight reduction in the percentage of grid-connected homes.

mained a closed system. Energy, food, clothing, and education were domestically produced from local resources; goods packed from afar were precious and rare because their supply was unreliable. In the twentieth century, the word troglodyte, which originally meant cave-dweller, has become a pejorative term for hermit, one who rejects modern life and embraces the oldest, most primitive values. Just in the last century, humans have moved unimaginably far from their cave-sheltering past, pretending to control the world for human benefit, unceremoniously taking its resources, bending all other creatures to human use. Accelerating mobility has enabled us to satisfy our needs from sources farther and farther away from our homes. My father vividly recalls the luxury of the first dandelion-green salad in the Colorado spring. Before 1920, lettuce was available for most of the year only in California and the far southern parts of the United States, but now it can be found beside peaches from Peru and melons from Mexico in any season. Before 1930, getting energy from far away was impossible; factory and town sites were chosen first for local availability of necessary power. During the Great Depression, great public-works projects and a frantic undervaluation of resources made it reasonable to waste prodigious amounts of potential energy in order to deliver a trickle of electricity, oil, or lettuce to faraway places. There were so many trees, so much falling water and buried oil, and human yearning for prosperity was so immediate, that our predecessors never reckoned the eventual cost. Half a century later, most of us scarcely notice that such dependences utterly determine the way we live. But economic and planetary realities have changed, and once-abundant resources now dwindle because of popu-

lation growth and heedless waste. Dependence is becoming horribly costly and even life-threatening.

A dependent house is a forlorn extension of the global scheme of exploitation and dominion, cut off from its immediate surroundings, irrelevant or damaging to the local ecology, crowded onto convenient tracts, scarcely more than a place for laborers to recover from their exertions in service of the global consumption

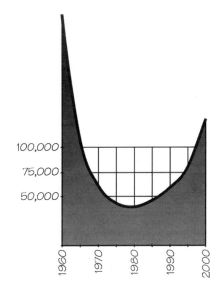

The number of off-the-grid homes dwindled, due to rural electrification, until the 1970s, despite increasing numbers of remote second homes. About 1980, a number of influences, including the availability of low-cost independent energy systems and the end of government-subsidized line extensions, reversed this trend. In 1993, when The Independent Home *was published, improved renewable energy systems allowed many off-the-gridders to electrify. The passage of net-metering laws in many states also increased the popularity of self-powered on-the-grid homes. This graph represents a very rough estimate of actual numbers, since no agency or organization has a good census of independently powered homes.*

machine. "Civilization," in the form of television, commuting to anonymous jobs, the bombast of sensationalistic meaningless news, conceals from us the cost of lettuce in February: our independence. Dependent householders often have no idea of the origin or true cost of their food, their goods, or their power: At any given moment, the electrons coursing through their walls may have come from Quebec or El Paso, Diablo Canyon or Three Mile Island, generated with oil shipped from the Persian Gulf or coal stripped from Appalachia. The emissions from the power plants mingle with the exhaust of millions of cars to pollute the air we breathe, to dull once-clear skies, and to poison the rain.

Today, many of us are striving to reduce our dependence on fragile and distant resources. We worry because we see that our sustenance is extracted and brought to us from far away by means that are mindless, abusive, and likely to crash. We are loathe to surrender our luxuries,

In summer, more of our necessities can be produced nearby—as much as half within fifty miles of Caspar, California and a third within the same distance of Exeter, New Hampshire. While our grandparents simply did without, we now throw energy and money at scarcity, importing luxuries we have come to consider necessities without regard for cost. In thickly settled areas of the nation, no amount of roof-top and vacant-lot gardening will balance population density, and this forces longer supply lines, particularly in winter, when half of Exeter's supplies travel more than a thousand miles.

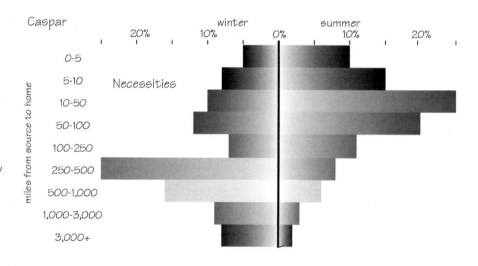

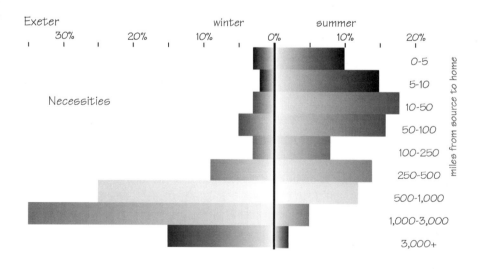

but we seek a happier, sustainable balance between dependence and independence. My life is enriched by chocolate and oranges, neither of which grow well where I live, and both of which I suppose I could live without. Energy, on the other hand, is essential in my life. Although energy falls freely from the sky here in Caspar, just as it does in Alaska, Vermont, Mississippi, Hawaii, and wherever the sun shines, most of my neighbors get theirs without a thought for the inefficiencies and inequi-

ties of its provenance. Why, since energy is so abundantly present in our environment, as evidenced by the rich natural bounty of life, should we be so eager to import it? Even when sunlight is feeble or absent (as it is north of the Arctic Circle in winter) the atmospheric phenomena the sun stirs, winds and falling water, provide clean and unfailing sources of energy. Do we have a right to live outside the budget provided by our local solar income? Through careful reduction and management of my

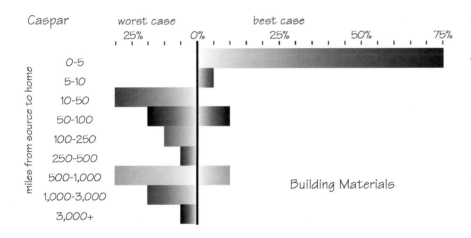

Increasingly, building materials are transported from great distances by large multinational corporations, with the transportation contributing significantly to their profit. In the days of early independent homes, most materials were found on the site or nearby, and local resources determined building techniques.

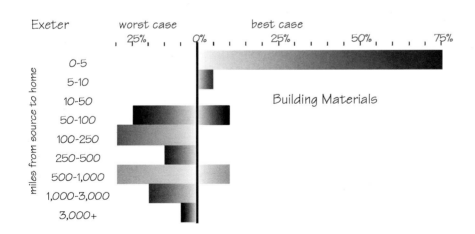

needs, and a comprehensive energy-harvesting strategy, I can produce enough power for all of my household uses—heating, cooling, lighting, pumping, washing, working, playing music—with energy generated in my own yard.

An independent home is a very hopeful political statement, declaring that we believe it is possible to live comfortably and responsibly on the share of energy that comes to us naturally. Many who live in independent homes on the North American continent might also suggest that we have grown uncomfortable being counted among the most wasteful and selfish people ever to have lived. The global economic system under which we are situated so fortunately may persistently fail to redress obvious global inequities, but we pioneers have unilaterally elected to lighten the load by disconnecting from the system, providing for ourselves instead.

Along the way, we have rediscovered lost delights and an invaluable connection to the wonders of living in harmony with the Earth.

The Present State of Energy

Before plunging into stories of intrepid energy pioneers and state-of-the-art hardware, electrons, kilowatts, and phantom loads, I must ask you to review with me the present state of energy.

Books, articles, speeches, television specials, movies, and every other known medium of communication document our planet's dire energy straits. (A list of current literature I have found valuable may be found in the bibliography at the back of the book.) Awareness of our plight has dawned in quite a short time, even in terms of human change. In barely half a century we have evolved from the necessary heed-

lessness of the Depression's grand public works to the present growing care and attention to pollution and scarcity. This reformation has motivated the pioneers in this book, and most independent homesteaders, to achieve astonishing transformations in consciousness. Our species, the most adaptive this world has ever known, has survived, evolved, and thrived by devoting ourselves to finding solutions. Of the many complicated problems, which do independent homesteaders consider most urgent?

— Our planet is under assault for her mineral riches. Many of these are limited and difficult to extract without damaging life. Our fossil hydrocarbons (flippantly referred to as dead dinosaurs) are quickly becoming resistant to retrieval.
— Inattention to what should be obvious, and short-term thinking including a persistent perception of plenty, has led to wasteful energy habits. While accelerating the depletion of resources, this sloppiness has led to serious pollution.
— We have come to expect a life of full employment and comfortable, wasteful consumption based on the assumptions of perpetual growth and unlimited resources. As these assumptions are called into question, we fear for our comforts and habits. When change is in the air, our first response is to fight it.
— Changes in the way we work, learn, and live have accompanied changes in the way we use energy, to the detriment of values that have long been held important; family, home, work, and nature have all been damaged by precipitous change. While most of us are concerned, few have good, relevant, constructive ideas about

what to do. Good news at last! Our wastefulness and superficial attention to the quick energy fix offers a shining hope. Because we waste more than half the energy we use, and new technologies are available that enable us to reduce our consumption and provide for our needs locally, we can preserve our comforts even while reducing our reliance on heritage energy sources.

The best energy sources, the miraculous flow of light from our sun and its natural effects, the planetary flow of water and wind, are relatively untapped and are more readily useful to us now than ever. Employing safe, proven technologies, we can preserve our planet's creatures while harvesting plentiful renewable energy to sustain our own comforts and share them with all the peoples of the world.

Petroleum has literally greased the skids of the industrial age. Oil enabled expansion and technological mastery beyond the wildest dreams of the most far-seeing futurists a century ago. Petroleum was always considered a bridging fuel by experts; it enabled many of us to emerge from the cave and enjoy unprecedented levels of freedom and independence. As industry abandons primitive means—"heat, beat, and treat" technologies developed during Queen Victoria's reign and not much improved to this date—in favor of clean, energy-efficient methods such as molecular assembly and bioengineering, the age of easy petroleum also grinds and rusts to a close. We can stop relying on the convenient but dirty energy content of heritage fuels—just in time! as dead dinosaurs are

Petroleum supply graph, showing various analysts' projections of future supplies. When half of a given resource is used up, as is the case with petroleum, the price inevitably starts to rise; if demand is undiminished, cost skyrockets. Experts believe this "Hubbert's Peak" was reached with petroleum in about 1998. Although scarcity and increased cost may motivate us to seek energy sources other than fossil fuels, the best arguments in favor of efficient use of current solar income are reduced pollution and preservation of heritage materials that may prove essential to future generations.

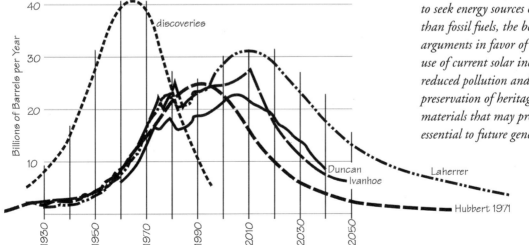

quickly being used up. We can preserve remaining reserves of petrochemicals for production of durable goods that require its peculiar molecular constitution, or for efficient combustion in the few applications that require gasoline's portability and energy density. Having left the cave and crossed the bridge, we turn now to less primitive means of generating the power and manufacturing the goods we need.

The sun is our original power source, and most terrestrial sources of power are merely translations, in time or form, of this ultimate resource's output. Fossil fuels are the compressed, carbonized remains of life forms that sought the sun more than 100 million years ago. Woodstoves and biomass generation use the same raw material without waiting through the millennia; sugar mills burn their waste, bagasse, to generate so much power that they supply electricity to the mill's company town as well. Wind turbines ride rivers of air powered by the sun and the earth's rotation. Hydroelectricity takes energy from the downhill side of the water cycle: Water, lifted by the sun from the ocean and forests to the clouds and then blown to the mountains by winds and precipitated, drives the turbines that spin the generators. Nuclear power, which imitates the sun's ability to convert mass to energy, is clearly a good idea for powering a solar system, but it is supremely messy, and should be done at a comfortable distance from living things, such as ninety-three million miles, then transmitted benignly as light before use. All our best power comes from a sun-driven, atmospheric engine.

I come to this inquiry, I must confess, as a life-long foe of disfiguring powerlines. As a child on my family's long summer travels, I imagined my eyes could decapitate roadside powerlines scarring the scenery, and enjoyed thinking about the consternation in my wake as people's electric toasters stopped toasting, their incandescent lights stopped lighting. I still hate the visual smog of powerlines, but have abandoned my childhood of terrorism. I recommend that we abandon the technologies of ooze and smash, now that we have found better ways.

I began the research for this book three decades ago, when penury forced me to conclude that I could and should change my own energy habits. Waste, where I find it, offends me; at first I suspected, and soon confirmed, that my own comfortable lifestyle, much of it learned in my small but comfortable family home, was in fact destructive and selfish, especially when contrasted with that of others in my California hometown, Oakland, and around the globe. As my eyes became accustomed to new ways of seeing, I began inching toward more rational use, where I made as much of my own energy as possible and was as careful with imported energy as I could be. My efforts met with unexpected help: The Arab oil embargo of 1973 and the recognition that pollution was threatening some of my favorite cities coincided with the successful introduction of small, efficient cars, and, because they were also inexpensive, I got one. I started planning my dream house at about the time my age broke into double digits, but I never thought to examine critically other homes and the home I was then living in, to evaluate the features that worked well and to abandon others that were no more than artifacts of empty habit. I realize now that few people do. Recognizing that I knew next to nothing about how a home should *really* work, I decided I could replicate the functional and exclude the habitual. I was blessed at the time by having (along with a natural curiosity and stubbornness) only enough money to build small and with my own hands. Friends and family were puzzled that I seemed to be turning away from their comfortable life paths. I was puzzled, in turn, by their insistence on hiring laborers to do the most rewarding work, on growing only inedible flowers, on smoking, on commuting for hours a day, on being so stubbornly resistant to the messages their bodies were sending them. I did not know what was right, but I had a hunch about what was not right for me, and that provided enough direction to start with.

When asked in 1993 if I could shape my experiences into a book, I wondered if my seemingly random walk contained anything useful to others. I was struck by anecdotal information: The average American family changes its address every 42 months, hardly long enough to put down roots, so their houses must not be working very well. In the forty years from 1949 to 1989, the average size of a new-built American house mushroomed from a modest 1,100 square feet to 2,300 square feet, even as the average family size decreased. At the rate real estate prices were escalating, would my children ever be able to afford to buy property and build even a modest home? If other pioneers could be found, who had traversed some of the same broken terrain by other, surer paths, perhaps more knowingly, I imagined their stories might offer some valuable clues. This book tells what I found.

Rediscovering the land: O'Malley Stoumen's story

Many stories started when mine did, in the late 1960s and early 1970s, as a "back to the land" movement drew a whole generation's attention to their estrangement from nature: a reaction to the rapid urbanization beginning in the 1930s and 1940s. O'Malley Stoumen found her place in the verdant, deeply indented land of southern Humboldt County in the heart of the redwood forest along California's northern coast.

In a way, her story is a romance, in that Jonathan Stoumen (whose own voice you will find in chapter 14) enters O'Malley's story as the trained architect and voice of practicality in her first summer back on the land. Together, they work the land to habitability.

I went camping when I was fourteen years old with my family up to the redwoods in northern California. Every summer vacation we went camping, always to a different spot in the state. But I thought that I had never been in such a beautiful spot, and immediately told everyone that I intended to live there when I grew up. Somehow that always stayed with me, and when I was twenty-one I headed for New York City, saying that I was going there to earn a lot of money so that I could buy my land up in the redwoods.

Well, as luck would have it, that is exactly what happened. In 1970, with money in the bank, two friends and I headed up to northern California and deposited ourselves in a real estate office. I can't begin to tell you how naive I was about buying land, but the one recommendation I had was "make sure it has water." So I said to the agent, "I want a fairly large piece of land that is about half meadows and half woods and has a creek on it. It needn't be all flat as I love to hike." (What a dumb thing to say.) He promptly put us in a land cruiser and took us for an hour drive farther and farther into the boonies.

We stopped on the top of a ridge that was just breathtaking and he said, "We'll have to walk from here, there's no road." It was the first day of January and it was sunny and quite cold. This turned out to be good luck for me, because even though I had never heard of a southern exposure, I felt warm whenever we looked at land on the south side, and cold as we looked at land on the north. So I unwittingly ended up with 160 acres of land with a beautiful southern exposure, half woods, half meadows and a half mile of creek (with fish) going through the bottom. Nobody had said anything about gravity flow.

I was so anxious to be there that as soon as the summer came I took my sleeping bag and headed up. It wasn't until then that the size of the project I had undertaken started to hit me. There was nothing there! There was no house, no water, and no one around. I started thinking that if I was going to build something myself it

had to be something easy and preferably out of a book. At first I thought of an A-frame, but I had never really liked the feel of them and so I decided on a dome. I purchased *The Dome Book* and promptly started making my list of materials and cutting all the two-by-fours into their correct shapes at a friend's house.

I made weekend trips up there with anyone I could talk into going, and started on the foundation (if you could call it that). I picked a terrible spot right out in the middle of a clearing with a little spring running under it. I thought it would be so pretty having the water there. It never occurred to me that after it had rained the expected 120 inches in a winter, the tiny spring would gush out with power to move mountains. We poured tiny pads, and put pier blocks on each one, nailed some redwood posts to them and made the framework to hold the plywood floor. This took several months as we didn't know what we were doing.

And then along came my rescuer. I had met Jon in New York. He had graduated from Cornell architecture school, done some graduate work there, then worked for the City of New York. We had been writing, and he was coming to visit me, and was very intrigued with the idea of a dome. He arrived and the first thing he said was that we would have to tear everything down. He said the dome was in the wrong place and stuck out like a sore thumb and that it should be moved over next to the trees where it could still get the warmth of the sun but nestle in with nature. He said it needed a proper foundation and the ground was not stable where the spring was. I finally agreed, but just couldn't take part in the demolition since so much misguided love had gone into the dome. I sat in the woods and sulked for two days while he tore it down. We set up a small semi-permanent

The Stoumens' homestead ranch in Humboldt county.

camp in the trees, and he said first we had to find water, as it was too far to go to the creek every time. We combed the land and finally found a little spring, built a redwood spring box and piped the water over to our site. We were water self-sufficient! It was a good start.

Jon never had thought of himself as a solar or environmental architect before, but we were forced to live with Nature, and she is the best teacher. The first time it really hit him, we had the dome almost all sheathed and we were lying down for an afternoon siesta, trying to let some of the heat of the day pass. We had left an opening for this huge south window and we couldn't get away from the heat or the sun because of it. Jon jumped up suddenly and said we could put eaves there that would block out the high summer sun but let the light in during winter. It sounds so simple and sensible and obvious, but when it hits you for the first time it is like magic. So we started on our campaign to have our house work with nature instead of against it, to have it be a collector instead of a consumer.

I don't think we ever even sat down and said that we wanted to be completely self-sufficient; it was just that we live so darned far from town and the conveniences that go with it. We did know that to continue working there we had to earn money, and that Jon wanted to continue his architecture. We also discovered that it was awfully hard to do architecture by kerosene light, and knew that electricity was much cleaner and safer. We had a small windmill that our ingenious friend Jim had built us for a wedding present, but it was very small and we knew that we would need more for those dark winter days of drawing. Jim suggested and ordered for us a 4-kilowatt Electro wind generator from Switzerland. This meant we had to add a room onto our house that would house all the giant batteries. Fifty-six 2-volt batteries!

The windmill was a joy to our lives, and made us become more deeply involved with nature. Our lives seemed to revolve around it. When the wind was blowing I would vacuum, make butter with the blender, bring out the power tools and the hair dryer. When it stopped we would conserve. The batteries were powerful and we could go a long time without wind if we were careful. It was also a giant weather vane: When it would turn around and point south we knew to put away all the tools, as it would soon rain. The sound it made was the most pleasant of all. Usually a soft swishing sound as it danced gracefully around. It had its wild sounds, too, when a storm would come up. Before we hooked up the automatic shut-off, there were many trips up the hill in the middle of the night to crank her out of the wind. But those sounds became familiar, too: when to leave it on for maximum power and when to shut it down.

Jon somehow still got work out in the middle of nowhere and was dying to try out his new natural ideas. Some work we traded for. One fellow owned a portable sawmill and came and sawed up our trees into all the lumber we needed to build an enormous barn and architecture office. Jon drew him beautiful plans for his little house in exchange. We got our first real break when Jon entered two of the houses he had designed and built in a HUD-sponsored national competition. He won two first prizes and ten thousand dollars plus a lot of publicity.

As he was taken away more often from home, I found I had fewer and fewer reasons to leave. We had built a beautiful barn, now full of horses, cows, chickens, geese, ducks, turkeys, and so forth. We had planted fifty fruit trees and sixty grape vines and had perennial and annual gardens full of vegetables. The only thing we were really short on was water. There was plenty of it, but it was all down at the bottom.

We had originally dug a small septic tank and had a flush toilet but

O'Malley Stoumen and her children.

Jon Stoumen.

soon found out what a water-waster that was. One day when two other solar-minded architects came to visit, they all sat down under a tree to brainstorm the perfect compost toilet. Jon had always wanted to do a building with a sod roof for a client, but he usually liked to experiment on us first. We also needed a woodshed, so all of this was to be incorporated into the building. One complication was that the best place for it was taken up by a winter-running creek. So these three guys sat there for hours with all these problems and came up with a beautifully simple building that spanned the creek. You had to walk up a few stairs to the toilet but you were rewarded with a beautiful view. We used dry compost from the barn to cover after every use and it all dropped down to a three-tiered system below. The front of the building faced south and the big door to the lower chamber was really a window made from a piece of fiberglass. We called it the solar assist. The sun beating in would really speed up the process. And with the venting there was no smell inside. The finished compost, which was taken out only once or twice a year, was beautiful and sweet smelling. We never used it on our vegetable garden, but it was great for the flowers and fruit trees. All our greywater was piped to various places and we moved the shower outside, too, surrounded by a bamboo screen. For hot water we put up a couple of solar panels and a coil in our woodstove, both leading to a high insulated tank. It was beginning to feel just like downtown.

Jon's work was really picking up and we felt like we needed a phone. There was no such thing as cellular phones then, so we bought a used-model car phone and installed it in our truck. We drove to town and got our number and got it working. Then we headed home and installed it in our home and bought a couple of photovoltaics to power that and a small car-radio/tapedeck for music.

Still we had the problem of water. And we had planted an enormous number of things to be watered. So we started researching drip irrigation. We had heard it was used a lot in Israel but no one around us had ever heard about it. We took a trip to the Bay Area and I started calling everywhere on the phone. We kept getting referred from one place to another until we finally found a commercial flower grower in Half Moon Bay who said he had just been to a dripper convention and showed us a lot of complicated samples that he had brought home. Then he pulled out a very simple design. It looked like a little threaded pipe fitting with a hole drilled in it and a piece of black plastic tape wrapped around it. He said this was the simplest and cheapest of them all and that was what he was going with. You had to have very low pressure (which is what we had) and a filter, but if the drippers did clog you could just increase the pressure a little and the plastic would balloon up and clean itself out. He gave us his source and all the information and we ordered a lot of stuff. Jon maintained that the reason they worked so well was confirmed by the fact that every time we had a hole in a garden hose or water pipe, we would try to wrap it with plastic tape and it would always drip. I spent days cutting and gluing pipe but all the components were so easy to handle; the pipe was only quarter-inch and very flexible, so that during the winter I just rolled the sections up on my arm like ropes, tagged them to say where they went, and hung them in the barn. We are still using some of those original drippers even though they are eighteen years old. So absolutely everything went on drippers, things started thriving, and weeding became less of a problem.

I think my pride and joy in our solar world was my solar oven. I bought a book on solar ovens and made myself one from the plans. It was very crude in some ways. It was made out of plywood with a sheet-metal insulated box inside. It had masonite reflectors with aluminum foil glued on them and a glass door on the front. But boy did it get hot. On really hot days it would almost reach 400 degrees, and even on cool days it was never below 300. So my life was revolving around the sun in another way: If the day was really hot, I knew I could make things that required a hot oven, and on cooler days I would put in a slow cooking item. In the summer hardly a day passed that I didn't use it. I made all our bread in it and just about everything else you could think of. If it was a beautiful warm day and the oven was empty, I would start to feel guilty because I was wasting energy.

But the big problem, and a universal one for sure, always came back to water. One day when walking down by the creek we started talking about all the water down there and how we could get it up to the house area. It was four hundred feet up and a half mile away. We knew we didn't want to install a big diesel pump down below that we would constantly be fueling. We measured the amount of water going past just to see what we were talking about. Armed with that information, we wrote to a hydraulic ram company back east, the start of a lot of correspondence and the eventual purchase of a ram pump. It was a project that took up almost the whole summer and was mostly just grunt work. We carried 110-pound sections of pipe up this steep creek to be put together with pipe wrenches almost as big as I am. We had a big 19,000-gallon ferroconcrete tank built up above the house, and had a smaller tank down below to catch the water that would shoot fifty feet straight down to run the ram. We built a concrete house for the ram and installed it and then put a big pipe from there back down to the creek where the water used for the power would be returned to the creek. At the highest place on the creek we built a small dam, which we could slip the boards out of in the winter and put them back in for next year. It was incredible sweat and toil, but when we finally got it going it was so awesome and unbelievable to see the water pouring into the tank that far from the creek.

There were lots of other things I did with my time. We had sheep, so I carded and spun and made warm hats and sweaters for the family. I took a four-day course at the local college on cheese making, and since we had two milk cows I was soon pressing, waxing, and storing cheese.

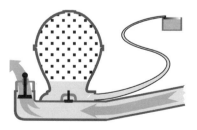 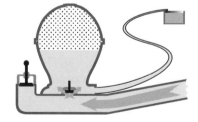 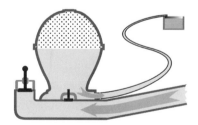

A hydraulic ram pump.

Of course there was always canning, drying, and putting up food. We had become almost self-sufficient and that meant putting away for the times that were leaner or out of season.

Suddenly we were confronted with a problem that we had never really considered in our plans. It was time for the kids to start school. The school down the hill would require four hours of driving a day, and the reason we were doing all this was so that we wouldn't have to drive down the hill all the time. Another problem that was bothering us was the dope growing. We were probably the only people on the whole mountain that weren't growing it and we worried about the kids again. For all of their friends it was a perfectly acceptable thing and even part of family life. They all helped in the gardens, sharing the harvesting, the cleaning, the profits, and the smoking. It didn't seem like what we had in mind. I hadn't ever thought of home schooling, or that might have been an option.

So we rented our farm and bought an old house in the little town of Healdsburg. But we still have our farm and it is still waiting for us to return, which I'm sure we will do. Meanwhile the whole experience there is carried with us every day. Jon's architecture was deeply affected, and he will always be doing environmentally friendly work. We still use drippers and conserve water in every aspect. When we remodeled our in-town house, we put in a separate line for all the greywater, and used all natural materials. We use compact fluorescents and low-E windows. I think it is probably more of a challenge trying to accomplish all of this in a city than it is in the country, where you may have no other choice. So that is the challenge we are working on now until we can get back closer to the land.

Looked Like Movement to Me

O'Malley's story may seem to end in a retreat from country life, but she assures us that her return to the city is temporary, for the sake of her children and her husband's work. She has given up neither the farm nor her dream of returning there, and both she and I believe she will return. I expect her to work out an up-to-date version of an age-old pattern of town house and family farm. The farm in the Eel River's basin and the town house a watershed south, along the Russian River, are both within the coastal redwood bioregion; between them runs a scenic railroad just now being refurbished and returned to passenger service. On horseback it would be a wonderful two- or three-day ride between house and farm.

We can also be sure that those who move back to the city after living at the end of the road have been changed, changed utterly by the country experience. Their new ways of seeing, of responding to challenges and caring for the land, alter and enrich their responses to the needs of home, even in an urban setting. Many of the storytellers in this book truly found their voices in the late 1960s and early 1970s. At that time, my decisions and those of many of my generation looked to our elders like a movement, a political reaction; at the time, neither we nor they could really say what we were reacting to. The Vietnam experience overwrought sensibilities already burdened with civil rights and freedom of speech issues, and so, for us, basic, earthy matters—nature, energy conservation, pollution—were negligible background elements. Stewardship permeated every current and backwater of my life, but at this time such a gentle notion could not catch and hold my attention.

Bullheadedness and myopic self-reliance led me to make many mistakes. I have come to identify a pattern in my responses:

1. I would encounter a stupidity, an inflexibility in the building codes, for example: "You can't let your greywater just run out into your garden, it's got bad chemicals and germs in it." Excuse me, but how could that be? If I am healthy, and am careful to put only good things—biodegradable soaps, toothpaste, organic food particles—in one end, how do bad things come out the other? After all, I did the plumbing, and knew there were no secret trap doors to the evil empire. "I don't know," says the sanitarian, "that's what the code says." So what happens if I ignore the code? "Maybe, you get sick. And for sure, I shut you down." What a challenge for a political malcontent like me!

2. Generalizing from this, I would decide that the whole code was stupid and all inspectors inflexible and powerless. My father calls this throwing out the baby with the bath water.

Years later, I have learned what the sanitarian should have told me: Some pathogens, like hepatitis, survive the dark ride from the bathroom sink to the puddle in the garden quite nicely, and a gust of wind just might loft one from an infected neighbor's greywater puddle in through my open window. Where houses are close together, and because hepatitis patients are usually discharged from hospital before their infectiousness has ended, this is a public health nightmare. However, injecting the greywater beneath a layer of mulch ends the danger, because the subsurface biota just love to munch those pathogens. It is difficult at times to tell whether codes and inspectors, like so much else in our consumerist civilization, mainly serve the public good or those who profit from the approved means. When these means, which by the nature of codes always lag behind necessity, are faced with the urgent need to reform, a better way is always found. To this day I still impatiently press authority to find the better way first.

Reinventing the Cave

Re-evaluating all the assumptions made by humans, bankers, insurance underwriters, architects, and interior decorators back to the cave to unpack the truth about functional shelter is at times discouraging and frustrating, but that is what I tried to do. I felt isolated, like the Lone Ranger, asking questions that felt basic to me and being assured by Authority that it would be best not to ask but to comply. Only when I finally started gathering stories for this book did I discover that I was not the only one who chose this demanding and eclectic quest.

One well-learned lesson: Initial cost is only a part of the whole cost. We think of ourselves as super-shoppers, but we seldom look beyond the sticker price or original cost of equipment to the whole cost, which includes maintaining the purchased appliance or device through its useful life. When energy is concerned, these operating costs are commonly the largest fraction of the whole-life cost. A wall is an appliance for keeping weather out and warmth in; if it leaks energy, initial cheapness is likely a false economy.

For me then and for many upstart home builders still, initial cost is often the only consideration: Building could proceed only if I had enough in my pocket to buy materials. The core of my new house grew slowly, and so walls that would one day be inside the house survived battering winters exposed to the cold and the rain. In this abrasive environment, I quickly learned which of my reinvented economies held up, and which failed. My waterproofing scheme, using methods recommended in books but meant for more clement climates, was repeatedly breached by the aggressive winds and waters of my coastal site . . . and continue to fail, though not as spectacularly as in the past.

In most climes, water is the most cunningly invasive element. As I struggled to keep water outside my walls, I learned that I had omitted a key ingredient, the impermeable membrane, which should have been the conceptual and mechanical heart of my wall. Building out from this heart would have been simple; adding it after the fact was between difficult and impossible. No amount of money, after the fact, could put an everlasting waterproof layer between stud and sheathing. Eventually I will rip out whole walls and rebuild them right, and the whole historical expense of the resulting wall beggars the cost I avoided at the beginning. Lesson: If you cannot find time or money to build it right, how will you be able to build it over? So do it right the first time.

While extending the margins of habitation, we find that only time and repeated experiments show us the proper materials and techniques. Meanwhile my home is, and will probably always be, an experimental shelter. Before the first room was closed in, and ever since, I knew how to build better next time.

On my travels, I have discerned two building styles: My own, which might be called organic, wherein much remains unfinished until inspiration comes and the builder can envision the completed detail; and a wiser, more architectural approach wherein the vision is elaborated from the broadest strokes down to the smallest details before the first spadeful of earth is turned. Architects may disagree with me, but I think the organic method produces a more comfortable, successful home.

Sticker Shock and Whole Cost

Energy-consuming products, from light bulbs and clock radios to refrigerators and automobiles, usually cost more to operate over their lifetime than they do to buy, and so purchase price is a deceptive yardstick. Suppose you are an apartment dweller whose electricity payments are included in your monthly rent. Would it be foolish to buy a light source that will outlast your intended stay? Here the "true cost" reckoning is complicated by hidden social costs. If you can demonstrate to the landlord a pattern of reduced electrical costs, might he not offer a rent reduction? Might you be wiser to separate utilities from rent, so that any economies might accrue directly to you, the economizer? The landlord stands, in this case, for society: By burning fewer kilowatts to generate the light we need, thereby reducing energy consumption, slowing the race to exploit energy sources, and decreasing emissions associated with their extraction and incineration, we benefit humanity. Even though a cost may be concealed, it is still real, and somewhere, someone pays. There is still no free lunch.

As the distance between original source and manufacturer and consumer stretches, some of

18

WHOLE-LIFE COST ANALYSIS

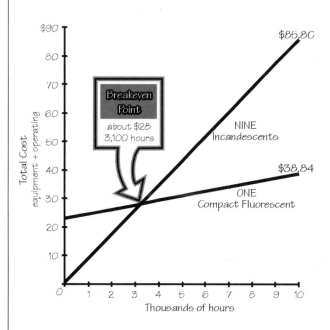

Hours	Incandescent	CF
0	$0.60	$23.00
1000	$9.12	$24.58
2000	$17.64	$26.17
3000	$26.16	$27.75
4000	$34.68	$29.34
5000	$43.20	$30.92
6000	$51.72	$32.50
7000	$60.24	$34.09
8000	$68.76	$35.67
9000	$77.28	$37.26
10000	$85.80	$38.84

Start with a simple whole-life, cradle-to-cradle cost comparison: in this corner, the champion and most-favored light source, the incandescent bulb. In that corner, the challenger, a first-generation compact fluorescent (CF) lamp. Considering only the standard economic consideration, equipment cost, the champion wins the first round easily: Who would buy a light for $25 when you can buy a package of four equally bright champs for under $3? But wait! the cheaper bulb lasts only 1,100 hours, one-ninth as long as the challenger, and more than two packages of four are required to compare fairly, so the challenger appears to be four times as expensive past the checkout counter. If you mean to buy light with only $3 in your pocket, study no more: Initial equipment cost decides the contest. For the longer, whole-life view, look again! The CF package bears a startling claim: The new-fangled light source consumes only a quarter of the electricity drawn by the conventional bulb. Doesn't someone pay for all that extra electricity? Over the whole life of the competing lights, which costs more? Assuming that electricity and lightbulbs will not increase in price—an exceedingly unlikely assumption—incandescent bulbs (it will take about nine) will use so much more energy over the life span of the compact fluorescent that the champion's overall equipment plus operating cost will be almost twice (188%) the challenger's. CF costs are dropping while incandescents are going up slowly, but judging by recent energy trends, a 10% annual increase in operating cost is reasonable, and the champ's whole-life cost may approach two-and-a-half times (236%) the challenger's by the time the CF burns out.

the real costs disappear into the infrastructure of modern life. In some cases, these deferred costs are greater than what we pay for an item; for an extreme example, could the people of the Ukraine add enough to the price of their electricity to pay for the clean-up of Chernobyl? Especially considering that the long-term effects of the Chernobyl disaster are global and incalculable?

There are at least two factors in the whole cost of any item: the cost to buy it (equipment cost), and the cost to maintain it (operating cost). To those costs, we should in most cases add two more: the true cost to create it, and the true cost to dispose of it at the end of its useful life. Both of these are usually absorbed by the environment or borne by the society in hidden ways. The accompanying sidebar may illuminate this idea.

Compact fluorescent bulbs continue to improve

Thrifty technology: Fred Rassman's story

Once a person is in the habit of buying goods based on the whole-life cost of each purchase, many assumptions about thrift come up for re-examination. Like reinventing shelter, this can become an engaging pursuit, yielding gratifying results.

Fred Rassman lives with his wife Linnea and his children Dylan and Alyssa in an off-the-grid A-frame in the Genesee Valley of northwestern New York State. He sought a life where he was responsible for dealing with the forces that acted on his family and their lives. As you will see in his story, this trained engineer wants to know the true cost of whatever he does.

I grew up in Pleasantville, down near Brooklyn, and we had a few trees and a pond or two, but it was city life. My old man took us camping for two weeks every year, but I always wondered, why do this two weeks a year when you can do it all year long? So after I studied engineering at Alfred University, I decided I never wanted to go back to the city.

I bought the land, ten acres for $1,500, in 1970. I guess I got it because it was pretty swampy then. I built my A-frame on the best spot without thinking about solar or anything: It was just going to be a summer cabin, and I was twenty-one, so what did I know? When I decided I was going to stay, I went to the power company, and they wanted $10,000 to bring power in, so I went without. Didn't make sense, spending ten times the value of the land.

I'm the quality control inspector for a plant that makes blacktop. I just fell into this job when I was looking for work one summer, and I've been real happy working there. As long as they're building and patching roads, I'll have a job. I work about as many hours in a year, thirty weeks at sixty hours a week, as most folks, which lets me stay home in the wintertime and

*The Rassmans and
their trackers.*

keep the fire burning. I get my head back together in the winter. I chose
this spot because I love the winters. I've always wondered why you build a
house in the middle of the woods if you work away all the time.

I'm a hands-on kind of guy, I work on my own cars, and I've built and
maintain everything here: the house, two sheds, and a pole barn. We
garden, more as a hobby, and have friends who have a roadside stand. I
raised pigs for awhile, and it was satisfying knowing it was my meat I was
eating, but I needed the space, and the feed got too expensive. Like I said,
watching the sun shine is more fun. I've added two wings to the A-frame,
so now we've got about 1,600 square feet. The house is grandfathered, and
I guess the county doesn't know I added on. Up in the peak I've got a batch
solar water heater, so I preheat my well water, then I've got two super-
insulated water heaters that run off my excess solar electricity, a 20-gallon
for showers that runs about 160 degrees even on a rainy day, and a 40-
gallon for the kitchen, that runs a little cooler. Right out the big octagon
window I put in a half-acre pond, for swimming after work in summer, ice
skating in winter. We've got a composting toilet too, just because I hate to
waste anything, you know?

Reclaiming this land has taken years. My land was covered with thornapple, kind of like hawthorn, with thorns that stick out an inch or two. Thornapple is called a "pioneer species," the first thing to grow on damaged land, which described my property well enough. Over the years I've gotten rid of those. When second growth comes in, the maple, oak, and other hardwoods, the thornapple gets shaded out. I've reclaimed the forest too, in pines, and it's too pretty to cut, so I pick up scrap wood at work and buy the firewood I need. I start my woodstove in October and it keeps going right through March or April.

In 1978 I came across an old Jacobs windmill, and built a tower for it. I'd never seen anybody else's towers, so I built what I thought would work; turned out, it's much stronger than it needs to be. I just took the Jacobs down two years ago. It was still running fine, but, with twenty-four panels on two twelve-panel trackers, I was getting more electricity than I knew what to do with, and wanted to save the Jacobs from wear.

Anyway, when I got electricity, it started to make things a little homier. I just got married in 1986. I guess Mama checked me and my place out pretty close, and she's pretty happy out here. My kids enjoy living out here; they don't seem to mind that there's no powerlines marching down the driveway. Sometimes a kid at school will pick on them, but that's the way kids are. Dylan knows that when the wind's blowing or the sun's shining he can play more video games. We live alone in the woods, and we love it. I'm fifty now, and I can't imagine that part being any better.

When I need to put something together, it's easy enough to get a book and read how to do it. Then I can pretty much make it work. My batteries are the best part of the system: I got them from the phone company back when they'd just let you haul them away or charge you for scrap. I had a contact at Western Electric, and hauled some batteries in from Indiana, some from Pennsylvania. Some of them are thirty years old, but they're in great shape: six tons of telephone batteries set up in several banks, with big old Frankenstein knife switches to control them. Since then I got some used nickel iron batteries from the railroads, must be fifty or sixty years old, and they're working just fine, too, 225-amp-hour Edison cells. I tried some nicads, but I like the nickel-irons much better. I'm a lead-acid man; I've got ten tons of them by now. I started out with golf-cart batteries, and I guess it's true that you've got to ruin one set to learn how to use them, because those only lasted me a year. With a wind generator, it's harder to get that last volt.*

---

*Getting "that last volt" of charge into a battery bank requires an energetic source typically producing about 30% more voltage than the battery bank's nominal voltage. A 12-volt battery will be thoroughly charged at about 13.5 volts, but will require a charging voltage greater than 16 volts to reach this state of charge. A 32-volt Jacobs windmill is mismatched to any battery bank made up of 6-volt golf-cart batteries, because a series of five batteries is nominally 30 volts, and six are 36 volts, and so one risks over- or under-charging the battery bank. Fred wisely opted for six batteries in series, but was never able to charge them fully, which decreased their life drastically. When Fred switched to strings of sixteen 2-volt cells, he solved the match problem . . . but getting "that last volt" into batteries is a perennial problem. Photovoltaic modules generally produce plenty of voltage, but it is uneconomical to have enough modules to generate the amperage required to charge the battery bank fully, and so many homesteaders use a fossil-fuel-fired generator or the grid for "topping off" their battery and getting "that last volt" of charge.

My original system was designed for 32 volts, because my first energy source was that old Jacobs. There are still a few folks doing 32 volts, but if I were doing it again, I'd probably go with 24 volts. I've got my PVs—photovoltaic modules—on a tracker wired to produce 38 volts. The only real sunny spot is 250 feet from the house, so I ran some big copper wire down there, and still lose a couple of volts, but it doesn't matter so much at the higher voltage.

In the house, we've got 32 volts and 12 volts DC and 110 volts AC. I like to keep everything I can—TV, lighting—on DC. I don't use my inverter very much, because I set everything up back before they were very reliable.

In the winter I help friends put together systems one piece at a time; windmills are my specialty. I tell them to get someone local to put the tower up. I've helped a lot of friends put up Jacobs-type mills, done it in all kinds of weather, ten below and windy, so I'm on borrowed time now, and I'd just as soon stay down off the tower and watch the trackers follow the sun. I'm scared of heights, always have been. Sometimes it takes me quite a while to get up the tower. But once

Fred's battery of batteries.

I'm up there and strapped and belted on, I forget where I am and get the job done. Some guys will put a $15,000 PV system together, a couple of them even got loans from the bank. Around here it's about $10 a foot to bring in the power, so it makes a lot of sense to be your own power company. I still like to use some of the DC lights and water pump so if they get struck by lightning or something, we're not sitting around in the dark.

Linnea got back into the garden full-time this summer. I tilled up a bigger area behind the barn, and we got a lot of corn, tomatoes . . . I've never seen such big pumpkins. It'll be expanding a little more every year. I think we'll all be looking to survival—the more you can produce yourself, the better off you're going to be. I don't buy into this whole "global everything" deal.

Our compost toilet is a Bowli, which they now call SunMar. It's gotta be ten years old, and it works great. I have to work on it a little more than they say, maybe a five-gallon bucket of nice-looking compost four times a year instead of twice like they say. The compost has no smell, although I'm a little leery of putting it on vegetables; it goes on flowers.

My driveway's a mess this winter (1997-98) because it's been so wet and warm. Linnea's been driving the school bus for three years now, and so she gets the kids to school and home. Before that I used to ride them out to the school bus in a garden cart wrapped up in a blanket on a beanbag chair.

My impression is that winters are getting warmer. I remember when I was at Alfred, the first snow would come in October and would still be on the ground in April. Now, it's that way maybe one winter in five.

I guess I could get grid power easily now—it's just a quarter mile away. But what a great feeling, making my own power. If you've never done it, you just don't know. We get ice storms here, and folks around us lose power for several days, and have no water, and can't cook, while we never have any problem. And the things they've got to do to generate electricity, nuclear and what-not, make you sick. Here in Allegheny County, I guess we're the poorest county in the state, and so they're trying to dump their garbage and their nuclear waste on us, and it makes you think.

I just like knowing this is all mine. ⟋·

The Hidden Costs of Electricity

In terms of hidden costs, those associated with the kilowatts we pump through our appliances are probably the most expensive and best disguised. Let us enumerate briefly here, identifying the issues without trying to quantify.

1. Costs of extraction and transportation. These costs are considerable, and largely hidden. Most power generation is presently fueled by natural gas and coal. Both resources are still reasonably abundant on our continent, and supplies should last well into the next century. Since both are extracted, given up freely by the earth, their true cost is impossible to reckon. The "monetized costs" (which we pay with real money) are arbitrary, although based on financially accountable extraction and transportation costs, plus a "reasonable" reward for ownership. Hidden costs, called externalities because they are external to the cost calculations used by the industry, are found all along the way: health risks and damage to the environment from extraction through

There are many costs associated with any mine, well, farm, or other resource-extraction activity; each industry and site differ. This generalized representation of relative costs includes those we pay for (the dollar signs) and "externalities" which are paid for indirectly, or not defrayed at all.

For heritage fuels—millions or billions of years in the making—the cost of reproducing the resource is larger than all other costs, accountable and external. Because scientists have not been able to produce synthetic petroleum at anywhere near competitive cost, the actual replacement cost is unknown.

delivery; emissions and losses during transportation; and (hardest of all to put a price on) the possible extinction of life forms, cultural sites, and artifacts, to name only a few. It is of course unthinkable that we might also pay the cost of replacing the resource.

2. **Waste created when electricity is generated.** Conventional power plants create megatons of fly ash and scrubbed sulfur; belch tons of noxious fumes (although, compared to other dead-dinosaur burners like automobiles, they are quite clean), and are, at the end of their relatively short life spans, monstrous stacks of used, useless, and often toxic junk. Until recently, the standard utility response to this problem has been to bury or dump the pollutants and let the machinery rust in place. In the extreme case, nuclear power plants bring a whole new meaning and time frame to the problem of defining and paying the costs of disposal. Twenty years after the first commercial nuclear plant was turned off, we have yet to decommission a nuclear plant. To do so will require that many of its parts and ingredients be encapsulated for periods up to four times the half life of plutonium, nearly 100,000 years at the very least. This is, above all, a building and communications challenge far beyond anything ever undertaken: Our oldest structures are barely a tenth that old, and how can we be sure that our warnings—"Keep out! Radioactive waste!"—will be intelligible in twenty centuries? (See the Radioactivity Time Line on page 26.) No scientist denies that a nuclear plant's whole cost should include funds for site and material decontamination; although never admitted into consideration during the proposed new plant's cost/benefit analysis, this cost may exceed the nuclear plant's construction cost by a thousandfold or more.

3. **Inefficiencies in generation and distribution.** Some of the extracted material goes awry, occasionally spectacularly, as in the case of the Exxon *Valdez*, but more often invisibly. Experts disagree about how much electricity is lost between generator and plug, or where it goes, but lost electricity amounts to about half the total generated. "Stray electricity" causes deformed livestock and decreased dairy production, but claims are quietly settled out of court by utilities when proven undeniably. Health experts speculate about the effects of electromagnetic radiation (EMR)

emanating from powerlines on humans, animals, and plants, but there is little certainty; we are not surprised that industry-supported research denies the existence of risks due to EMR while anecdotal information suggests there is something ugly going on. Can we even begin to evaluate the harm done to our magnificent scenery by powerlines draped insouciantly over hill and dale?

By improving the efficiency of North American and European homes, we could use half as much energy as we do now without sacrificing a bit of comfort. We could then rely on locally-produced electricity, reducing electricity consumption by three-quarters, and eliminating our need to burn fossil fuel for electricity.

Perverse incentives govern the massive, self-interested power structure already in place, and so our best efforts to identify and reduce energy's true whole-life, whole-system cost to living humans and to all future life on the planet are frustrated. Long-term survival ought to be a greater concern than short-term cost, but in a corporation-dominated economy motivated by quarterly profit, environmental poisoning and resource depletion are denied until they are overwhelming. The fact that our infrastructure and installed equipment wastes three of every four units of electricity generated may have a fiscal meaning, but it is also threatens the continuity of our species. Nature simply does not allow such heedless wastefulness to continue.

If we are to survive, we must hasten to institute sweeping changes, so our individual efforts will compound themselves rapidly—before the petrochemicals are used up and our environment becomes inhospitable to all but the insects and genetically engineered microbes that love to eat toxins.

What is true for electricity also applies to every other use of energy, from the cars we drive to the way we heat our hot water, the way we keep our homes comfortable, and the way we refrigerate our food. In the 1950s and 1960s, decisions about energy-consuming devices were made assuming that energy costs were then and always would be a minor component of overall costs. In 1973, Americans received a wake-up call we are just now beginning to forget, but the 1973 and 1978 oil crises sounded the death knell of easy energy. A 1998 *Scientific American* article written by two senior oil company scientists assures us that we have seen the last of cheap oil. Energy costs are highly politicized, and I believe that America's distorted energy market deliberately and short-sightedly prices energy at only one-third of its real value to stimulate its growth-based

If we are to survive, we must hasten to institute sweeping changes, so our individual efforts will compound themselves rapidly— before the petrochemicals are used up and our environment becomes inhospitable to all but the insects and genetically engineered microbes that love to eat toxins.

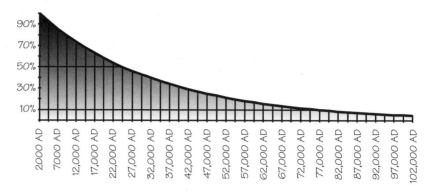

Plutonium Radioactivity Time Line. Plutonium is one of the most toxic substances known; a single atom can inflict fatally carcinogenic, mutagenic, and teratogenic harm if it lodges within a life form. Plutonium—element Pu—is astonishingly persistent: its virulence is reduced by half every 22,500 years, its halflife. Since Pu is so toxic in such small quantities, it is reasonable to classify it as a "zero tolerance" pollutant, which means waste plutonium sequestered for more than 100,000 years—twenty times longer than the oldest known buildings have been standing—may still be unsafe.

economy. Another third is collected indirectly, through personal taxes converted into tax breaks for energy corporations, subsidies, and gunship diplomacy. The remainder is borrowed from the future: a mortgage on our grandchildren's quality of life, which they will be forced to repay through reduced quality of life, diminished natural diversity, and staggeringly expensive necessities. Resource costing based on an economic model called "Hubbert's Peak" (see page 7) demonstrates that production starts to fall and prices start rising sharply when half of any resource is extracted. The evidence suggests that oil, our most important energy resource, was at or beyond that point before 1998, with no credible replacement in development. All petroleum demand projections, based on a burgeoning population, show a steadily *increasing* demand, with no end in sight. The longer oil prices are artificially depressedthrough political means, the steeper the eventual upward oil-price slope will be; it may look more like a wall than a slope: the Last Drop of Oil Wall. Therefore, as caring parents and stewards of the earth, or prudent managers, or wise investors, it makes sense to reduce wasteful consumption, and adjust our personal budgets and energy-use plans in anticipation of rising prices and disrupted flows of extracted energy, even while enjoying the last years of life in an energy fool's paradise.

In order to use less, we must first examine what we use.

CHAPTER 2

HOMEWORK: OUTGROWING DEPENDENCE

FOLLOW THE MONEY: GOOD ADVICE FOR SOLVING ANY PUZZLE ABOUT costs and benefits. The best opportunities for lightening the weight we add to the planetary energy overload are likely to be found right in the areas where we spend the most money on energy.

In every climatic regime, from Alaska and Finland to Ecuador and Singapore, the primary energy consumers in a "modern" house are, in descending order by energy consumption: space heating and cooling, hot water, refrigeration, lighting, cooking, and various plug loads—entertainment and office devices, tools, and the plethora, one might even say plague, of small appliances that burden our electrical systems. By conscientiously examining every energy use starting with the largest, we will find ways to lighten the load and simplify our lives.

Future historians may well call the 20th century "the uprooted era" because so many of us lost or abandoned our ancestral homes. At the start of the 21st century, many people yearn to end this state of homelessness. In the year 2010, a decade into the next millennium, demographers suggest that four of every five presently existing houses will still be in use, and many of us hope to be living comfortably in them. Considering the irreplaceable "embodied energy" already bought and incorporated into this housing stock, and the generations of emotional investment in these old buildings, we cannot simply "write them off." Nevertheless, due to bad energy decisions made from the 1950s onward, most of these houses are energy pits into which we throw money.

Luckily, we are also a nation of tinkerers and do-it-yourselfers, and when we have time, our homes are our favorite hobbies. This is especially true for independent homesteaders: The home becomes the *magnum opus*, the Great Work, and living becomes the ultimate form of art. Independence requires dedication and discipline. We have always heard that constant vigilance is the price of freedom; this chapter tells what we must know and do to secure our energy independence.

Cost-Effective Home Energy Systems

Many homes, as well as the major appliances in them, were designed and manufactured between 1945 and 1978, before the public awakened to the harsh light of energy reality. Before 1978, our massive manufacturing apparatus was hustling to cut as many trees and cover as much acreage as possible with pavement and commodity housing to fulfill the American Dream sold by profiteers and earth rapists. Then and now, the housing market rewards fast, light, and cheap construction practices and high turnover of properties, not householder satisfaction, and so the system remains

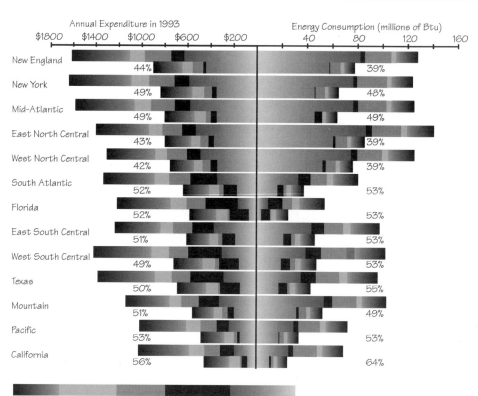

REGIONAL ENERGY ECONOMIES IN THE U.S.

out of control. Because an industry can be expected to respond in proportion to the directness of its exposure to the effects of spiraling energy costs and increased pollution awareness, some industries are already energy efficient. The energy industry itself has been quick to adopt the best available technologies. The American automobile industry noted the need for energy efficiency, but sidestepped government fleet mileage requirements by marketing bulky urban assault vehicles to a gullible, bigger-is-better-brainwashed public; the same obsession with features over function can be seen in refrigerators. Reformation of the housing industry has come about slowly because the homeowner, not the manufacturer, pays for

inefficiencies after the sale. Building codes; architectural, engineering, and production practices; and choice of materials all lag a decade or two behind the best of known options. Commodity home-building, bound by fashion as well as by building codes from a more profligate era, will probably be among the last industries to change, because consumers are uncritical and poorly prepared to make good housing decisions, even though they bear all the risks and costs. Books such as this, and smart solar architects, will help a lucky few escape the onrushing consequences. Unfortunately, the best solutions, like most benefits of our lopsided economic system, accrue to those who can afford high-end custom homes.

This chart shows regional opportunities for energy savings in the U.S. Homes consume differing amounts of energy depending on many factors—home size, family size, orientation, construction type, to name a few—but local climate is a major determinant. The U. S. Department of Energy's statistics show the average quantities and expenditures for all forms of household energy except garbage and sewage. In this chart, each region is represented by four bars. Each pair of bars shows all 1993 household energy usage for a region as reported to the Department of Energy (reaffirmed by 1996 data). The upper bar is average, and the lower bar shows the energy mix and reduction possible with efficiency measures. The left side of the chart shows energy cost; the right shows actual consumption in kilowatt-hours. Lighting and cooking energy is included with plug loads.

Across the country, the easiest opportunities for energy- and cost-reductions are more efficient hot water and refrigeration. While plug

loads consume a small amount of energy, they are quite expensive—they use electricity, our most costly form of energy—and so small economies in this area result in good returns.

New England's housing stock is the oldest, so its heating costs are disproportionately high compared to regions with similar climates. New England's energy also costs more, so a perfect opportunity for dramatic improvement exists: a 39% reduction in consumption produces a 44% decrease in payments.

By comparing BTUs of heating and cooling with corresponding costs, we see that heating is more cost-effective than air conditioning, in large part because heating can be done more directly, with oil or wood, but air conditioning must be done with electricity. This shows that better hot weather insulation measures would be most cost effective. In Florida and most of California, well-designed passive solar homes require virtually no heating.

ELECTRICAL SYSTEM COMPONENTS

Power systems are composed of four parts: source (where the energy starts), distribution (how the energy is moved from the source), control (that measures, directs, and adjusts the flow), and load (that converts the energy into power). These words are most often used to describe electrical systems, but they apply equally well to any energy system. For example, a woodburning stove is the load at the end of an energy chain that starts when the woodlot is planted, because trees are the source.

Electricity is a flow of electrons which is generated at a source, moved and controlled with conductors and switches, and converted into work by loads. The form of electricity called "Direct Current" can be stored in a battery. For more about Electricity, look at "Describing Electricity" on pages 64–65.

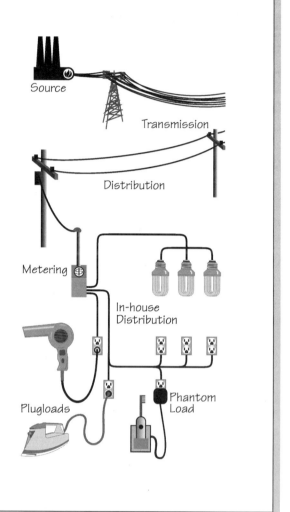

Source

Transmission

Distribution

Metering

In-house Distribution

Plugloads

Phantom Load

For an energy feature—a brighter lightbulb, a better door seal on a refrigerator, a high-efficiency zero-pollution automobile engine—to be cost-effective it must repay its added cost within the feature's effective life. This standard, called *payback*, can be calculated and applied quite rigorously if one is of an accounting turn of mind; an example of the necessary calculations will be encountered later in this chapter, in our discussion of refrigeration. On the energy market's "uneven field," we must often consult our conscience as well as our wallets when making hard decisions. Coal power, for example, is known to be the dirtiest and most dangerous, as well as the most abundant and profitable for its providers. Living in the shadow of the upper Midwest's coal burners, hard-nosed Yankees of Vermont have agreed that health aspects alone justify assessing new proposed fossil-fuel-fired electric plants an "externalities adder" of 30% when reckoning overall long-term plant costs. Even if unmonetized external costs such as the health effects of breathing coal-smoke-bearing air are not reflected by energy prices, a conscientious homeowner might easily decide to apply her own personal "adder" when accounting for the whole cost of contemplated energy features. I suggest that the proceeds from this adder be directed into a fund for pre-emptive energy conservation measures.

Traditional economic analysis uses a "steady-state" assumption—energy and replacement equipment costs will remain constant over time—which simplifies calculations but understates operating costs and predicts eventual overall costs poorly. To lay a firm foundation for my economic analysis of energy features, I researched the best available projections for the cost of energy. Amidst the hype, I found expert opinions suggesting that "scarcity pricing" of energy could easily increase at a rate of as much as 20% per year through 2020. In calculating payback periods, I have found it useful to do so twice, once using this unlikely "steady state" assumption of straight-line energy costs, where the amount paid per kilowatt-hour or therm does not increase from present levels, a scenario preferred by appliance makers, government energy planners, and other dreamers; and once with a compound increase model of 10% per year which, we can only hope, is the worst case we will face as we approach the Last Drop of Oil Wall. The real payback should be somewhere between calculations.

Many of our dollars-and-sense decisions are warped by traditional means of accounting, which separate bricks-and-hardware capital costs from operating costs. This practice can be traced back to a time at the beginning of the industrial age when capital was scarce and resources plentiful. Because our whole economic system, including the amount of disposable cash we have in our pockets at any given point in time, is determined by these outdated assumptions, it is easy for us to continue making decisions using a 19th-century model. Capital costs are usually buried in mortgage or consumer indebtedness, while operating costs come out of the petty purse, our paycheck-to-paycheck budget. Sticker shock at the higher capital cost of more efficient equipment too often motivates cheap-to-buy, expensive-to-operate decisions. The purpose of the longer-view, whole-life cost assessment is to recognize all costs no matter when they

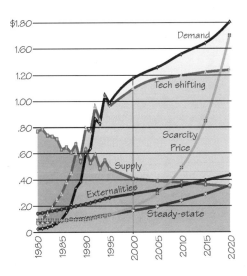

According to classical economics, price is a function of supply and demand. This graph shows projected electricity price based on cost per kilowatt hour in California. The lowest curve shows the "steady-state" price quadrupling in 40 years. This projection takes dwindling supply and increasing demand into account. Supply dwindles less sharply due to assumed technological breakthroughs (shown as "Tech shifting"), but experts seldom predict that demand will decline. I have optimistically accepted the "Tech-shifted" demand line, based on my assumption that rising prices will motivate social reforms and increased applications of efficient technology, somewhat reducing the forces that could raise the projected "Scarcity Price" tenfold in the next twenty years. Data for 1995 and before are actual, showing the jittery uncertainties of resource supply.

must be met. If an energy feature is worthwhile, even if it requires more expensive equipment, we may expect to enjoy reduced energy costs over its lifetime.

The basic principle of payback is nothing more than common sense, belatedly applied to an area where we enjoyed the brief and foolish luxury of ignorance brought on by a surfeit of cheap energy. In 1960 we knew no better. Now we do.

Some of us willingly pay more in partial restitution for past energy depredations. After a year of deregulation, commercial energy providers are finding a brisk market for "renewable" electricity, including old and new wind, PV, hydro and geothermal electrons, costing a premium of 15%. Sacramento's publicly-owned utility SMUD allows conservation-minded customers to pay the same premium for their electrons in its "Give Us Your Roof" program, which sells out every year.

Retrofitting: Steven Strong's story

Steven Strong designs solar electric systems and environmentally responsive build-ings with Solar Design Associates, his architectural and engineering firm in Harvard, Massachusetts. Here are some strong ideas about bringing energy effi-ciency home.

Retrofits are no architect's passion, but we must face the fact that the majority of the buildings for use in the next generation are already built. It's sometimes difficult and often unglamorous work but there are far too many existing buildings to ignore. What can we do with these structures? After increasing their thermal integrity and optimizing their energy efficiency, we must retrofit them to produce some or all of their own energy on-site from renewable resources. It turns out to be reasonably easy to address these problems, and most of the solutions are equally appropriate to new con-struction.

The basic hierarchy of options can be prioritized in terms of cost-effectiveness and return on investment.

The first priority is thermal integrity. Start with the building envelope. Upgrade the insulation and gaskets, and eliminate infiltration. Consider controlled, heat-recovery ventilation. Check out the windows. Regular double-glazing has an R-value of one and seven-tenths; today's high-tech window's R-value is nine.

It is simply inexcusable not to employ the very best windows you can buy. Just in terms of resource allocation: microns-thick, low-emissivity (low-E) surface coatings and inert-gas fills justify themselves. Glass itself will last more than the lifetime of the building and, with the investment of less than an ounce of additional high-tech materials, you can cut energy waste by a factor of two, three, or more over the building's entire life.

In the last eight years there has been a full-scale revolution in the windows you can buy. For example, Andersen no longer sells standard windows. When you buy any Andersen product today, you get argon-filled, low-emissivity windows whether you ask for them or not. This is a major, major advance. By effectively doubling the R-value of their entire product line, Andersen has, almost on its own—because of its volume—set a new industry-wide standard for energy efficiency. Other manufacturers have then followed suit.

As the race to develop "super windows" continues, new technologies recently introduced provide almost three times the R-value of the standard low-E Andersen units which are themselves about twice as good as ordinary "standard" double glazing. Let me sound like Amory Lovins for a moment: If we retrofitted the glass in all existing buildings in this country using today's best-available technology, we would save more energy than flows from the Alaskan Pipeline. That may sound like a formidable task, but it's cheaper and easier and "nicer in every way" than another war over oil.

Next, install a state-of-the art heating plant. Many state-of-the-art plants burn conventional fuels at close to 95 % efficiency, and easily pay back their cost even when they're used just half the year.

Then, look at your lighting, appliances, and plug loads. Shopping for the most energy-efficient appliances is a critical (and simple) thing each of us can do to move the country toward greater energy efficiency. It may even make sense to change our appliances before the end of their design life if you gain significant efficiency. It's fairly easy to figure. Amory calculated this years ago and the utilities are just now catching on.

It's technically possible to build a refrigerator that uses one-tenth the energy of the average unit out there in American households. There's an energy revolution in refrigerators on the way as evidenced by Pacific Gas & Electric's "Golden Carrot" program in the U.S. (See page 154.)

Refrigerator energy consumption will likely fall by at least a factor of two or three over the next ten years. Other major appliances are undergoing similar redesign.

Finally, after you've tightened up the building and reduced the loads, look at the possibility of producing your own energy on-site. Solar water heating is very effective: It's off-the-shelf technology, well-debugged, supported by a good service infrastructure; it's easily interfaceable with existing systems and works equally well in retrofits and new construction.

Further up the return-on-investment curve, you'll find solar thermal systems for combined space heating and domestic hot water, and, ultimately, on-site generation of electricity with wind or hydro power, where available, and photovoltaics anywhere. It is possible to achieve energy independence with today's technology even in the northern tier of states, where degree-days are high and sunlight resources are modest.

Like high-tech glass, the ultimate solutions will be invisible, with more benefit for less money, in the long run, than we would expect . . . and you won't even see the difference. These changes are coming about, first, because of the energy market, but second, because it is not that hard to do much better. If we, as a society, invest our talents and resources wisely to develop the right set of technologies, austerity and deprivation will not be necessary. If we fail to do this, they most certainly will be.

For new buildings, which is what all architects like to build, we must make sure our philosophy is in tune with what we're trying to accomplish on the global scale. In our firm, we will only work with clients who have a high degree of concern for environmental issues. At the same time, our clients want a high-quality living environment. Our houses reconcile this conflict by making it possible to live lightly on the earth without taking a vow of deprivation to do so. Many of these houses are energy independent and only a few might be considered "inexpensive." That's a difficult issue for me, that our work is looked at as just for the rich. But, it's a fact of life in the profession—whether or not they want sustainable buildings, only people with reasonable means hire architects.

I'm in love with photovoltaics. It is the most compelling technology I have ever encountered—silent, nonpolluting, self-renewing—simply elegant. We powered our first building with PVs in 1979. It was an eleven-story mid-rise and cost a truckload because it was twenty-five or thirty years ahead of its time. We designed and built the Carlisle House in 1980 and the PV system for it also cost a truckload.

When I'm criticized on the cost of our PV-powered houses and buildings (usually by the press), I point out the twentyfold reduction in the cost of pocket calculators in the last fifteen years.

It costs real money to bring a new technology from concept to a mature commercial product and early demonstrations are going to be expensive. If the calculator industry had been shamed into giving up in its infancy by complaints that their simple four-function model was "too pricey!" we'd all still be using those clunky, mechanical adding machines. Predicting the future is like taking a snapshot through a keyhole, but I feel our clients are beginning to invest in photovoltaics because they see the watershed of change coming. We've got the technical solutions now; all we need is the political will to implement them.

I realized many years ago that we simply can't change the world, one custom house at a time—they may be fun to do but there just aren't enough to make any difference to society. So we became politically active, working with state and local policymakers and utilities to find ways to restructure capital resources and introduce energy efficiency and renewables. Our private clients have supported our explorations of solutions that, in the near future, can then be shared with working-class homeowners who can't afford to be early innovators.

Off-the-gridders are true pioneers who have created a significant base

The Gardner homes are unremarkable in every way—that was the point of the experiment—except for their solar orientation and the PVs on their roofs.

marketplace and proving ground for PV technology. But they are not, by themselves, going to make the changes required in society—there aren't enough of them, either. It's a nice solution for a very small minority of folks, but we can't take all of our buildings off-grid—nor should we want to. If photovoltaics (and other renewables) are going to make any significant contribution to society, then the utilities must become active partners in the process.

In 1984, I helped plan and execute the world's first PV-powered neighborhood in Gardner, Massachusetts. Working under contract with New England Electric, we found a subdivision with good solar exposure: thirty existing, modest, working-class ranch houses, classic post–World War Two housing stock, various sizes from fifteen hundred to twenty-two hundred square feet with varying loads. We installed utility-interactive PV systems of 2.2 kilowatts on each roof, and we ended up with a win-win-win symbiosis.

A utility-interactive PV system uses the power grid in lieu of on-site storage. The homeowner gets free, one-hundred-percent efficient, no-maintenance energy storage, without batteries. The utility gets on-site power generation with a surplus coincident with their system's peak summer air-conditioning load; they give back the surplus during off-peak or baseload periods when they're hungry for consumers just to keep the generators turning over. Society wins because the renewably generated electricity displaces polluting "conventional" sources of power.

You ask the people in the world's first PV-powered neighborhood how it's been after more than ten years. The minority, perhaps 10%, really got into it—counting the solar kilowatt hours every day, and so on—but the rest

blissfully reported there is no noticeable change in their lives. Most significantly, no one is unhappy. Again, the ultimate solutions will be essentially invisible to the end user. Here we have a neighborhood where thirty new electric power plants have been installed in people's yards and no one has noticed any difference in their lives! Compare that to the siting process for a coal or nuclear facility and you can begin to understand what I'm getting at.

The Sacramento Municipal Utility District (SMUD) is presently leading the way for utilities worldwide in the use of renewable energy. They're a unique experiment in democracy, where the ratepayers are the shareholders, and they can thumb their noses at the state Public Utilities Commission, from which they're exempt. Earlier managers built Rancho Seco, a particularly troubled nuclear plant, right near Sacramento, but the people of Sacramento, with the support of the current managers, voted to shut it down. SMUD's new program goes something like: "You told us to shut down the nuke, now give us your roofs so we can install solar thermal systems and photovoltaics." In the program's first three years, SMUD installed 5 megawatts of distributed PV systems on rooftops in their service territory.

For utility-interactive PV to work requires a revolution in utility rate structures. The key element here is net metering, which is already the rule in California, Texas, Oklahoma, Maine, Massachusetts, Minnesota, Wisconsin, and several other states: Whatever you generate spins the meter in your favor, and whatever you consume spins it the other way. Kilowatt-hours are traded back and forth, one-for-one. At the end of the month, you pay or are paid for the difference. With such a buy-back arrangement established, you can address each site's energy

The Gardner "solar neighborhood."

If your goal is trying to build an environmentally responsible building, you're missing the whole point if you get all lathered up over a non-volatile natural finish on the handrails, while you're connected to a plutonium-generator power plant down the road.

resources. Photovoltaics are certainly the most elegant and universally applicable, but wind and hydro are much more cost-effective where those resources exist.

I like Denis Hayes's conclusion that it is impossible to get American society to give up cars willingly, so we must redefine them: Four-passenger vehicles getting sixty or eighty miles per gallon, recyclable components, automatic braking systems, airbags and the whole crash package, etcetera. Establish what the car of the future should be and then challenge the U.S. car companies to build it . . . with a big reward to the one that gets there first, a billion dollars on the table as a major carrot for the development of the next-generation automobile.

Such bold new vision is what is required to bring about the watershed changes that are necessary (and inevitable) in contemporary American society—called the "paradigm shift" by PG&E's former research chief Carl Weinberg. This is also the easiest, best, and least painful way to bring about effective change.

My message to architects and engineers is: Look at the whole picture. In the trade press recently, there was an article hailing a custom, 9,000-square-foot, architect-designed house as the latest in environmentally responsible design. Its principal claim to fame seemed to be the use of natural, nontoxic finishes on the woodwork. In the rush to commercialize "Green Architecture," no one noticed that this house consumes more energy than a small New England town.

If your goal is trying to build an environmentally responsible building, you're missing the whole point if you get all lathered up over a nonvolatile natural finish on the handrails, while you're connected to a plutonium-generator power plant down the road. It's the same old "out of site, out of mind" again, with a new face on it. "I'm doing all I can for the environment, my architect specified beeswax on my new woodwork—someone else will just have to figure out what we're going to do with all this radioactive waste" . . . and acid rain and oil spills and global warming and ozone depletion and unhealthy air quality and . . .

You hear a lot about sustainability these days. I've been at this since 1973, long enough to be certain that, without addressing the energy issues, you're in the weeds. All the fuss over "my milk-based paints transported in from Europe" is just a myopic distraction from the issues that really matter on a global scale. True, natural-based finishes are desirable, but they fall far short of the answer. Establishing an energy infrastructure based on renewable resources is a necessary and fundamental precondition to establishing a sustainable society or to achieving sustainability at any scale. If you are not addressing the energy issues, don't even pretend that your building is environmentally responsible.

One-time Apple CEO John Scully says the best way to predict the future is to invent it. I say, look for the "invisible solutions" that are already here, available today, and apply them. If we all just do this, the future we want will arrive before you know it. ↜

Heating and Cooling Space

Our costliest domestic energy activity is regulating the temperature of the space we occupy. Many of us remember President Carter's cardiganed appeal from the Oval Office to turn our thermostats down in winter and up in summer. Not bad advice, but it is not the first thing we should do. First it makes sense and saves dollars to eliminate waste, something energy companies have been helping us do for the last twenty years.

Designing and building a new house, a multitude of small energy improvements may be better and cheaper than brute force heating and cooling. In chapter 16, Amory Lovins describes a program in Davis, California, one of the most intense cooling regimes in the United States, where designers of a new building identified every possible efficiency feature, calculated a cost and benefit, and put them in a box. When the conventional design was completed, and the necessary air conditioning equipment specified and costed over the life of the building, they worked the box, comparing incremental capital (building) and operating costs. The designers were surprised to find that the building could be entirely cooled by their boxful of passive measures, solar-powered air-handlers and pumps, with a very attractive breakeven point. No secret design tricks or magic were employed, just bottom-up step-by-step attention to sensible design, from siting and orientation through materials selection, construction techniques, and building operation. If this approach works in Davis, it can work in any temperate environment.

The measure of a climate's intensity is described by "degree-days." One degree-day is one day with an average ambient (shaded outdoor) temperature one degree above or below optimal. Typical calculations define the optimal as 68° to 72° Fahrenheit. If the average outdoor temperature is 67°F for a given day at a given site, that represents one degree-day of heating; if the average outdoor temperature is 73°F, add one degree-day of cooling. A locale's degree-day figure is the total of all degree-days calculated for one year. As you may see in the accompanying map, most of us live in regimes requiring 2,000 or more degree-days of heating or cooling, and often both. Because heating can be done with a variety of competing fuels, degree-days of heating are less costly than the

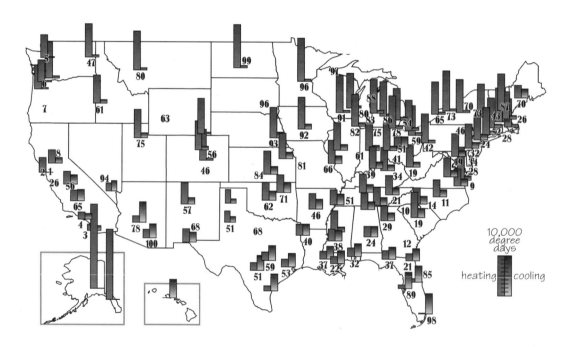

Degree-days of heating (left) and cooling (right) for cities in the United States, using optimal figures of 68 degrees Fahrenheit (°F) for heating and 72°F for cooling. Numerals are rankings of 100 cities, based on the energy required, using the CNG Energy Index (commissioned by Consolidated Natural Gas Company), a more sensitive measure of actual energy requirements. The CNG Energy Index considers temperature, insolation, humidity, and the previous day's weather in order to predict the amount of energy required to keep space livable. Oakland, California ranks number one, Phoenix, Arizona is number 100.

same number of degree days of cooling. Cooling is the monopoly of refrigeration manufacturers, and it is brutally inefficient. I suspect that modern buildings are designed principally to benefit the makers of air conditioning, since in most of the United States buildings can be built that require little or no air conditioning to be livable in summer.

The measure of an exterior surface's resistance to heat gain and loss is its R-value. Glass, no matter how thick, has an R-value of 1, as does any thickness of water, or an inch of wood. Since heat escapes to the outside sur-

Thickness	R-value
Water	0.00
Concrete 1"	0.08
Rammed Earth 1"	0.11
Plaster 1"	0.12
Rock 1"	0.20
Ground surface	0.50
Hardwood 1"	0.91
Glass any	1.00
Air gap non-reflective 1"	1.01
Plywood 1"	1.24
Soft wood 1"	1.25
Redwood 1"	1.75
Asphalt shingles 1"	1.76
Straw 1"	1.78
Double-glazing	2.00
Carpet and pad 1"	2.00
Sawdust 1"	2.44
Sheep's wool 1"	2.96
Cork board 1"	3.33
Air gap reflective 1"	3.48
Polystyrene 1"	3.57
Balsam wool 1"	3.70
Glass wool 1"	3.76
Best "super-windows"	8.00

faces of a building mainly by convection, the roof and ceiling are more wasteful in winter, because heat rises, and the sun-struck roof and southern and western faces are most wasteful in summer. Careful builders insulate these surfaces more heavily than other walls, and will also insulate floors, which once were not insulated at all. In fact, many houses built before 1980 were not insulated at all—remember, it was cheaper to use lots of cheap fuel. Since then, insulation practices have improved. Smart, future-oriented builders say there is no such thing as too much insulation and commonly build R-50 roofs and R-30 walls.

Building codes and design practices have improved, and many states now mandate energy efficiency. As noted by Steven Strong, buying a single-pane window is no longer possible. But conventional stud-framed houses still employ a corrupt structural strategy. The Forest Abatement lobby of large timber companies has institutionalized a pattern of using too much wood in houses. Architectural stress analysis suggests that typical framing practices—2x6s every 16 inches—puts 25% more wood in a house than stiffness and load-bearing requires. This joke on the proud new homeowner makes builders, suppliers, and providers happy, but the real punchline comes during heating and cooling season. Remember that wood has an R-value of one per inch? Studs are less well-insulated than the rest of the wall. "Thermal bridging" designed into this wall and roof strategy means that about 16% of every exterior surface has an R-value equal to the thickness of the wall in inches. Modern walls are often seven inches thick to accommodate thicker insulation between the studs (and to sell even more lumber) but more than a tenth of the surface area plus all of the windows has an R-value under ten. Advocates of steel and aluminum studs neglect to mention that thermal bridging problems are even more acute when using these materials. Fortunately, there are other, much more energy-efficient strategies, from straw bale and insulating foam to overlapping arrangements of studs to eliminate thermal bridging, but the building industry resists these innovative measures because they require craftsmanship and cost more to build.

For existing homes, the best advice can be summed up simply: Replace glass and other heating and cooling units with the most efficient equipment available. Yes, windows *are* the most abundant and significant heating, cooling, and lighting appliances in the home, a fact we too often ignore. Few of us drive cars made in the 1950s, but most of us live in houses with the largest energy appliance—the windows—made in that energy-impaired era. Single-pane glass has no place in an energy-

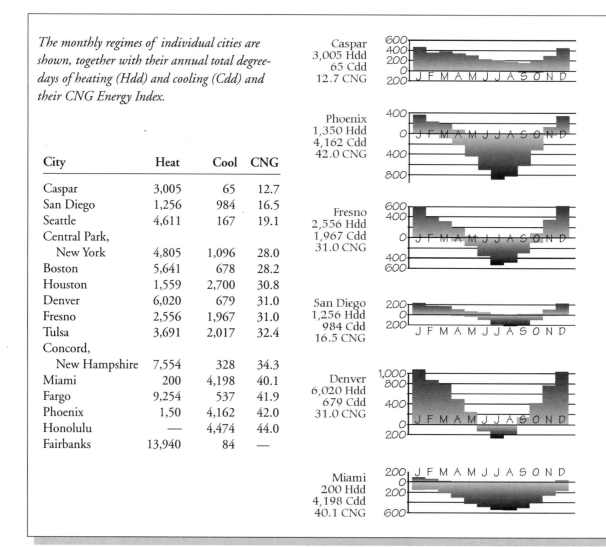

The monthly regimes of individual cities are shown, together with their annual total degree-days of heating (Hdd) and cooling (Cdd) and their CNG Energy Index.

City	Heat	Cool	CNG
Caspar	3,005	65	12.7
San Diego	1,256	984	16.5
Seattle	4,611	167	19.1
Central Park, New York	4,805	1,096	28.0
Boston	5,641	678	28.2
Houston	1,559	2,700	30.8
Denver	6,020	679	31.0
Fresno	2,556	1,967	31.0
Tulsa	3,691	2,017	32.4
Concord, New Hampshire	7,554	328	34.3
Miami	200	4,198	40.1
Fargo	9,254	537	41.9
Phoenix	1,50	4,162	42.0
Honolulu	—	4,474	44.0
Fairbanks	13,940	84	—

Caspar
3,005 Hdd
65 Cdd
12.7 CNG

Phoenix
1,350 Hdd
4,162 Cdd
42.0 CNG

Fresno
2,556 Hdd
1,967 Cdd
31.0 CNG

San Diego
1,256 Hdd
984 Cdd
16.5 CNG

Denver
6,020 Hdd
679 Cdd
31.0 CNG

Miami
200 Hdd
4,198 Cdd
40.1 CNG

efficient North American home, and the economics of replacing it are very favorable. Replace north-side windows and known energy-offenders first.

Rewindowing is straightforward compared to other remedies which may require serious house surgery. More radical but nevertheless cost-effective measures include improving insulation, plugging leaks, and installing zoned heating and cooling with set-back thermostats so rooms are heated or cooled only when they are in use. This advice applies to new construction as well as retrofitting houses built after 1910 and before 1990, during the time when it was assumed that energy would always be cheap.

People Cookers

You may be surprised to learn that right across the United States, the days when our electric supply runs closest to its capacity are not the

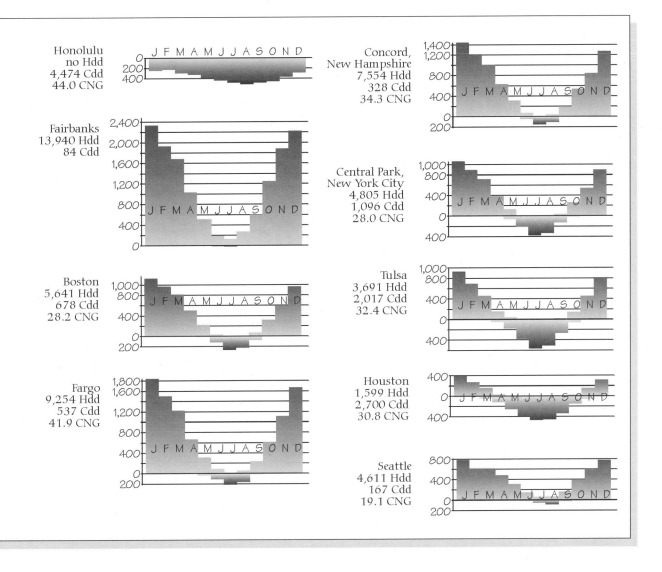

coldest days of winter, but the dog days of summer, when the air conditioners are set at maximum and it still feels too hot. The worst offenders, and the greatest energy consumers in this stressful time, are office buildings built—one might be tempted to say designed, except that wrongly implies intelligence on the part of the architects—to stack workers in configurations that could scarcely be better conceived to waste coolness and trap heat: solar ovens for cooking people. Energy theorist Amory Lovins has called these curtain-wall monstrosities "all glass but no windows."

The same vain and sloppy architectural practices, applied to residential construction, produced a generation of once-fashionable, cheaply built, flat-roofed, poorly insulated, glass-walled buildings habitable only because of their massive heating and cooling systems. Owners of these artifacts are reminded of their misfortune every month when they open their heating and cooling bills. Little can be done: increase and cherish shade trees, abandon and close off poorly ventilated southern rooms in summer, seek and encourage any nighttime cross-ventilation opportunities. Costlier remedies include replacing old glass with high-tech glass, installing or extending overhangs to block summer sun, increasing interior thermal mass in the cooler and more easily ventilated areas of the house, replacing old floors with massive hydronic floors, adding a second roof or a cooling tower, and retiring the original air conditioning unit in favor of the most advanced and efficient model available. With energy trends as they are, you may expect any of these measures to pay you back within your lifetime.

The gracious houses built before architects and mass-production housing spoiled the American vision of a sensible dwelling may handle summer's heat well. In areas with hot summers, stately deciduous trees may already be thriving; whole neighborhoods often enjoy temperatures several degrees below those reported at the treeless airport. Remember that it is best to keep the heat where you want it, indoors in winter, outdoors in summer. For older homes, employing the common-sense measures intended by the original builders and the best modern means can be very effective, will be quickly repaid in energy savings, and will improve thermal performance in summer and winter:

— Install radiant barriers and beef up attic and, if possible, in-wall insulation.
— Add or increase attic ventilation with a solar-powered fan.
— Replace single-glazed windows with the best possible modern fenestration; do not buy cheap windows. Minimizing infiltration of outdoor air with weather stripping and caulking windows is a stop-gap.
— Maintain storm windows and sashes, cross-ventilate in summer, close off unused rooms in winter, and use door-snakes (those long, bean- or sand-filled socks that block the crack below the door).
— Renovate and use the shutters provided by wise old-timers on the south side in summer.

    Improve window shades, especially on the north side, to retain winter heat and reflect summer heat.
— Install efficient new cooling and heating units before the old ones are worn out. If you have all the preceding measures in place, you may be surprised to find you

can get by comfortably with a smaller unit than you had before.

If in doubt about where your home is hemorrhaging energy, a home energy audit might be warranted. In addition to documenting all energy expenditures—follow the money—comprehensive audits employ physical measurements of the home's energy integrity using blower doors and infrared cameras to identify leaks and weak spots in your home's defense against the frigid outdoors. Since most of us must heat as well as cool, improvements in wintertime performance often apply as well in summer, when the house's thermal behavior must be inverted to keep heat outside. The results of such diagnostic tests may help you decide which measures will be most cost-effective. In many states, audits are now being used to "energy rate" homes for bank financing and government-sponsored renovation programs. A typical finding will be that heat loss is occurring primarily through single-pane windows, and therefore all the caulking and weather stripping in the world will make a very small impact. Based on this finding, it may be possible to justify, and perhaps even finance, rewindowing a whole house, or at least the most wasteful windows. (You are invited to conduct your own home energy audit in chapter 5.)

Whether planning a new structure or a retrofit, consider the dwelling's entire heating envelope. Here is a nightmare scenario: two o'clock on an August afternoon in a Lodi, California subdivision, where summertime electric bills for a normal dwelling average more than $200 a month. Or worse, if you are very brave, imagine a house near Houston, where cooling bills are typically in excess of $300 a month. The sun boils down on the shallow-pitched roof, and attic temperatures (if there is an attic) get high enough to bake bread. The single-pane windows, especially in the late afternoon, are a misery, permitting solar gain as well as radiating the heat from outdoors, which may easily reach 120°F, and so the tenants have covered the windows with aluminum foil. Of course this renders the indoor space dark, and so incandescent lighting adds to the cooling load which has long ago overwhelmed the inefficient air conditioning unit roaring like an injured bear in full sun on the roof. Just home from school, the senior child dutifully washes the dishes and starts preparing dinner. Every time he runs the hot water, he unintentionally heats the slab beneath him as the water travels in transit through pipes buried in the slab from the water heater located conveniently (for the original builder) in the service room at the opposite end of the house; the heat radiates gently into the interior space, further taxing the air conditioner, and wilting the people. Were any of these perversions anticipated or considered by the house's builders? A more urgent question: What can we do to salvage this valuable housing stock and make it more livable?

For new commodity housing, improved techniques are just coming into use for collecting hot water and storing the heat for future use—please see the case study on radiant floors in chapter 8. For some existing homes, like the tract home described in the last paragraph, we may reasonably conclude that no remedies will suffice, and the building should be demolished.

Passive ways of controlling temperature extremes inside the house—insulation, ventilation—should have been built into your house,

THE CASE OF THE OVERHEATED ATTIC

Your house creates its own microclimate, and on a summer day the extremes of this climate can be quite surprising. Around on the sunny side, air temperatures soar, while on the shady side the temperatures may remain quite comfortable. The roof turned perpendicular to the sun is an efficient solar panel, harvesting the sun's rays and turning them into thermal energy—heat. Too bad it isn't covered with solar panels to harvest all those energetic photons! Roofs oriented south and west are particularly productive, because they harvest the most heat in the afternoon when it is already hot. Many American homes capture and store this unwanted heat in their attics, and it is not unusual for temperatures in that dark, still space to exceed 160° Fahrenheit. Because of convection—heavy cold air falls, displacing lighter hot air—the ceiling temperature just below the attic will lag behind, but the whole ceiling is apt to act like a radiator, transferring heat into the living space. If your house is air conditioned, this extra heat makes the refrigeration system work harder, and increases your utility bill. If you rely on nature's air conditioning, you may be uncomfortably warm.

Conventional insulation above the ceiling helps in the winter, when heat is meant to be kept inside, but is less helpful in the summer, because conduction—heat traveling through a solid like the ceiling—works well in any direction, unlike convection, which works best when moving cold down and heat up. Eventually, the attic's heat penetrates the insulation and warms the living space below.

The simplest, least costly remedy for this problem is an attic fan.

Properly installed, this fan will push the hot air out of the attic, so that cooler air can come in and replace it. If this "make-up air" is pulled from underneath the eaves on the shady side, for example, it is quite possible to lower the attic temperature below the sunny-side air temperature despite the ongoing harvest of inbound solar energy by the roof.

Designers try to avoid active solutions like fans and air conditioning because they consume energy, but the overheated attic presents an opportunity to convert the problem into its own solution, by placing a small photovoltaic module (PV) in the sunniest spot on the roof and using its output to drive the attic fan! Our objections to active systems usually center on the complicated control equipment, probes, relays, thermostats, and logic that tell the business end of a system what to do and when, but the beauty of a PV-direct attic fan system is that the energy source is perfectly linked to the need: The attic overheats when the sun shines most brightly.

By harnessing sunlight to spin the fan, we use the source of the problem to generate its own solution.

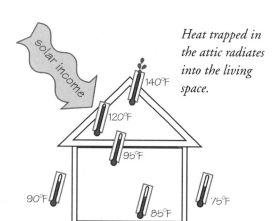

Heat trapped in the attic radiates into the living space.

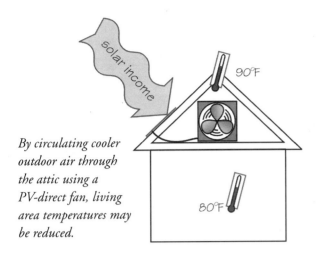

By circulating cooler outdoor air through the attic using a PV-direct fan, living area temperatures may be reduced.

Sticker shock is a real danger when planning a system like this; too often, we forget that the operating cost of many types of equipment is often much higher than the original equipment cost. Photovoltaic modules are made of the same stuff as transistors and computer chips, a highly refined and treated form of silicon. Think of a PV as a bit of blue metal, which it is, and the price may seem staggering; think of it as a nearly perpetual source of electricity, and to many people it seems nothing less than miraculous. A discount-store house-current fan will be cheap, but it will use considerable electricity over its life, while the PV-direct fan works for free. An AC attic fan project will include the cost of extending a house-current circuit into the attic and connecting a thermostat or other controller to make sure the fan only runs when it should. An accurate reckoning suggests that the simplicity of the self-contained system, a PV module connected directly to the fan, may be a bargain over the life of the system.

Using a low-voltage DC fan adds two other features that are impossible to evaluate from a monetary perspective but that should be considered. Low-voltage electricity is inherently

much safer than the AC (alternating current) we generally use in our houses, so you can safely do this simple wiring job yourself if you'd like. Controls for low-voltage DC fans and motors are also simpler and more reliable, and so if you would like to add a thermostat or a switch to control the direction the fan spins, you can do so cheaply and with no fear of causing a fire or damaging your new equipment.

An attic fan is a perfect introductory project for the do-it-yourselfer who wants to learn about renewable energy. Summer is AGES away, you say? Well, maybe, but it sure is nicer to work in the attic when the sun is less ferocious. The old pro says, do it NOW.

The resource list at the end of the book includes sources for attic fan kits.

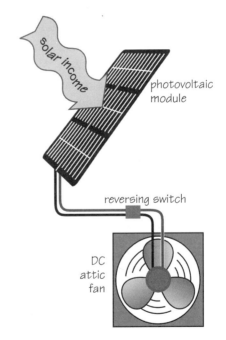

When sun shines on the PV, the fan spins. By reversing the current with a switch, the direction of air circulation can be reversed.

but an entire generation of commodity housing relies instead on costly heating and cooling equipment. Adding passive measures later may help improve performance and cut costs, but may not be wholly successful. Attic heat buildup does not yield easily to conventional methods of insulation. Without resorting to active, costly measures such as air conditioning, heat must be kept entirely outside to preserve comfortable space inside the home's envelope. In Florida, experiments have shown that a bright-white-painted roof cuts attic temperatures by 25% over a conventional white pebble roof. A radiant barrier, made of construction-grade aluminum foil hung just below the rafters, can turn some of the radiant heat around, but only after the heat is inside the perimeter, and by then the struggle is already partly lost. Many classic old buildings, constructed before this century's love affair with the kilowatt began, successfully employ shading, thermal mass, and ventilation to maintain a comfortable interior climate through the hottest days.

If you live in a climate with cool nights, using a "whole-house fan" may make sense. By closing the house tightly during the heat of the day while controlling heat infiltration aggressively using reflective roofing, overhangs, deciduous plantings, reflective window coverings or shutters, and, as a last resort in truly unpleasant climes, air conditioning, the interior can be kept livable. During the cooler evenings, the whole-house fan comes on to replace the air heated during the day with fresh, cool air from outside. If your house contains thermal mass—masonry floors or walls, masses of water, or other materials with high specific heat—these masses are recharged with the cooler evening air and act as batteries for storing the cool until it is needed the next day. Thermal mass can have its bad side: If interior temperature gets out of hand during the day, thermal masses store daytime heat instead and can make nighttime unbearable, so managing heat by opening and closing doors and windows becomes very important.

One final note: Advocates of healthy houses and workplaces assert that we define our "comfort zone" too narrowly for our own good. "Air conditioning flu" is a real hazard for those of us who must go in and out between superheated urban reality and supercooled shops, offices, and homes. Even the best air conditioning systems drastically reduce the humidity in the air, which puts added stress on our respiratory systems. As energy costs continue to spiral ever upward and fossil fuels become scarcer, it makes sense for us to broaden our notions of comfort. We survive quite nicely at 85°F if the air is moving enough for our personal evaporative cooling systems to operate.

An American Fetish: Hoarding Hot Water

Installing a solar hot water system is probably the best resource- and money-saving action you can take. Choose one that is appropriate for your climate, and it will pay for itself in five years or less. But our modern-day hot water behavior is quite strange, and bears close examination.

Let me ask some personal questions. Are you a hot water hoarder? Do you have, somewhere in your abode, a tank full of hot water patiently awaiting your call? Most Americans do; most everyone in the rest of the world does not. It's one of our strangest cultural quirks.

How long does it take from the time you turn on the hot water tap until water at its hottest comes out? One energy theorist pointed out that having a closetful of hot water is like leaving the car running in case you want to take a ride.

There are several remedies for this cultural aberration of ours. The simplest and most surprising, I find after years of teaching this subject, is to change our habits. Here's another personal question: When you go to the sink or bathroom lavatory and turn on the tap for a quick rinse, which tap do you turn? Most of us turn on the hot, and cheerfully wash our hands in the cold water that comes out. A stubborn few of us wait for the hot water; I have a bit more to say about them in a moment. Friends, the water from the cold tap is the same temperature: Discipline yourselves to turn that one instead. You do not suppose, I am sure, that the water heater, knowing your intentions, simply neglects to heat the water that will be orphaned in the walls or under the slab. Generally, by the time you have finished rinsing your hands, the hot water will have reached the tap, and when you turn it off, all its costly heat radiates into the indoor space so you or another member of your family can be treated to another cold rinse later in the day. Not even perfect insulation will fix this problem; self-discipline and energy awareness are the only remedies. I urge you to observe your behavior, and accept this small exercise: Get your cold water from the cold faucet!

In the most extreme (one might even be tempted to say obsessive) manifestation of energy abuse, many luxury homes employ a supremely wasteful practice called "circulating hot water," in which the house's entire hot water system, albeit usually well insulated, is connected to a pump which keeps hot water instantly available at any hot tap. Only the most stubborn, impatient, and I might say, greediest of American energy hogs would insist on hot water "in waiting." If you have circulating hot water, please turn off the pump—the system will still work fine, and your water heating bill will cut itself by three-quarters.

Another simple remedy is to make sure that your water heater is set to a reasonable temperature. Because we have agreed to give up the sporadic hot rinse, our heater should be set to deliver a large amount of perfectly tempered water; for most of us, that would be at or below 120°F. Hot water heaters are commonly delivered set to 140°F or hotter, so, since each degree Fahrenheit costs about a dollar a month, you can save quite a bit by adjusting the temperature so hot water out of the tap never scalds. If you have not already done this, please do it right now: Patiently adjust your hot water thermostat so that at its hottest, tap water is as hot as you

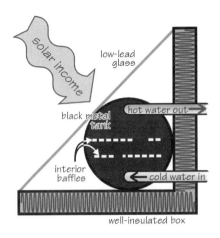

A "breadbox" or batch solar water heater consists of a black tank in a well-insulated box. Cold water enters the tank near its lowest point and is removed from near the top. Baffles keep the incoming cold water from diluting water already heated by the sun. At night, the insulated box keeps the tank from losing its heat. The tank holds a "batch" of water, usually between 40 and 80 gallons. Because batch hot water heaters are passive and plumbed directly into the house system, they require no pumps or controls. Because they hold a large volume of water and typically a "pillow" of air, freezing is seldom a problem except in hard freezes experienced in the far northern tier of states.

48

ever want it, but no hotter. One hundred twenty degrees Fahrenheit is a good maximum. One hundred thirty degrees Fahrenheit, only ten degrees hotter (and hot enough to scald) costs up to 25% more.

The next remedy, as already pointed out, is to install solar water heating. By using solar energy to heat water, we replace a highly significant energy load with a free source, and we are put in touch in a very important way with the day-to-day state of our planet. If we rely wholly on the sun for our hot water, before every shower we will devote just an instant's thought to our home's recent sun and water history. Has it been sunny? If so, there will be an abundance of hot water, but if not, it would be wiser to conserve. Has there been much hot water use since the last sunny period? If so, we would be wise to use sparingly or defer our water use, but if this is impossible, we must at least exercise our normal, water-conscious discipline. Independent families do not quarrel about this; they seem to take care of each other. An occasional tepid shower helps refine our estimating skills.

The solar industry has been severely criticized because of poorly installed water-heating equipment hustled into service during the OPEC oil crisis in order to take advantage of tax credits then being offered by a frightened government. The technology is much more mature now, as designers have learned many lessons about the need to plan for repairs necessitated by the corrosive qualities of hot water and the devastating effects of freezing. Opportunistic developers and other get-rich-quickers who flocked to the government trough have left town, and solar water-heater installers still in business are likely to be professional and competent, although you should still check references.

Most solar hot water systems are assisted by conventional water heaters. As the graph shows, my well water runs within a degree or two of 55°F, but I like my hot water at about 115°F, which means I must somehow get about 60°F of heat gain. The solar fraction, in winter, may be as low as 25%, because my solar collector supplies 70°F water even in a storm, and provides an average 15°F gain during our coldest month, February. On a typical Caspar summer day, the heating coils never see an electron, because the sun provides our total requirement of domestic hot water. I know this because I like to measure results, and have added three thermometers to my system; now I can watch the water temperature rise on a sunny day, monitor hot water availability, and gloat over my energy savings.

Although such systems still hoard hot water, in sunny times their electricity consumption is reduced to nothing. Some systems, which

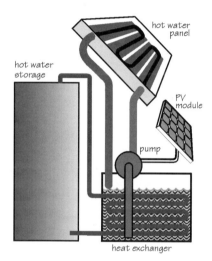

Because the hot water panel in this system only holds fluid when the sun shines on the PV module and the pump runs, this is a reasonably freeze-proof system. Heat is collected in the panel, and raises the temperature of the heat exchanger, where it is transferred to incoming cold water. This is not a passive system, but it is PV-direct, meaning it operates only when the sun shines, which coincides with the time the panel will heat the transfer fluid. In cold climates, the transfer fluid often contains glycol-based antifreeze.

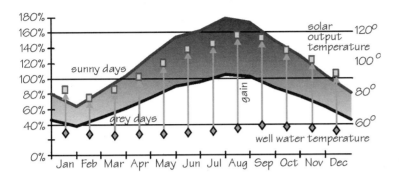

In Caspar, the fraction of hot water energy provided by the sun shining on our pair of batch water heaters varies with the seasons, from 100% in August to 40% in February. These heaters paid for themselves in four years.

have little capacity for storing hot water in the collectors themselves, circulate heated water by thermosiphon (or, more commonly, by means of a pump) to a heat exchanger or storage tank, which partially justifies the large tankful of hot water we Americans like to hoard. In more gentle climates, a "breadbox" water heater like ours in Caspar, assisted by a propane- or natural-gas-fired "on-demand" water heater fitted with a sensor that suits the flame to the temperature of the incoming solar-preheated water, provides unlimited hot water while consuming the minimum amount of fuel possible. This is the most energy-efficient remedy. In especially cold climates, elaborate heat-exchanger systems are necessary to transfer and retain heat while avoiding freezing, but even so these will pay back their costs within seven years or less.

By putting a timer on the fossil-fueled backup we can avoid electrically heating water that will not be used until a little later in the day after the sun could do the job. For example, morning showerers cause the hot water they use to be replaced by cold water, which, being heavier, stays on the bottom of the water heater; the thermostat senses its presence and starts expensively heating it even though we may not need more hot water until evening, by which time the sun would have heated it for free. Most conventional water heaters recover in an hour or less, so allowing an hour's lead time is adequate. My system turns itself on once a day, late in the afternoon at the end of the solar day so that, if there has been sun, the water heater's thermostat allows it to use just enough energy to bring it up to my preferred temperature. Morning ablutions are more expensive because the solar-heated supply is at its lowest ebb then, and grid energy will almost always be needed for hot water, so it is reasonable to consider moving the habitual shower to the evening.

Even without the help of the sun, adjusting the water heater's performance to suit our particular life, so that it heats water no hotter than the

On a typical Caspar summer day, the heating coils never see an electron, because the sun provides our total requirement of domestic hot water.

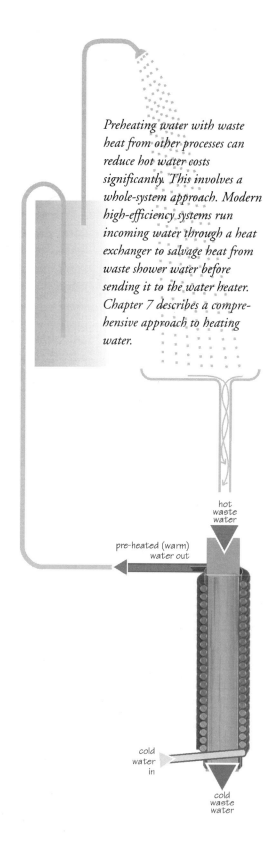

Preheating water with waste heat from other processes can reduce hot water costs significantly. This involves a whole-system approach. Modern high-efficiency systems run incoming water through a heat exchanger to salvage heat from waste shower water before sending it to the water heater. Chapter 7 describes a comprehensive approach to heating water.

hot
waste
water

pre-heated (warm)
water out

cold
water
in

cold
waste
water

temperature at which we will use it, and only during times we habitually require it, can save as much as a third of our present water heating bill.

The final option, if we are unable for technical, aesthetic, or historic reasons to install solar hot water, is to disconnect the energy-guzzling hot water storage-tank heater and install local hot water sources like the ones used everywhere else in the world where hot water is served. These devices, called demand or instantaneous water heaters, heat water only when it flows through them. By placing an instantaneous water heater close to the point of use, within a few feet of the sink or shower, minimal water will be wasted while waiting for the hot stuff. Even if the new heater is installed at a distance from the kitchen sink, shower, and bathroom sink, for instance in the closet or basement where the old hoarder used to be and where plumbing already exists, hot-water energy expenses will decrease because keeping a standby quantity of hot water is costly.

The Outrageous Cost of Refrigeration

Why does a home refrigerator cost so much to operate? Familiar answer: The manufacturer pays to build it, but you pay to run it. If you, like the average appliance shopper, base your purchasing decision primarily on sticker price and features, you deserve what you purchase. In private, refrigeration makers are baffled and cynical about the decisions made by what they call the "American housewife," who seems not to notice the cost of the electricity gobbled by the behemoths they build, but demands ice cubes, water, and chilled beverages coming out of little ports on the door. Laughing, they call this "terminal featuritis." If you are working toward energy independence, you buy a small, quiet, efficient cold box. On-the-grid, the true costs and benefits are as usual disguised: Buy the cheap model, which costs much less, and pay more for electricity over time, or buy the more expensive, more efficient unit and within seven or eight years, comparing equipment cost (original purchase price) plus the operating cost (price paid for the electricity consumed), you will break even and be a resource hero. If you keep the refrigerator for twenty years, your total lifetime cost will be half that of the inefficient unit.

Refrigerators and freezers work better when they are full, and this is mostly true, but is not a justification for a large appliance. If too full, the circulation required to keep the whole box cool is defeated, or the fan runs too much. A *chest* freezer works best if it is packed full, except that it may be slow to reach full cold if a large amount of new food is added at one time; upright freezers lose less cold air every time the door is opened if they are full. But efficiency is not served by buying a unit too big, and then putting things in it that do not require refrigeration—see the Refrigerator Challenge in the sidebar, and then read about Berta's Monster in chapter 5. It is much cheaper to buy a small efficient unit and use refrigeration intelligently. We get along nicely with an 11-cubic-foot super-efficient unit, and all our brooms, mops, and other long-handled riding equipment fit nicely into the space saved.

In the late 1960s the lifetime cost of operating a new refrigerator was expected to be about as much as the unit itself originally cost, a reasonable tax to pay for twenty years of cold. Compared to features, energy efficiency was unknown. In the late 1970s, we started buying refrigerators with big yellow stickers on their

REFRIGERATOR COWBOYS

Ask a dozen engineers why refrigerators are built with their compressors and heat exchangers underneath their cold chambers, and get a dozen answers, but not one of them will say, "Because it works better that way." Because it doesn't. Some say it moves the unit's weight lower, thus making it more stable so it is less liable to fall over on people. Maybe it makes the unit easier to ship. It probably looks more pleasing to the householder's eye. But it makes the refrigerator inefficient.

Conventional refrigerators work by compressing a normally gaseous compound, a chlorinated flourocarbon (CFC), until it becomes liquid, a process that uses electricity and gives off heat. The liquid refrigerant is kept compressed in the radiators, where it dumps the heat of compression into the room, then is allowed to expand in coils inside the refrigerated chamber. Expansion requires heat, and as the heat-absorbing coils in which the refrigerant expands draw heat out of the chamber, the air in that enclosed and insulated space is cooled. The vapor circulates back to the compressor, where it is compressed again, using more electricity and giving off more heat, and around and around the cycle goes. Remember that heat rises, so if all this compressing and radiating takes place below or beside the cold part, the heat will interfere with the cooling, making the whole operation less efficient, particularly if the cold part is imperfectly insulated, as it generally is. Add a few more tricks, like a defrost system that periodically heats things up in the cold part, heat tape (also in the cold part) to keep the door seals free of frost, an incandescent lightbulb, and other filigrees that sacrifice energy for convenience and easy production, and you have an energy hog that ranks just below the hot water hoarder for energy wastefulness.

THE GREAT AMERICAN REFRIGERATOR CHALLENGE

An iconoclastic colleague at one of my seminars challenged our adult students to take "The Great American Refrigerator Challenge":

Get two large cardboard boxes. Open the refrigerator. (Might as well turn it off, since this may take a while.)

Phase one: Take out everything that has been napping undisturbed in the refrigerator for some arbitrary time (shall we say three months?) and put it in one box. Do the same for the freezer, but perhaps we can agree to a longer threshold, possibly six months. The rules of the Challenge do not provide for leaving time capsules longer than a year in either case.

Phase two: Look quickly in the freezer to see if anything there does not require refrigeration, but there shouldn't be much. You can close the freezer now. While it was open, could you feel the cold pouring out of the freezer door onto your toes? That is why intelligently designed freezers are chests, not upright. Now, remove from the refrigerator everything that requires no refrigeration. If the label says "Refrigerate after opening" assume the makers know whereof they speak, but otherwise, into the second box it goes. If not yet opened and not marked "must be refrigerated," it too goes into the second box.

My friend reports that in the majority of Challenges he has administered, more than half of the contents of the refrigerator are now in the boxes. Why should this be? Could it be that we grossly overestimate our need for refrigeration? Could it be that our refrigerators are so big, we have no other place for most of these foodstuffs?

doors; we began slowly awakening to energy consciousness. Today's feature-laden refrigerator humming away happily in your kitchen can be expected to consume energy costing four to eight times the purchase price before the unit is replaced. Despite the yellow stickers, consumers continue to buy without regard for operating costs, and refrigerator designers persist in adding conveniences. Ice boxes were fairly efficient, as were the first refrigerators, but during the decades following their introduction to the consumer marketplace, efficiency decreased spectacularly. Refrigerators became the household's showpiece appliance—and an energy and ecological horror. For example, although the refrigerator's primary mission is keeping things cold, the door seals and internal circulation patterns of most standard refrigerators are so poorly designed that they must be fitted with heaters to keep coils and doors frost-free. The waste heat dumped into the kitchen, already the steamiest room in the house, further burdens domestic cooling systems, consuming yet more energy. Chlorinated fluorocarbons (CFCs) in the refrigerant and embodied in the insulation of older refrigerators endanger ozone and therefore decrease our ability to enjoy the sun. Research is not yet complete on the new coolants called HCFCs, but if our track record is any guide, these, too, will be found to have some detrimental effect on the environment. There is only one remedy for such appliances: Replace them with efficient models which use half the energy.

After the foregoing advice was published in the 1993 edition of this book, I was accosted by an unhappy pair of homeowners who claimed their electric bill had gone UP when they brought in the expensive new super-efficient fridge. "A lot?" I asked. No, the couple admitted, scratching their heads. Puzzled, I asked several questions about brand, open-door food-gazing policies, and other possible causes. Finally, I asked, "What did you do with the old refrigerator?" "Oh, we moved it to the garage for soft drinks and beer." Moral: If you want to enjoy the reduced cost of a new, efficient refrigerator, you'll have to unplug the old one!

Another word about replacing old, CFC-filled refrigerators. Environmentalists worry about the latency of the past's bad decisions. Because a single CFC molecule can wipe out tens of thousands of ozone molecules, the installed base of millions of old CFC-filled refrigerators in use and in junkyards means that even though the manufacture of these chemicals has been banned, releases of CFCs will continue until the last old refrigerator vomits its poisonous blood into the atmosphere. If you have one of these horrors, please ensure that it goes directly into the hands of a responsible licensed appliance recycler so the CFCs may be removed and destroyed properly.

Many old-timers remember the spring box, the icebox, the root cellar, and the unheated pantry. Refrigerators are another one-size-fits-all energy-intensive solution that suits manufacturers and merchandisers, but serves people poorly. Several tinkerers have come up with homemade schemes employing thermal mass, cold extracted from the north side of the house, and other strategies. Michael Reynolds, a man who makes friends with what he needs, then asks for some, playfully designed an unconventional thermal mass refrigerator. The mass is located in the chamber's concrete walls, in the concrete freezer and box shelves, but mostly in the refrigerant stored in cases at the bottom where the coolness pools. During the night in cold months, which in the mountains of New Mexico go from January to December, when outside temperature is at or below the target temperature for the freezer, the operator opens the box's skylight. In places where nighttime temperatures seldom fall below 40°F, the device could be fitted with a small refrigeration unit and thermostat to keep the temperature low. As already noted, this Reynolds design employs a static refrigerant in the form of cases of beer or soft drinks instead of the toxic CFC used in active refrigerators. So far the primary development problem has been refrigerant loss to thirsty lab colleagues. (More will be heard from Michael Reynolds in chapter 13.)

Standard refrigerators have been inefficient for decades. The present generation does away with some of the more wasteful tricks and improves the seals and insulation. Many large utilities have run campaigns to get the really inefficient refrigerators, those made before 1980, out of

Mike Reynolds's thermal mass refrigerator is made out of concrete. Cool stored in the case of "refrigerant" is recharged by opening the trap door at night.

54

HOMEWORK: OUTGROWING DEPENDENCE

But Where's the Stick? Refrigerator Breakeven

In the early 1990s, several electric utilities offered the refrigeration industry a Golden Carrot: one million dollars to the company that could build a really efficient refrigerator. Sun Frost and others already manufacturing units that met the performance criteria were unable to compete because they did not make "enough" refrigerators in the qualifying year. The problem is, building a really efficient refrigerator costs quite a bit more than a feature-laden mass-market box, and American buyers are unwilling to pay in advance for efficiency, so Sun Frost produces for a coterie of pioneers with limited sun and renewable energy sources who simply must have the most efficient appliances. Now that efficiency has become a mainstream motivator, the big manufacturers beat Sun Frost to the breakeven point by a decade.

The comparison of refrigerators is straightforward. If two refrigerators, an efficient (Effy if you know her well) and the cheapest-to-buy (Cheapo), are tested side by side, we will require answers to four questions: (1) Do they work equally well? (2) Do they last equally long? (3) How much do they cost to buy? (4) How much do they cost to run?

Suppose we establish that they work equally well and last equally long. (That's not really true; the pricier efficient models are usually better made.) The answer to our third question is, Effy now costs nearly twice as much to buy as the Cheapo unit. This difference is enough to

Cost of refrigerator plus electricity over a twenty-year lifetime for three different 19-cubic-foot refrigerator-freezer units. In each case, the lower line represents "constant cost" electricity at today's rates in Caspar, and the upper curve represents the costs with 10% compound rate increases.

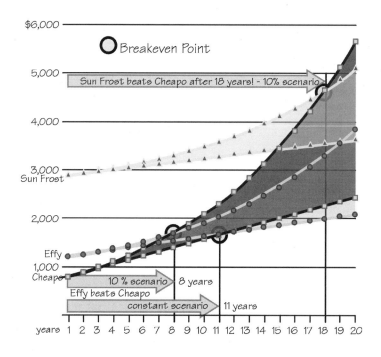

stop most buyers from considering further; sticker shock dulls their wits, and another Cheapo unit is sold. But wait! Unit Cheapo costs almost twice as much to operate. (That's progress. In 1993, there was no mass-manufactured Effy, and Sun Frosts were six times as efficient as the best mainstream model.) Said another way, if we graph the whole package cost (refrigerator price plus electricity used to date) over time, Effy's total falls below Cheapo's somewhere between the eighth and eleventh year of service, depending on actual electricity costs. If our calculations assume that energy costs keep going up at about 10% per year, compounded annually –which accounts for recent increases fairly accurately—the payback period, the time it takes for the whole Effy package cost to be less than Cheapo's, is about eight years, or about 40% of the appliance's presumed life. After that, the efficient unit is substantially cheaper to own.

We bumped up against this issue in our earlier discussion of lighting, and we will encounter it again. Pay a lot now, or pay even more later. Better equipment usually costs more to buy and less to operate. By paying more now, we pay less in the long term and have the satisfaction of knowing that we have made a positive decision for the planet. Do consumers pick up the big ticket? In 1993's first edition, I said "Not enough of us—at least, not yet" but now the consumers have waved the biggest stick of all, their buying power, and the manufacturers have responded with affordable efficiency. Well done, consumers!

service by offering to get rid of them in favor of newer models, even subsidizing the purchase. Leave it to high-end consumers to subvert progress: The Sub-Zero refrigerator, darling of kitchen designers, is to the world of conscious refrigeration what the urban assault vehicle is to sensible transportation. Efficiency has not yet invaded the pricey realm of custom kitchens, but it has slightly reformed the mass-production boxes most people buy, although they are still not efficient enough for independent homeowners. We know that even these new models could be improved dramatically, because a few small manufacturers build units that are six times as efficient using the same technology, units that (guess what!) have their compressors above the cold part, have well-designed doors, a minimum of energy-wasting features, and are extremely well insulated.

For a number of years, independent homesteaders had little choice of refrigerators. A small company in Arcata made the only reasonable option, called the Sun Frost, which used between a seventh and a tenth of the energy used by commercially available refrigerators made by big manufacturers. Sun Frosts accomplished this frugality with precious electrons by clever door seals and solid hinges, well-conceived interior design, the compressor and radiators on top (or even in an adjoining room is you wanted) but above all, massive insulation. Recently, super-efficient European appliances have also appeared on the market.

*Proudly
standing
alone:*

*Catherine,
Eva, and
Lilia's
story*

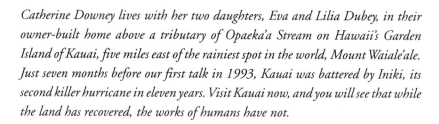

Catherine Downey lives with her two daughters, Eva and Lilia Dubey, in their owner-built home above a tributary of Opaeka'a Stream on Hawaii's Garden Island of Kauai, five miles east of the rainiest spot in the world, Mount Waiale'ale. Just seven months before our first talk in 1993, Kauai was battered by Iniki, its second killer hurricane in eleven years. Visit Kauai now, and you will see that while the land has recovered, the works of humans have not.

Catherine: I changed the brake shoes on my car once, and decided I didn't want to be mechanical. As a naturopathic doctor, I'm a scientist only by default. As a midwife and naturopath, I'm interested in natural living.

We try to live naturally. The bugs bug us, especially the big flying cockroaches we call B-52s. Roach hotels work— *Tora! Tora! bes' kin', bra'!*— and they get their feet stuck. I won't use poisons, except for boric acid traps for the ants. Ants are really bad when it rains. We're trying to get away from using plastic bottles because I just read that Americans are throwing away two and a half million a day.

I believe in midwifery, in its spiritual basis. When white man and red man met, the patriarchy was supposed to meet the matriarchy and create a more intuitive culture, but somehow we blew it . . . and that's why we still worship a god up in the sky. In my work, I bring together modern things, IVs and medications, and ancient midwife traditions. I have a doppler that I use to hear the baby's heart; I imagine that ancient midwives would have loved to have that, to be able to say to the mother, "Roll over on the other side; you're pinching the cord."

For me, this consciousness started when I became a vegetarian when I was twenty-one. (I'm not strict; I'll eat fish sometimes, and a little chicken. . . .)

Eva: You don't eat much chicken. But Lili eats hamburgers!

Catherine: Yes, she does.

To the best of my ability and knowledge, I built this house as well as I could. I moved very little earth, and the lumber came from farmed trees. The house is built of untreated lumber, cement, and steel. If I did it over, I would use more cement. It was important to build on marginal land, and let the good land alone for farming. We have the technology to build on hills, and we should.

I left the trees—Java plum, Formosan koa, volunteer bananas—by the stream. There are probably few native species, but the birds love the plums. My neighbors cleared all their trees. They say to me, "When you going to clear your land, make it useful?" I think I will leave it. Now, I get all the birds.

Catherine and Eva checking their batteries.

Except for the koa, which broke in half, Iniki (the hurricane) left nothing but sticks. They looked like they were dead. "You just as well bulldoze it now, it's dead, and take to the dump," the neighbors said, but look at it now, in less than a year, it's grown back.

During the storm, we stayed downstairs, stood on the porch and watched. It was quieter there. At first the storm came from the east and broke a few windows upstairs. I think the fact that there's so much screen, and the wind could blow in and out again, saved the house. It sounded like a big train coming, and the trees in the distance were bending over, going *Eeyah*! I could tell from that palm which way the wind was blowing. It shifted more south, and then at the end came from the west. Pieces of neighbors' roofs were flying over and settling into our valley where the wind was not so strong. I remember seeing birds blown in by the gusts and taking refuge down by the stream; I remember Eva saying, "Mom, there are still birds!" Right across the stream, that house moved fifteen feet off its foundation with the people still in it. Whew! We just had some water damage, and lots of leaves in the house.

After Iniki they wouldn't let us scavenge at the dump. I needed a second toilet, and there were people whose houses had been blown apart—they'd

It was important to build on marginal land, and let the good land alone for farming. We have the technology to build on hills, and we should.

thrown everything into dumpsters, and there were lots of good toilets, but I had to buy one new. We had just started recycling on Kauai. Before, they said there was no way to do it. I say, you live on an island, you better recycle! The island generated five years' worth of garbage in that one day. They could have called it the Garbage Island.

I've lived on Kauai for thirteen years now, and both my daughters were born here. When I bought this land, close to where my ex-husband lives, both he and my boyfriend were upset, but I did it for the girls. This way, their dad is close; he's got the piano, so Eva can go up there to practice. They each have one bike, one boogie board, and they get to have their dad. It works out well.

We lived in a yurt on the land while we built this house—I loved it! The electric company charged three thousand dollars to install our electricity, and they made me give easements to my two neighbors before they would do it. I was pissed, because I wanted to spend that money on solar. When the electricity went down after Iniki, the insurance company had a loss-of-use clause that let people buy generators so they could stay living in their houses, and there were all these noisy gas generators around. I told them I wanted to buy a "solar generator," and they said okay, just show us the receipts. That's how I got my system.

I built this house as the owner-builder, but I had a lot of help; I didn't pound very many nails. We had the usual inspections, foundation, framing. In Hawaii, you have to have a licensed plumber and electrician do those parts. We never mentioned to the inspector the house was on solar, but we didn't hide it either, and of course he could see the modules, and he didn't say a word. The AC smoke detector required by

the building code didn't like the electricity out of the inverter, so I disconnected it and put in one with a battery after he left.

Now, I use the grid as my backup, and we run everything on electricity from the four modules on the roof. I watch the meter, and when it says my batteries are 90% full, and I know it's going to be sunny, I do a wash. Sometimes, but seldom, when somebody turns on something like the vacuum in the morning when the batteries are below 85%, I can see by the meter lights that we're using the grid. I don't know how much grid power we use, but I can tell you it isn't much.

Eva: Because Mom's nagging me all the time to turn off the lights!

Catherine: We're adjusting to living within the amount we produce. I don't know too much about how these things work, so Ross, the installer, wrote everything inside the box, and gave me lessons.

What got me interested in solar power? I thought it would be so neat to refrigerate with the sun. I finally got my Sun Frost, and we had to remodel the kitchen because it's so big. I want to add at least eight more modules and four more batteries; if I add more batteries, I'll have to build another battery box. Next, I'm interested in solar water heating. I have an instantaneous water heater, but I'm paying too much for gas. And an electric car . . . I'm committed to being petroleum-free by the year 2000.

Hawaii is one of the few states that licenses naturopathic doctors. There are two others on the island. My specialty is women and children, and I'm the only one that does home births. I've done, maybe, four hundred, and the one this morning was so beautiful! But the more you do, the more you see problems, and appreciate that it's life and death. Here's what I did so far today:

I was at that birth all night, and the baby was born at 4:33 this morning. At nine I was in a canoe race; we came in third overall and won our division—the masters, women over thirty-five. I guess we're masters because we've mastered the art of being athletes. We're not as strong as the twenty-year-olds, but we're better paddlers. Look at this silver medal from the 1991 statewide Hawaiian Canoe Racing Association competition . . . Now I'm paddling with the Kaiola club down in Nawili'wili. We have a smart coach, and this year we'll win! ⟿

What Part of OFF Don't You Understand?

Phantom loads are small but constant energy drains—clocks, transformers, and other miscellaneous energy guzzlers. Felicia Cowden calls these phantom loads the energy criminals of the small appliance world. (You will find Felicia and Charlie Cowden's story in chapter 4.) If you have office or entertainment equipment plugged directly into your wall plugs, you may be wasting quite a lot of power even when the equipment is not in use. Instant-on televisions and other gear with remote controls are always on, even when they appear to be powered off, because they are always poised for the call of the remote control unit. The little power cubes that power calculators, clocks, rechargers, and all manner of other small appliances are also horribly wasteful. If your goal is energy independence, you will not buy this kind of device; if you must, you will plug them into plug strips or switched outlets so that they may be completely turned off when not in use. (This strategy does not work well with clocks. Rechargeable battery-powered clocks are recommended.)

Small electronic devices that require transformers are themselves often quite efficient, especially because they use low-voltage current. The real thief is the little power cube, the transformer: This cheaply made inductive energy-hog wastes as much as 80% of the energy you feed it. "Aw, but they're cute little piglets and don't eat much," you might say, but you would be surprised. People who serve as their own power companies tell me that without vigilance, as much as a quarter of the energy in a well-equipped home with a home office can go to power thieves like these.

Phantom loads are *plug load non grata* in any energy self-sufficient house, but we benefit by considering them just as offensive on-the-grid, since the only difference is that the offense has its impact in somebody else's backyard. Beyond the big three of potential energy savings discussed above (space conditioning, water heating, and refrigeration) and lighting (discussed in chapter 4), eliminating unneeded phantom loads, reducing plug loads, and buying small appliances that are designed for efficiency can save the most domestic energy.

To get a useful grip on what your phantom loads add up to, try the Unplug Your House Challenge. Choose a comfortable, bright, quiet day. Go around and unplug or throw the breakers on your major appliances and loads but be sure to leave all your outlets and lighting circuits "hot." (Children, warn your parents first.) Turn off all obvious loads like lights, heaters, fans. Leave microwave ovens, VCRs, and other things that are theoretically "off" plugged into live circuits, along with plug-in alarm clocks, clock radios, and anything else that loses its grip on time if unplugged. Now, go to the meter and record your starting point. (Your electric company should be happy to teach you how to read your meter.) Run silent for an hour, again note the reading, and calculate the difference: the kilowatt-hours consumed, electricity stolen from you every hour of every day by your household energy criminals. Think of it as a voluntary tax you pay to support convenience over efficiency. It will undoubtedly make you feel good to know that one trophy home on Northshore Kauai holds the phantom load record of 1.5 kilowatt-hours per hour.

Off-the-grid purchasers of large appliances must always be aware of hidden energy thieves, but anyone seeking to live an energy-aware life can learn from their choices.

Obviously, an electric stove is out of the question—heating anything with electricity is out of bounds for all but those with a year-round excess of hydro-power. Yet you might think that a propane stove with an electric clock would be fine. Not necessarily: Make sure that the stove of your dreams will start its burners or ovens with some means other than glowplugs, which are exceedingly wasteful. A typical oven glowplug consumes 30 watts all the time the oven is in use. In some jurisdictions, it is illegal to sell a stove without glowplugs, purportedly for safety reasons. Piezo-electric ignitors, which require house current, consume a tiny amount of electricity.

The road to town: Noel Perrin's story

No discussion of energy dependence can neglect our love for the motor vehicle. Compared to the burden imposed on the planet by our vehicular misbehavior, the energy we waste at home is small. Internal combustion engines are still one of the dirtiest power sources ever devised, despite our recent (and halfhearted) efforts to clean them up. (Chain saws, lawnmowers, and other two-cycle engines are even worse.) This issue overwhelms most of us. We are infatuated with mobility, surrender to it unquestioningly, and make no effort to cope. A few brave souls, among them Noel Perrin, have tackled the issue head on.

Noel Perrin lives in a Federalist farmhouse near Thetford Center, Vermont. He teaches literature and environmental studies at Dartmouth, where his students and subject matter have led him to think deeply about energy questions, and suit his actions to his conclusions. His experience with solar electric vehicles has prompted him to take the next step: building an off-the-grid home.

We are a restless folk. A rich old woman of my acquaintance wants to be taken for a daily drive in her car—often she's out for two hours. She has a perfect right to like motion. But wouldn't it be better, using the same amount of money, and without environmental damage, to get a nice matched pair of greys, a comfortable rig, and perhaps an Amish driver-keeper to maintain horses and rig, and take her for her daily drive?

We need to get real. It's as John Jerome says in *The Death of the Automobile:* Everybody knew the automatic shift wasted power, so we just had to add more power. Now we must learn that power is not inexhaustible, and is unacceptably costly.

I came to live as I do because I love this old house, and because of an accidental character trait—being frugal. I came in 1959, with no prescience or thought of the environment, to teach English Literature. Living here, and heating with wood, I became aware of a productive closed loop, wherein I was clearing my woods, and saving money, and getting firewood. By Earth Day in 1970, I was selling excess firewood off the back of my pickup in Hanover to collegians interested in practicing seduction by firelight.

The wood harvesting could be cleaner, and will be when a friend develops his portable electric chainsaw system. An unfortunate fact of internal combustion engines is that the smaller they are, the dirtier. We bring the wood in, and perform other farm tasks, with a diesel farm tractor. We may lie slightly more lightly on the earth than a typical middle-class family, but we are still harder on the environment than two or three families in India.

That first Earth Day was another major force in my life. Regular classes at Dartmouth were cancelled in favor of a special curriculum, and I attended a class, "Pollution in Dartmouth's Backyard," which revealed that we were burying low-level radioactive waste on college-owned land. To this day, there is a striking discrepancy between the preachings of the Environmental Studies faculty, and the practices of the college (and I daresay this would be true of any college with such a faculty). Why not follow the advice of the environmentalists on campus? And commence behaving as we know the First World must? Or, if we are not believed, why are there no firings or gaggings? In fact, what we say is believed, but only as dry facts. Even the best education moves us only a little; events, like Three Mile Island, move us a lot.

Does that mean we must have more and better disasters before we commit to a more rational approach to energy? I hope that an abstract understanding of extinction will prove enough. If I'm wrong, I'd rather not know it. My wife asked her fourth- through eighth-graders at Lyndonville to write an essay about "the Earth our Garden" and they all wrote about toxic waste dumps. To be totally environmentally aware as a child would be

I came to live as I do because I love this old house, and because of an accidental character trait—being frugal.

Noel Perrin.

too awful, as was the idea of "mutually assured destructive capability" for children in previous generations.

One must present the information without preaching; most of us prefer not to be preached to. Since television is the source of half the world's information, Population Communications International has been financing sitcoms and soap operas seeded with population and energy awareness messages, which have been produced by *O Globo* in Brazil and syndicated internationally. I think the best operating principle is, whatever works, works.

Most of us would rather *do* something (like burn down a billboard, and there are plenty of small wooden billboards for those who wish to amuse themselves that way) rather than meeting to legislate against billboards. I wouldn't claim that my work, exploring Edward Abbey and his peers with my college seniors, is more powerful than teaching environmental science, but it should be an important way to spread awareness. A student brought me a letter from his father in response to my assignment to conduct a family environmental assessment: "Son, may I remind you how we came to own our fourth car . . . ."

We need not look too far, yet, for things to improve upon. Nor, I think, need we look too far ahead; it works best for me to look forward no more than two or three years.

A few years ago I was taken to task by my students for commuting thirteen miles from home to work, and I decided to correct my error. Pollution seems the most threatening risk, so I decided on an Electric Vehicle (EV) recharged photovoltaically by a grid-backed system. I wrote about finding and retrieving the EV in *Life With An Electric Car*, but the story about the rest of the system may be more

interesting here. I wanted enough pollution-free electricity to recharge my car, and I preferred not to buy twenty or thirty extra batteries besides those in the car. Since I wanted to connect my photovoltaic array to the powerlines, this caused some initial difficulty for the utility, which they were quick to share. They wanted me to carry five million dollars worth of liability insurance, which was more than I wanted to carry, and they required absolute assurance that my little system would shut down in case of grid failure, so that linemen would not be at risk. (I wanted that assurance too, because I know some of the linemen and women.) The utility executives were at last convinced that a standard homeowner's insurance policy would suffice, and that my inverter's fail-safe mechanism would offer protection. As it has turned out, it has been grid power that has caused the problems: Twice, line surges have blown up my inverter.

I have two meters; theirs records the electricity I consume, which I pay for at the standard rate, and mine shows my generation, for which I am paid a lower rate (which is fair; I don't

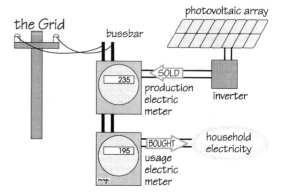

Noel Perrin's dual-meter intertie system. In areas without "net metering," small power producers buy energy at a high rate, and sell it at a lower rate, necessitating two meters.

maintain a distribution system). On a winter day, I may generate between 800 and 900 watts in the four peak sun hours between ten and two. That comes to about 3 kilowatt-hours a day. My EV has a range of forty-five to sixty miles on level ground, and so in summer, in theory, I might be self-sufficient for transportation. You may have noticed that around here it's mostly not level, and mostly not summer.

It can get very cold here, so some winters the pipes freeze. I can close off all of this house but the stove room, where the cat and I stay comfortable, close to the woodstove.

1999 Update from Noel:

My PV array and system has gotten much more reliable. Each of the first three years I had it, the inverter got blown by a line surge, presumably caused by lightning. Each year we added more protection after the inverter was repaired. This was new territory for Leigh Seddon of Solar Works as well as for me; it was his first interconnect.

The third time we obviously got things right. There has been no failure of any kind for four years now. Nor are the panels perceptibly less efficient at age seven. Several times last fall, on a clear day around noon, the system, twenty 75-watt modules on the barn roof, produced 1,530 watts. Considering that it's rated at 1,500, not bad. ⌐

A Cleaner Obsession

Electricity is a clean and well-behaved form of energy that we can harvest ourselves with almost no effort. The forces underlying many other forms of energy are uncertain and obscure, but electricity is elemental in its simplicity. By mastering its measurement and man-

agement, we become entranced by electricity's proximity to some of the most basic properties of matter and life.

In the building trades, domestic electricity is a relative newcomer, and as a result its practices and procedures have not completely settled down yet. Household electricity is not sneaky, taking advantage of the tiniest flaw in the conduit to leak the way water does. Likewise, the materials needed—wire, boxes, outlets, switches, and meters—are straightforward and comprehensible, and the tools—hammer, drill, stripper, nipper, knife—are easily managed. Working with electricity does not require strength, but we have to be smarter than electrons. Even people whose hands are "all thumbs" can put electricity to work for them. Our efforts in improving our homestead electrical systems also remove a weight from the overburdened planet.

Yet people have been encouraged to feel ignorant and fearful about electricity, which has become mysticized and mythologized. The Electrical Priesthood, abetted by the High Priests of Tort and Insurance, have exerted much effort to instill awed respect in all lay users of electricity. Anyone who has ever gotten hold of both sides of a live AC circuit can tell you the respect is deserved. There are even a few who, due to unusual circumstances at the time of their encounter, can no longer tell us anything. This chapter should begin to demystify electricity, a process that the next chapters carry forward, until you will feel yourself joining the electrical cognoscenti. Having looked closely at the ways we use energy, and some of the techniques we can employ to use it more efficiently, we must now ask: If we are ready to harvest electricity ourselves, what will we need to do?

DESCRIBING ELECTRICITY

In all but the simplest of homes, electricity is everywhere, our energy source of choice. There is a wholesomeness and purity about electricity that appeals to people, and its purposefulness and intensity makes us happy to surround ourselves with it. Electricity is invisible, undeniably modern, a pervasive and reliable energy source which reaches into every corner of our homes.

Warning: There is a mystique about electricity promulgated by a group who keep the arcane knowledge to themselves. We call them the High Priests of Electricity, or, more familiarly, electricians. Electricity, particularly the low-voltage kind harvested locally and renewably by the people celebrated in this book, is actually gentle, benign, and eminently logical. There is no need for fear or mystery. Sit still and let me anoint you: Read this, understand, and you become members of the Guild of Electricians, with all the rights and privileges appertaining thereunto.

The basic unit of electricity is the electron. Unfortunately, an electron is only a theoretical beast. Physicists say it has virtually no mass, no permanent location . . . in fact, an electron's whereabouts, Heisenberg assures us, simply cannot be known with certainty. In subatomic theory, an electron represents the smallest persistent packet of negative charge, but the theory further speculates that since the "negative" assignment is arbitrary, it might just as well be positive, a positron, moving in the opposite direction. In which case, I suppose, what follows could be a short treatise on positricity, and you would be studying to become positricians. Enough theory.

Electricity is flowing electrical charge—whether electrons flowing one way or positrons the other, it matters not at all on our practical level. To describe the flow, we require units more our size; we cannot count electrons any more easily than we could describe the flow of water in a pipe by counting water molecules.

We routinely work with electricity flowing in two fashions: direct current (DC), the kind of electricity that comes out of batteries and flows in one direction; or alternating current (AC), our common house current, which flows back and forth through the wire, changing its direction 120 times per second.

Direct current (DC) is easiest to understand. It is frequently compared to water flowing: You open the valve (flip the switch) and water (electricity) flows at a rate (current) determined by the pump (battery or generator) through the pipes (wires) in gallons

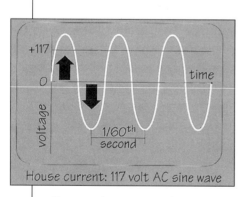

House current: 117 volt AC sine wave

Common house current is alternating current, varying sinusoidally between +170 volts and -170 volts 60 times per second.

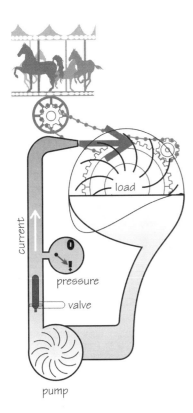

The water-electricity analogy. A pump (the power source, corresponding to a generator) induces a current of water (distribution) through a pipe (wire) and control valve (switch) toward a waterwheel (the load). As it flows in the pipe (wire) its pressure (voltage) and volume (amperage) may be measured. At the load, the energy from flowing water (or electrons) is converted into work, in this case to turn the carousel. Because of conservation of matter, the water (electrons) circulates back to the power source.

per minute (amperes) at a given pressure (volts). A light bulb or other electrical load is like a waterwheel, converting the flow of water (electrons) into work (light, heat, spin . . .).

Note that electrons are not created by sources, but sources induce or push existing electrons into motion. Batteries and outlets seem to produce electricity, but they are simply agents: Batteries store potential energy, and the electrons flowing through an outlet may have been pushed into motion by sources anywhere on the North American continent. Because electrons (and all matter) are always conserved—they cannot be created or made to go away—electricity always flows in a circuit.

Loads consume energy. Anything that uses electricity to perform work, from a battery-powered wristwatch to the winch that lifts a drawbridge, is a load on some energy source while it is in operation. Electrons are not "used up" by loads, but the potential energy of their motion—their voltage—is reduced.

The amount of electricity that sources produce and loads consume is measured in watts; a kilowatt (abbreviated kW) is one thousand watts, and a gigawatt (gW) is a billion watts.

In the United States, the electrical grid supplies house current as a complicated electron flow: alternating current somewhere between 110 to 120 volts. The voltage alternates sinusoidally sixty times per second between two peak voltages, somewhere around plus and minus 167 volts. This means that the voltage at any given instant can be measured as +167 volts, -167 volts, or anywhere in between. The nominal voltage is the root mean square (RMS), or the effective average potential received by a load. On an oscilloscope, a device that depicts electrical changes over time, house current produces the smooth sine wave shown in the diagram. Electricians like to talk about how loads "see" the energy: Most loads "see" the RMS voltage; a few, including early core/coil ballasted compact fluorescents and laser printers, "want to see" a peak voltage up in the mid 160s or they will not work.

By comparing the energy used to generate electricity with the work performed by the device the electricity drives, we may calculate the efficiency of the process. Some tasks can only be accomplished with electrical power—most electronic devices fall into this group—while others are extremely well suited to electrical power, and still others use electricity very inefficiently. For example, heating anything electrically is inefficient, because grid electricity is almost entirely

DESCRIBING
ELECTRICITY,
CONTINUED

heat-produced, by burning natural gas, coal, or biomass, or fissioning radioactive materials; because any conversion is inefficient, it would be more efficient to burn the original fuel where the heat is needed. Therefore, an electric water heater uses about twice as much energy as a natural gas water heater, because the energy is converted twice. We overlook this inefficiency because electrical heating is so convenient; imagine a propane-fired fuser in a copying machine or a nuclear hair dryer!

Electric motors can be made to use either AC or DC, but a DC motor is more efficient because it will consume less energy than an AC motor to produce a given amount of rotational power. Most electronic devices, which neither heat nor whirl fast—clock radios, televisions, computers, anything rechargeable—convert all or part of the alternating current they draw to direct current before passing it along to the parts that work for us, and this conversion is inefficient. Quite often, we can confirm this inefficiency directly by checking the heat emanating from a device—for example a television set—that is not really meant to be a space heater. The best uses of electricity are efficient. A compact fluorescent lightbulb emanates about a quarter of the heat of an incandescent lightbulb of the same brightness, and is, not coincidentally, about four times more efficient.

Alternating current was invented for convenient distribution over distance. (In 1890, at about the same time, Nicola Tesla suggested it would be good to avoid the vibrational frequencies to which humans are sensitive, from about 50 to 150 cycles per second, but you can guess what happened.) Electricity can be generated either as direct or alternating current, depending on the configuration of magnets and coils in the generator. Alternating current's advantage is that electricity packaged that way moves more manageably over great distances. On the home level, except for wasteful inductive loads (heating or transforming), it is not particularly convenient to have alternating current. For most of the best uses of electricity, house current's alternations are a nuisance that must be converted by each appliance to a more suitable form, usually wastefully. That is what those energy criminals, the phantom loads, are doing underneath the sofa. (For more about phantom loads, read Felicia Cowden's story in chapter 4.) Anyone who has tried to get the buzz out of a stereo or phone set can testify to the nuisance of the pervasive sixty-cycle hum. We are also becoming aware of possible health risks associated with constantly bathing ourselves in this electrical pollution, called Electromagnetic Radiation (EMR).

CHAPTER 3

HARVESTING POWER

IMAGINE: SOMETHING AS ESSENTIAL AS ELECTRICITY COMES FROM simply basking in the sun! Gathering energy is one of my most rewarding pursuits. Anyone who delights in using a light-powered calculator, solar flashlight, or stand-alone walkway lamp understands. In less than a decade, photovoltaically powered call boxes have sprung up along the nation's highways. In addition to their life-saving mission, they are prominent reminders that the energy needed to do important work can be harvested anywhere the sun shines.

The best energy path is always the most direct. If light or heat are required, get them with as few transmission and conversion losses as possible—use daylighting and wood heat. If mechanical energy is needed, get as close as possible to the source: The flour mill's waterwheel is a picturesque example. Every time energy changes form or direction, passing through gears or around a pulley, a certain amount is lost to friction, usually converted to unwanted heat and wear. Each of these little "conversions" costs energy, so simpler systems are more efficient. Since the sun provides both light and heat, solutions that use this resource without any intermediaries are most efficient.

As we noted in the preceding chapter, electricity is a remarkably clean and protean energy source, but applying it to our lives has not added unalloyed beauty. Most of our electricity comes unsustainably from the combustion of fossil fuel. The commonest renewable electricity sources

DAVE KNAPP

Off-the-grid and loving it near Fairfield, Iowa! This passive solar straw bale house has a thatched roof made from locally grown long-stem straw. Earthen mixtures are used for wall plasters and floors. The PV system is a collection of used Arco PV modules with a simple 12-volt system and small inverter. The owners catch their water in a small pond and filter it before use. They are debt-free and working toward self-sufficiency.

are photovoltaic (PV), micro-hydro, wind, and biomass. Schemes for exploiting other potential renewable energy sources—geothermal, ocean waves, tides, and even the thermocline (the temperature difference between the ocean's surface and its depths)—are being explored. So far, most of these schemes have been disappointing because of their ecologically sensitive, remote, or unusual locations, because of difficulties in making them large enough to be economical, or because the sources are intermittent or difficult to work with. Equipment at pilot geothermal plants fails much earlier than originally predicted due to corrosion, and the ocean proves to be persistently destructive.

Flat blue, black, and brown objects passively converting sun into energy look better all the time.

Harvesting Sunshine: Photovoltaic Cells

In the 1950s, when the photoelectric effect was explored, it was an expensive laboratory curiosity, yet scientists saw that its promise, a local, lasting, and trouble-free supply of electricity, was very important. The world's finest technology was applied to put this scientific breakthrough to use, with a clear objective: develop a reliable energy source (1) that liberates as many electrons as possible (2) using the cheapest and most abundant materials (3) manufactured and operating in a nonpolluting way. This last consideration was new at the time; humanity was just beginning to admit that its works were having unintended effects. The manufacturing problem for semiconductors, the stuff of photovoltaics, is not trivial because their processing, purification, and assembly uses toxic chemicals and prodigious amounts of energy. Semiconductors produced for electronics are noteworthy principally for their smallness. This parallel development made semiconductor use in photovoltaics affordable, but PVs harvest electricity from sunlight in direct proportion to their size, and so bigger is better. Materials scientists have dramatically decreased the costs and increased efficiency at every step in the production of silicon-based devices, growing purer and larger boules in pressurized ovens, slicing them with ever-thinner diamond saws so less material is lost to the saw kerf, doping them with other materials, and devising robots for assembling them into finished cells. Amorphous and multicrystalline cells are now grown in thin sheets, eliminating the boules and wasteful slicing; this process continues to allow decreases in cost per watt. Other materials besides silicon, and many doping and fabrication strategies, are being evaluated, but there appear to be no profound breakthroughs close at hand. Costs will continue to decrease because of the technological learning curve, as materials and techniques are fine-tuned, and modules are produced in ever-larger volumes. Photovoltaic technology is mature and ready for deployment.

A good way to measure the cost-effectiveness of a solar cell is in dollars per peak watt: that is, the system's whole-life cost divided by the cell's highest energy output. The first PV cells were astonishingly expensive, costing more than $1,000 per peak watt in 1958. The first customers worked, of course, for the space program. Since then, cost has decreased by roughly an order of magnitude (a multiple of ten) every fifteen years. As with many new technologies, the price has dropped so dramatically in such a short period of time that it is impossible to foretell what may

Costs will continue to decrease because of the technological learning curve, as materials and techniques are fine-tuned, and modules are produced in ever-larger volumes.

THE PHOTOVOLTAIC EFFECT

By far the most magical energy source uses the photovoltaic effect to turn light directly into electricity.

Billions of inbound photons (light) enter the cell past a conductive collecting grid. Somewhere in the thin cell, each photon gives up its energy. In the diagram, the leftmost photon sails through the whole cell and encounters the opaque backplane, where it gives up its energy as heat. Photons two and four make it part way through, before whacking one of the electrons in the outermost layer (the valence band) of the silicon atoms with sufficient force to liberate an electron-hole pair into the silicon lattice. Photon three hits an impurity or silicon nucleus and also dissipates as heat.

The liberated electrons, encouraged by a charge gradient created by doping the upper part of the silicon lattice with a rare earth such as phosphorus, wander from silicon atom to silicon atom generally "downhill" (for them, due to the charge gradient) away from the positive doping of the p-type silicon, toward the conductor on the cell's face, where they flow along the circuit to the battery where they are stored. The holes (spaces in the valence band "wanting" an electron) do the same random walk in the opposite direction "downhill" away from the negatively doped n type silicon.

Photons falling on a photovoltaic cell can make electrons flow. In this schematic, the first and third photons strike atoms in the silicon lattice, liberating outer valence-band electrons and corresponding "holes" that begin their random walks down the electrical gradient created by the dopant in the semiconductor. The electrons in this semiconductor migrate toward the collection grid on the cell's surface while the holes move toward the backplane.

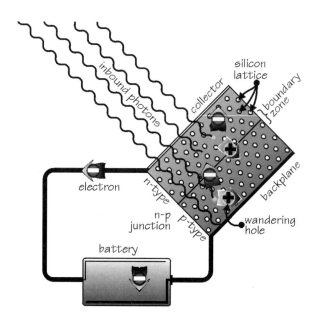

happen in the future. Extrapolating from this trend, it might seem that photovoltaic modules will sell for one dollar per peak watt by the year 2000, but recently the curve has flattened. In five years from 1993 to 1998 no real decrease took place, and so new home systems driven by four-dollar-per-peak-watt cells still produced a kilowatt of electricity for about thirty-two cents.

As we will see in the next chapter, different families use varying amounts of electricity. To supply a household with photovoltaic power, one must first know how much power is required. Despite careful conservation, my house consumes an average of eleven kilowatt-hours of energy every day, and I generate barely a third of this with eight photovoltaic modules. A typical photovoltaic module produces about 50 watt-hours of energy for every hour it spends in the full sun, and few sites are favored with more than six hours of full sun

per day, averaged over the year. If waste is eliminated, a 2-kilowatt array provides abundant power for most families at an installed cost of something like $20,000.

Experts nevertheless consider PV the most easily fielded, most reliable, and most cost-effective renewable technique for most installations, and most systems utilize PVs.

Evolving Modules

While the physics of photovoltaics has not changed in five years, the packaging has. In 1993, PV modules were long, narrow, glass-topped, aluminum-framed collections of large blue silicon cells. Some manufacturers were toying with concentrating the sun's power with mirrors and lenses onto smaller bits of purer silicon using elaborate trackers to maintain perfect alignment with the sun, hoping to collect more energy by increasing light falling on

ARE PVS TOXIC?

Manufactured amorphous or crystalline silicon-based PV cells are about as toxic as the beach sand they are made from, but their production involves halide gases and aggressive toxics that must be very carefully handled and meticulously reclaimed and recycled to avoid damaging the environment. PV manufacture consumes an enormous amount of energy; at least a year in the sun goes by before a module harvests as much energy as was embodied in it during its manufacture. Rigorous whole-life cost analysis shows that material extraction costs, substance risks, and production energy, if charged at fair, whole-cost prices, might add as much as 20% to the cost of each already expensive module we buy. Even so, this is a bargain, as externalities go, compared to burning up nonrenewable, therefore precious, heritage fossil fuels. And at the end of their thirty- to fifty-year useful cycle, when their output will have decreased so much it will be economical to replace the PVs with new modules, the ingredients can be recycled very cost-effectively, whereas the burnt fossil fuel is gone forever.

the costly active ingredients. Recently this strategy has not advanced, but thin-film PV technology has. Thin-film fabrication lines produce a flexible band composed of thin layers of doped semiconductor interlaced with the necessary wiring, yielding a more pliable, cheaper, but much less efficient photon harvester. This black, brown, or rainbow-colored band is cut into tiles which may be affixed to many surfaces. The roof of Stephen Heckeroth's barn (described and pictured in chapter 5) is made from "integral" plate-glass panels coated with thin-film material, which allows a pleasing but not oppressive amount of light to pass through while converting some of the rest to electricity; using a second plate of glass and inert-gas-filled super-window technology, this roof has an acceptably high R-value. In Europe and to a lesser degree in America, thin-film technology is being used to moderate light in glass-walled buildings, transforming the roofs of atriums, entrance canopies, and south-facing curtain-walls into electricity-producing surfaces. In northern latitudes, because the winter sun comes from a low angle, south walls may be more productive than roofs.

Using the top of a building for energy harvesting is an increasingly sensible idea. New materials, including roof panels that look like either metal roofing or shingles, can turn a roof into an energy producer while also serving to shed weather and even catch potable water: three functions, where in the past we have been content with just one. Integrated solutions muddy the waters of traditional economic analysis—how are such a roof's various functions evaluated and compared with other products performing less comprehensive service?—but general agreement is emerging that these synergies add good value. At Kahikinui, above

the southern tip of Maui, for a project that promises to resettle a population of Native Hawaiians in their original pre-contact village while restoring the sandalwood forest, all roofing materials must be barged and trucked to an extremely dry region far from the powerlines. In terms of appropriate resource use, it makes perfect sense to employ lightweight, multipurpose materials, and so an integrated thin-film PV roof has been identified as the best solution for shelter, energy production, and water catchment.

Single-cell modules are still the most cost-effective choice when the area available for solar energy generation is limited or must be created just for the purpose, but when a large area is available—the entire properly oriented roof of a home or energy shed—to be covered with PVs, thin-film technology may be better. There are unanswered questions: Thin-film modules are known to degrade. Different manufacturers make differing claims, and actual guarantees are shorter than experience will probably prove. Single-cell modules are likely to produce useful power for fifty years, but thin-film modules may not be worth their place in the sun after thirty years. Since thin-film module performance decays most quickly during the first few months, their capacity is often understated, confounding system designers. In the full sun of Kahikinui, the first test array produced 15% more than the rated maximum, threatening to damage system control devices, and so we chose to disconnect half of the array until the PV shingles had "settled" closer to their rated capacity.

Innovation, improvisation, and constant, insistent change inspired and bedeviled the aboriginal independent homesteaders until they abandoned their caves, and these forces continue to induce modern homesteaders to enlarge and upgrade their homes and domestic systems. In the decades since PVs became available, innovation in renewable energy technology has followed an unusually orderly development path, ironically due in part to its ancestry in the automotive and computer industries. "Legacy" PVs happily copopulate an array with the latest modules, and veteran batteries hold their charge well enough to power state-of-the-art inverters. With solar electricity, the future retrofits more smoothly with the past than we might have a right to expect.

As simple as possible: David Katz's story

David Katz lives off-the-grid on the hillslope above Briceland, California, in southern Humboldt County. He presides over an anarchic band of alternative energy equipment makers and distributors, most of whom also live off-the-grid. An engineer by training, he is often involved in the discovery, rediscovery, or introduction of appropriate alternative technologies.

I had a Swiss Family Robinson idea: We wanted to see how much we could get for ourselves, instead of from outside. We got some cheap land, and started from nothing, with little knowledge. For example, I built the house facing magnetic south, which turns out to be sixteen degrees west of south, but what did I know? As you see, I built a rack to reorient the PV array to true south, and it works fine. Part of my life is always doing things wrong the first time. It usually works out good by the end. My house is pole-frame construction. The poles are set in six-foot-deep holes dug to bedrock, with cement footings and gravel backfill. It's a very good construction technique for earthquakes. We had the poles set plumb, but by the time we attached the roof, they had shifted and there was nothing we could do, so the whole house is a bit out of square.

The little village of Briceland's water supply used to be a couple of parcels below me, and not too reliable. They'd always wanted to use my spring, but hadn't been able to get access. At the driest time, in fall, there's at least five gallons a minute, more than enough drinking water for me and for gardening. In winter, there's thirty gallons a minute, enough to run a micro-hydro. The sun picks up the slack in the summer. I've got a generator, but I doubt I run it four times a year.

The spring is several hundred feet below us, but there's so much water that two High-lifter pumps can pump enough to the gravity-feed tank up above the houses. We started with two rams. They're very satisfying, the way they pump water using a water hammer, but they don't restart themselves the way the High-lifter does; the High-lifters are altogether more elegant.

It's Humboldt County tradition to tax your property for the value of the buildings, even if they are built without the Building Inspector's blessing. They started a program a few years back to let the Building Inspector write tickets for building infractions, but three thousand people, mostly straight, showed up for the hearings, so the idea was dropped.

David Katz surveying his winter vineyard.

I had a business down in Briceland installing oversized alternators and batteries in cars, so you could drive your car around to generate electricity, then park at home and plug your house in. It was a sensible solution, in that automotive lights and devices were all that were available then, and gas was cheap. My partner Roger and I went to a trade show and saw some kind of panel parked in the back of a booth, and the guy knew enough to tell us it was military, and expensive, and did something with sunlight and electricity. When we finally tracked the product down, we found out they were perfect for remote generation, and that we could buy them by the hundreds. We had our first order for two hundred in very soon.

A while later, a guy in a tie showed up at our workshop. He was from Photowatt in Arizona, and had been looking for us for two days. They'd somehow found out that some folks in southern Humboldt were buying more modules than anybody else except the military, and he was supposed to find out why. We ended up buying from Photowatt, too.

The less electronics, the better, is my theory. Photovoltaics are just so simple.

I don't make decisions, so things just happen. When I make decisions, I probably do the wrong thing. My business just happened. I'm trying to keep it the size it is, because I'm managing as many people as I can. I am getting better at delegating work, trying to make it more fun and less work for me. I like to go to conferences and hang out with guys in suits. I enjoy traveling, and I like talking on the phone.

We manufacture a lot of the parts we sell, like the High-lifters. It's like the Japanese mini-business system, where we have a network of cottage industrialists working for us. We're building a new building with a research and development lab in it. Besides the things we make and our dealer sales, we support a very active solar community with service and installation. That feels very important, that we can help our neighbors get their equipment installed and working right.

My house's main problem is that nothing is finished. I need more light, a new inverter . . . I'd like to do a radiant floor, and I want hydrogen. . . . ⌐

Generators: Turning Spin into Flow

PV is magical, but most electricity is generated comparatively simply. Because electrons roam relatively freely in a conductor, large groups of

In this schematic of a generator, the squirrel spins his cage, which cranks the spinning magnet. The magnet's force field, shown as the figure-eight shape, "chases" electrons around in the conductive coil, inducing them to flow as electricity.

them can be induced to flow by exposing them to the flux of a moving magnet. A generator is simply a magnet and a coil of wire, one held steady while the other is rotated, inducing electrons to flow through the coil. When the coil's output is connected to a circuit, electricity flows.

Wind power is a perfect example of transforming a linear flow—wind—into rotational power using a rotor. Hydro power transforms linear stream flow into rotational power using a turbine. Dirtier techniques

WE NEED IMPLEMENTATION, NOT A BREAKTHROUGH

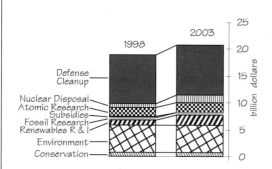

Research and investment into renewable energy sources gets a small but growing share of the Department of Energy's annual budget.

The obstacles to developing photovoltaic power as a competitive energy source are formidable, and unfortunately include political and military as well as technical complications. The Department of Energy, which supervises and funds research into new energy sources, has devoted a small portion of its research funds to renewable sources, preferring to spend its funds cleaning up after earlier mistakes, investigating potentially strategic energy sources, and supporting glamorous Big $cience such as nuclear and plasma research—projects with military promise. The National Renewable Energy Laboratories (NREL) research program, sponsored by the government, has focused its solar-cell research on five out of hundreds of possible semiconductor strategies: amorphous silicon, copper-indium-selenium, indium phosphide, gallium arsenide, and cadmium telluride. Amorphous silicon, refined from sand, consists of a chaotic tetrahedral lattice, but PV electricity emerges only from areas where the lattice is orderly. All amorphous semiconductors have problems with age and large-area uniformity—the lattice tends to revert to chaos—which means it is hard to make them big, cheap, and tough enough to compete with crystalline silicon in efficiency. Amorphous silicon is a reasonable strategy to

convert combustion directly or indirectly into rotation using a piston-crankshaft arrangement. Rotational power is then harnessed to a generator, which produces a flow of electrons.

Wind power typically requires a "power wind" with minimum average speed of ten miles per hour (quite a windy site) to be economical. Wind and photovoltaic often make good partners, because wind frequently blows when the sun is not shining.

develop because we have acres of roofs and plenty of sand, but this is not a good argument for government involvement. Reasons for continuing research on the other four strategies are not purely scientific; in fact, there are reasons why none of that research should be carried forward. Apart from technical problems, Copper-Indium-Selenium (CIS) cells consist of a strategic metal, a rare metal, and a poison; in fact, indium is so rare that world supplies, if entirely exhausted in order to manufacture CIS cells, would produce only a gigawatt of cells, enough to provide just over one percent of the energy shortfall projected by the year 2000. Indium Phosphide (InP) shares the indium problem. Why should we spend money researching a technology that is sorely limited by global resources? With a government as upright and sensible as ours, the decision could not possibly be influenced by the fact that sources of indium are controlled by a few wealthy, politically well-connected profiteers. What if, having devoted the planet's whole stock of indium to this effort, we then find some other, essential need for indium? Gallium arsenide (GaAs) is comprised of a rare metal and a poison. Cadmium telluride is similarly forbidding; in this case the main problem may be the health risk: Cadmium is known to accumulate in human bones, and we have

no idea how much of it will find its way into our food chain. Does it make sense to repeat our history of broadcasting pollution when a well-developed, non-polluting resource is already available?

Researchers working privately with more common and benign materials and strategies have found NREL to be unhelpful, and suggest that NREL's decisions are too often based on the wishes of politically astute scientists. Which opens the question: Does it make sense for us to ask government to support research like this at all? Pure research is always valuable; particle accelerators, plasma generators, and exotic photovoltaic materials can all lead to an improved understanding of nature. Meanwhile, equal funding should go toward conservation, distributed power, wind, water, biomass, and storage technologies, because they are as likely to produce significant results. Should the Department of Energy continue to fund fossil-fuel combustion research, or are oil companies wealthy enough to "beat a dead dinosaur" on their own? DOE and NREL officials operate in a self-created vacuum, having isolated themselves and the research establishment they support from the people they ostensibly serve. I believe we need implementation of existing viable strategies, not a miraculous breakthrough.

Wind energy: Paul Gipe's story

Paul Gipe is an enthusiastic and influential advocate of wind power, which he has seen on a large scale in Europe, California, and the American Midwest. Considering its power and simplicity, Paul wonders why wind power is not more widely employed, and has devoted his life to changing that situation. He has written Wind Energy Comes of Age, Wind Power for Home and Business, *and* Wind Energy Basics.

I originally studied mechanical engineering at General Motors Institute of Technology in Michigan in the days when they had a dress code: no sideburns, no moustaches, a yes sir/no sir kind of place. After a couple of years of that I went back to Indiana. While there I began working with a student environmental group at Ball State in Muncie. We were fighting strip mining in the southern part of the state. I eventually got a degree in Natural Resources, but it was secondary to the environmental work.

During the early 1970s I was studying the geo-hydrologic impact of strip mining in over 18,000 square miles of the southwestern corner of Montana. I kept coming across these abandoned old windchargers that were used on the Great Plains prior to the Rural Electrification Administration (REA), which spawned the electric co-ops. I bought and sold windmills that were relegated to the junk pile after REA came through. The co-ops wouldn't allow ranchers to use both. They forced the ranchers to take down their windchargers. Finally I took the back seat out of my VW bug and began dickering with ranchers for what to them was just junk. By the end of the summer I had a tractor-trailer full of scrap windmill parts.

It was a wonderful experience. I'll never forget it. We renewables advocates were all so innocent then. We were out to save the world. Those windchargers fit perfectly with what I wanted to do.

One day I was testifying at a hearing when some canny old pol said, "Son, those alternatives you talk about, they just don't work." At the time he was right, so I decided to put my life where my mouth was and help make renewables happen. You can't stop strip mining on dreams alone.

There were plenty of people already working on solar and it seemed, well, a little boring. After all, solar panels just sit there doing what they do without any fuss. There was also far less attention paid to wind's prospects than solar. It was the renewable underdog. Proponents would say, "We want solar energy," and then as an afterthought, "Oh yeah, wind energy too." That's still true today.

All in all wind energy simply appealed more to me. It was mechanical, and used the aeronautic arts. I grew up in the Sputnik era and was always fascinated by aircraft. Wind best fit my combined environmental and technical interests. I found operating wind turbines entrancing—still do. I felt I could make a contribution in wind.

So, I shipped those junk Montana windmills back East with every intent of refurbishing them for some working demonstration of renewable energy. While networking with environmental and energy groups in Pennsylvania, I called a dealer who rebuilt used windchargers. He showed up on my doorstep the next day, money in hand, wanting to buy my windchargers. I soon found that I'd become a wholesaler of junk windmills. And, as they say, "It's been downhill ever since."

When that lode played out in the mid-1970s, I began writing about wind energy. In 1988, the American Wind Energy Association named me the wind industry's man of the year. I guess they figured that anyone who has survived that long writing about wind energy deserves a medal.

Wind still offers tremendous promise. We estimate that there are enough resources for wind energy to easily meet 20% of the nation's electrical needs, even after excluding unsuitable areas like parks and wilderness areas. We often say that North Dakota alone is a potential Saudi Arabia of wind energy. If we get to just 10% during my professional life, I'll be happy. That's one hundred times what we have today. Technically we can do it. The question is, do we (the nation) have the will to do it?

The public seems willing, according to opinion polls. But the nature of utility regulation and entrenched interests make it problematic. Utility regulations, which vary from state to state, effectively limit what can be done. Wind development could follow three paths here in the United States: wind farms for bulk power; individual wind machines for homeowners, farmers, and businesses; and small wind turbines for off-the-gridders. Today, at least, the off-the-grid market is doing well.

If the wind is available, then small and what I call micro-wind turbines are a far better buy than PV. But there's no beating a hybrid system using both PV and wind. You get the best of both worlds because they complement each other. In late summer when winds are light, solar reaches its peak. During the winter, the wind system picks up the load. Advances in inverter electronics, compact fluorescents, and other energy appliances have revolutionized life off-the-grid. Anyone living more than half a mile from a utility line will find building their own hybrid power system a better buy than bringing in utility power.

But wind doesn't have to mean only small turbines off-the-grid and thousands of utility-scale turbines in wind plants. There are other ways. As the Danes have shown us, we can install small- and medium-sized turbines

We often say that North Dakota alone is a potential Saudi Arabia of wind energy.

Paul Gipe.

Paul Gipe and his wife, Nancy Nies, erected this mini wind turbine on a hillside in the Tehachapi Mountains of southern California.

for residences, farms, and businesses who use as much electricity as they need, and then sell the rest. In Denmark, Germany, and the Netherlands, homeowners and farmers install the same size turbines as we install here in our utility-scale wind plants.

The economics of individual turbines in Europe is more attractive than here. That's why single turbines are sprouting up all over the countryside. You can't drive down a road in Denmark and not see a turbine spinning in the distance. The same is true in northern Germany and in the northern provinces of the Netherlands. It's an impressive sight.

What makes this possible is the price

northern European utilities pay for wind-generated electricity. They pay enough to justify buying the most economic wind turbine available even if it will generate more electricity than needed by the owner. Because cost-effectiveness often increases with the size of the turbine, everyone wins. The farmer benefits by getting a good buy on the turbine. Society benefits by getting clean energy for use elsewhere.

Wind has a lot to contribute. Wind energy could revitalize the American Midwest. If we can provide farmers there with a new cash crop in the form of wind-generated electricity, they could stay on the land, tilling the soil like their forebears. Wind could provide manufacturing in the industrial heartland, and service jobs in thousands of struggling small communities. The technology exists, and the time seems right. We'll know by the mid-1990s if wind energy will blossom in the United States.

If you know the average wind speed your wind turbine will see, you can estimate its potential output by using this table. For example a Bergey 1500, which uses a rotor three meters (9.8 feet) in diameter, will generate 1,900 kilowatt-hours per year at an off-the-grid site with an average speed of ten miles per hour. Its big brother, the seven-meter (23-foot) Bergey Excel, will generate five times as much: 10,000 kilowatt-hours per year. Because the wind resource is sporadic, it is not meaningful to speak of daily output; at times, an adequately sized plant will give you more than you can handle, and at times the spinner will stand becalmed.

1999 Update from Paul:

I've been lecturing around the world about how to design wind turbines and wind power projects that are aesthetically pleasing. We should respect all landscapes, arid lands as well as forests. One bad example is the Altamont Pass, the first big development in California. It could and should have been done better, but it probably shouldn't have been done at all. I've come to the conclusion that when the developers sense a tax break or a subsidy, they stop listening, reading books, or using their heads. Altamont Pass was part of the first great wind rush, when wind energy was a very lucrative investment. Technically, the winds at Altamont aren't as good as elsewhere. Altamont is also a very visible location. Lots of people drive by on I-580. The first windmills went in near the freeway, and of course the first ones were the ones that failed first—they were junk, and they broke. It's taken the industry years to explain why the turbines everyone sees near the freeway don't work much of the time. And it will take a decade to get those junkers off Altamont's hillsides.

Then there's the golden eagle problem. Altamont may have one of the greatest concentrations of golden eagles in the United States. Golden eagles and wind turbines don't mix. They're a protected species. We have a moral and legal obligation to protect them. Most of us were unaware that this might become a problem when development started, but there were biologists who warned us to be careful. The developers didn't listen . . . until after installing 7,000 turbines. Today, in a calmer frame of mind, Altamont Pass would be one of the last places we'd think of putting turbines, after we had put wind turbines elsewhere, and had figured out how to build wind turbines that worked and how to install and operate them without killing golden eagles.

In the best of all possible worlds, I'd suggest we start in the upper midwest, in Minnesota, the Dakotas, and Iowa. You bet! Let's start at Buffalo Ridge, in southwestern Minnesota. We can easily fit more wind power capacity there than in all of California. One of the advantages of going to Minnesota is that the terrain isn't steeply rolling. We wouldn't gouge the mountainsides like we've done here in California, so we wouldn't have the erosion and visual scarring either. Secondly, the planning authorities in Minnesota are stricter than in California. Even after fifteen years of abusive wind development in California, operators can still get away with murder on the landscape. Minnesotans, for example, might require a wind farm to contain wind turbines that all look alike. They don't have to be alike, but they should look alike. That's the single most important factor in making a wind farm visually appealing. The beauty of the landscape in Minnesota is that you can do a lot there without making mistakes, because the terrain is more forgiving than in California. Now that we know how not to make wind turbines, and how not to develop wind farms, we can go to places like Minnesota and do it right.

When I talk about how much energy wind turbines can provide and how many people they can serve, I use two comparisons. I say, "These wind turbines can meet the needs of so many Californians, or twice that number of Europeans, or one-third that many Texans, because Europeans are twice as energy efficient (in terms of electrical usage) as Californians. And Californians are the most efficient users of electricity in the U.S. Texans consume two and a half times as much as Californians."

One reason we consume so much in the United States is that electricity is too cheap. It's heresy to say that in America, but electricity should be 10 cents per kilowatt-hour at the minimum . . . I think I hear the lynch mob coming now. We're told that Americans won't stand for expensive electricity. "It'll ruin the economy," they whine. But it's not uncommon at all for northern Europeans to pay 10 cents or more for a kilowatt-hour, and they have dynamic economies, and a higher quality of life than we do.

Deregulation is a disaster for decentralized energy and will set back the possibility of Danish-style clusters of owner cooperatives, which seem to me the best form of organization for wind plants. In California, I see very little positive value in deregulation either for ratepayers or for developers of renewable energy. It's an enormous bailout for California utilities, including $28 billion for stranded costs in bad energy projects, mostly nuclear. There are some thirty million people in California. You can figure it out. This is serious money!

Deregulation has been sold to the public as a way to reduce electric utility costs, but that's just the come-on. To make the sale, the legislature perpetrated a subterfuge, promising us all a 10% rate cut, but floating in the same law a bond to finance the 10% rate cut. We not only have to pay off the rate cut, we have to pay for the interest as well. What a deal! To top it off, now we're getting phone calls from companies claiming to sell green resources, and offering us the 10% discount as if it was something only they could offer! This is so despicable that words escape me.

What effect will deregulation have on the development of renewable energy? Some people say we'll see green resources. Maybe. But so far all I've seen is greenwashing! The green hype is so thick that some sensible organizations like the Sierra Club are planning to form their own

green utilities. Unless they do, this is shaping up to be a disaster for the ratepayers and the environmental community.

I've become increasingly outspoken about wind energy politics. Worldwide, wind energy is booming, with new capacity worth two billion dollars installed in 1997, including 1,000 new megawatts in Europe alone. Last year (1997) Germany and Denmark installed one billion dollars worth of new wind turbines. Germany now has more than 2,000 megawatts of wind energy in place, more than in all of North America, which has something like forty times the land area and six times the people. In 1997 Germany surpassed the U.S., and Denmark will surpass California by 2000 unless something major happens. In 1997 U.S. wind capacity *declined* by more than 20 megawatts, a continuation of a steady decline in U.S. wind generation, particularly in California, for several years. Either the Europeans know something we don't know, or they're all fools.

Unfortunately, the U.S. appears to be on the brink of another wind rush, with all the attendant ills. A lot of energy companies are trying to position themselves to take advantage of the newly deregulated industry. Wind energy production tax credits expire in 1999 and developers are tripping over themselves trying to get their hands on the money in time. When they do, there will be rush-rush slap-it-up development like we saw in California in the 1980s.

But these tax credits have obviously done nothing to spur wind energy development until now. The industry wants the tax credits renewed because in that game, only the big boys can play. I favor a solution that doesn't respond to market maneuvering and changing political winds, but instead is directly related to energy costs.

There is a much simpler, better way to develop renewable energy. I continue to travel extensively in northern Europe. Last year I spent four months studying wind energy in Denmark. We stayed in the township of Syd Thy where they are net exporters of electricity due to their wind turbines. From where I lived I could see thirty-eight wind turbines. Every one was owned by a farmer or cooperative of local people and every one was working—something you don't see in California. I concluded from that experience that the American wind industry's lobbying for tax credits goes in exactly the wrong direction. What we need is something simple and direct, an arrangement like they have in Denmark and Germany. And we'll only get it if we ask for it.

We need an "electricity feed law" like that of Germany: a Public Utility Regulatory Policy Act (PURPA) for the new millennium. The German electricity feed law says, like PURPA, that you have the right to connect to the grid and sell power. Unlike PURPA, the German law spells out what you get paid: 90% of the utility's retail rate. No endless meetings, no paper shuffling, no lawyers! It's as simple as that.

Net metering won't do it. Every state that has net metering limits it to either solar or small-scale wind generation. In Germany, there's no limit on the size of your wind turbine. You use what makes the most economic sense. That's the way it should be.

I'm for taking a radical step, adopting an electricity feed law. The electricity feed law in Germany is startlingly simple, it's only two paragraphs. Let the lawyers, utility apologists, greenwashing con artists, and beltway bandits getting fat on deregulation and tax credits designed for only rich corporations stew in their own juices!

This is a controversial stand. The wind industry says I'm being unrealistic, we're Americans, and it doesn't fit within our political and economic

ESTIMATED ANNUAL ENERGY OUTPUT
AT HUB HEIGHT IN THOUSAND KILOWATT-HOURS PER YEAR

Average Wind		*Rotor Diameter*							
	(m)	1.0	1.5	2.0	3.0	4.0	5.0	6.0	7.0
Speed in mph	(ft)	3.3	4.9	6.6	9.8	13.1	16.4	19.7	23.0
9		0.1	0.3	0.6	1.3	2.3	3.6	5.2	7.1
10		0.2	0.5	0.9	1.9	3.4	5.3	7.6	10.0
11		0.3	0.6	1.0	2.3	4.1	6.5	9.3	13.0

environment. I disagree. It fits with a Jeffersonian vision of the kind of America we want.

My vision of energy development in the best of all possible worlds would combine a healthy mix of wind farm development and distributed, decentralized ownership. Every farmer in Iowa, Minnesota, and the Dakotas ought to have his own turbine, just as they do in Denmark.

Of course developing wind energy makes no sense without a comparable commitment to conservation. They have to be undertaken side by side so we maximize the benefits of wind and other renewables. We have to remember that renewables have an environmental impact, too. We can justify wind development only when we engage in a major conservation program, and only when we design our wind turbines and our wind power plants "as if people matter." ⌁

Falling Water: The Best Source

"Falling water is a goldmine," I was told when I first saw a micro-hydro system in action. Micro-hydro is the most continuous and cost-effective renewable energy source. If you have a good year-round watercourse with adequate head, you may not even need to use the draconian measures of conservation that shape most independent power strategies . . . in fact, you may be looking for ways to use extra energy. Owners of hydro systems often share with neighbors or burn their excess to heat water and space, all unthinkable with more precious sources like PV. Year-round hydro systems need less storage than solar or wind, just enough to match peak usage with twenty-four-hour production and cover times when the turbine and generator require servicing.

Large-scale hydro has its unique terminologies: Water enters the forebay, then is directed through the headrace into the penstock which carries it down to the turbine generator and back through the tailrace into the afterbay and watercourse. Micro-hydro uses humbler hardware, but the idea is the same. Water fills a barrel connected to a pipe which runs a few hundred feet downhill, and, under considerable pressure due to the head, or vertical drop, squirts through one or more nozzles to strike the cups of a pelton wheel turbine, which spins a generator, typically a converted automobile alternator. Micro-hydro requires plenty of water falling quickly, so these systems are most often found in wet, mountainous sites. Flow rates are usually measured in gallons per minute (GPM). Head is the vertical distance between the intake and the generator site, and need not be a dramatic drop, as in a waterfall. If the horizontal distance the water must travel in the pipe, called the run, is long, the flow's energy will be decreased; a run greater than four feet of run to each foot of head will slow the water too much to spin the turbine effectively. Serious power production starts when

A two-nozzle and four-nozzle hydro-turbine (shown upside down).

Estimated Micro-Hydroelectric Potential in Watts of Output

Flow in GPM	Head (in feet)						
	25	50	75	100	200	300	600
3	—	—	—	—	40	70	150
6	—	—	10	20	100	150	300
10	—	15	45	75	180	275	550
15	—	50	85	120	260	400	800
20	25	75	125	190	375	550	1100
30	50	125	200	285	580	800	1500
50	115	230	350	500	800	1200	++
100	200	425	625	850	1500	++	++

the head exceeds fifty feet. For a small household requiring 500 watt-hours per day, a modest flow of ten gallons per minute and sixty feet of head provides abundance, and an average family would get by on fifty gallons per minute falling one hundred feet.

The relationship between flow and head is complex, and is usually calculated using a special computer application. Micro-hydro systems using relatively small amounts of water falling a considerable distance are simple to design and install, and run for years with little more attention than cleaning the leaves out of the intake grating and the salamanders and frogs out of the nozzles. Low-head, high-flow systems, usually involving river water with only a few feet of head and a flow reckoned in cubic feet per second—a lot of water!—can be incredibly powerful, but necessitate a dam or diversion and larger equipment. Some of the best systems use recycled parts. A system just below "the rainiest place on earth," Kauai's Mount Waiale'ale, diverts water through an abandoned irrigation ditch to turn a pair of homemade overshot waterwheels which grind an upside-down reduction gear salvaged from a sugar refinery to produce up to two kilowatts of electricity. Low-head high-flow potential is very site-specific, and requires expert planning (or lots of tinkering) and ongoing maintenance.

Every hydro project presents a unique combination of water source and topography, which can require more careful surveying and planning than an equally productive wind or photovoltaic installation. Developing a hydro resource is often complicated by environmental concerns, by neighbors, and by local regulatory agencies. Water quality, fish migration, and the effect on the portion of the streambed from which water is diverted are all legitimate issues to study and mitigate carefully. Critics are often confused to learn that hydro projects do not permanently remove water from the stream, they merely borrow it for a few moments and a few feet of stream, so hydro projects in no way diminish a water resource; infringing upon the water rights of others is seldom an issue except in the rare case where water is diverted from one watercourse to an-

A run-of-stream generator is like an underwater wind turbine: Held in the flow, its propeller spins a generator which produces power.

other. Nevertheless, regulatory agencies live to regulate, and private power generation makes them nervous. For all these reasons, many micro-hydro systems are undeclared, and micro-hydro is probably more common than we think. I visited one system in New England where the operator, a farmer whose family had held the land for four generations, installed his wheel as a defiant act of independence, and was thrilled to have a secret energy source. More power to him!

Because high-head, low-flow hydro equipment is inexpensive and reasonably trouble-free, seasonal hydro sites, which only produce well in the wet months when PVs are unproductive, are cost-effective contributors to a steady energy budget. Although studies suggest that only a tenth of our nation's hydro potential has been developed, ideal hydro sites are as rare as they are perfectly suited to independent power production. In many western states, despite regulatory resistance, hydro enjoys favored treatment: An undeveloped hydro site may be developed in some jurisdictions against its owner's will in a process quite similar to exercising eminent domain, because the law recognizes that the water isn't used, it is only borrowed, and the energy source is acknowledged to be of greater community importance than mere property ownership. In the West, at least, it is understood that water does not belong to the owner of the property it crosses.

Hybrid Systems: Multiple Energy Sources

It is a rare site where a single renewable source provides adequate year-round power through all weather conditions, and so use of multiple sources is necessary. If you are lucky enough to have access to wind or hydro resources, you will find they often complement the photovoltaic source well, because water flows and wind blows when the sun is not shining. Photovoltaic and wind generation seldom use the whole available resource, and so their production can be increased simply by installing more modules or another wind-spinner or two; unfortunately, both still produce relatively expensive electricity, and so we are tempted to look for a cheaper source. Off-the-grid power systems typically include some form of fossil-fueled generator for the "swing seasons," those awkward weeks in early fall and late spring between reliable storms that set the windmill spinning and steady summer sun to power PVs.

For the 99% of existing homes that are on-the-grid, grid power is a much cleaner and more economical backup and storage than backyard fossil-fueled generation. Utilities do much of their generating of electricity using thermoelectric generation: burning non-renewable fuels to boil water into high-temperature, high-pressure, superheated steam, with which steam-turbine generators are driven. As fossil-fuel-fired engines go, generators work fairly efficiently; the bigger they are the more efficient and cleaner they can be. Smaller gen-sets using internal combustion or diesel engines to drive generators provide backup during outages in hospitals and other places requiring absolutely reliable power, and are often used as the source of last resort in off-the-grid systems. In hybrid systems with gas-, diesel-, or propane-powered backup, the system's battery storage allows operators to run their generators at optimum efficiency and store excess output. This contrasts sharply with the generator-only practice of running the 2.5 kilowatt generator for a 25 watt lightbulb. Renewable energy users shake their heads at these lazy gear-heads: what sloppy wastefulness! If we must burn fuel for electricity, let us at least do so as efficiently as possible! Generator performance can always be improved by integrating their output into a battery-storage and power-conditioning system. Small gen-sets like these convert fuel to electricity inefficiently, particularly when running with light loads, and are environmentally dirty; smaller, in this case, is uglier.

Fossil-fuel-powered generation is the last, worst choice for backup power from the standpoint of cost and pollution. When pollution is considered, propane is the best of a bad set of options; if operating cost is the determining factor, diesel is best (and dirtiest and least efficient), and propane is a close second; for cheapness, local availability and service, and fuel convenience, gasoline-powered generators are the easiest and commonest choice. At the most successful installations, where householders manage their energy usage realistically and carefully with attention to weather and living patterns, generator use decreases as they learn to live within their renewable energy budget. Often the generator stands unused for months, and is started a half a dozen times a year to keep it limber or to power a large tool, while also topping off the batteries after a lengthy cloudy spell during swing season.

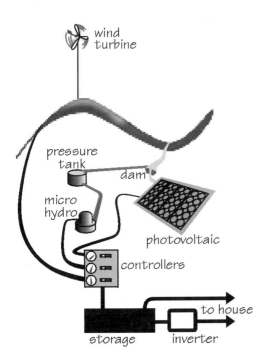

The more diverse the sources of renewable energy, the more reliable the system, and the less reliant on fossil-fuel power. With wind, water, and sun, this system requires little or no generator time to provide year-round energy.

Power's Dark Underbelly

I visited a friend's new farm in a pristine wooded valley where a strong year-round stream begs to generate micro-hydro, but a

The generator usually coughed to life in the pre-dawn cold, filling the still valley with its ghastly throbbing and a miasma of local smog . . .

huge 25-kilowatt diesel generator and industrial-strength battery bank allow for a no-load-spared, all-electric lifestyle. This generator, the size of a small tractor, ran an average of five hours every day. Pollution and maintenance aside, the day-to-day costs of electricity compared favorably with grid power. However, overall costs, including original equipment and periodic major engine generator maintenance, were staggering. The generator was set to start automatically to recharge the battery when system voltage reached a setpoint. Since the chemical reaction that produces battery voltage becomes less vigorous as temperature drops, the generator usually coughed to life in the pre-dawn cold, filling the still valley with its ghastly throbbing and a miasma of local smog, which, trapped by the protected little valley's natural inversion layer, lingered for hours after the generator had recharged the battery and shut down.

At an "eco-lodge" on beautiful Maho Bay, where the sun shines and wind blows steadily, I stumbled upon a dirty little secret: a huge buried tank of sump oil, the toxic, dirty byproduct of internal combustion-powered generation. Laughing, one of the lodge managers suggested that every departing guest should be given a souvenir gallon bottle of sump oil to take home. In practice, these arrangements reproduce in miniature what is wrong with all fossil-fuel electricity production.

And at both places, the owners claimed they could not afford renewable energy.

Dead dinosaur fumes! It is not hard to imagine the harm that comes to our bodies when we breathe the leftovers of burnt fossil fuels, but it is proving amazingly hard to get our medical establishment and multinational corporations to own this apparent truth. A researcher at NREL (the National Renewable Energy Laboratory) wondered what uproar might result if folks in a typical town were asked for permission for several truckloads of highly explosive, toxic, and flammable material to pass through their town every day. Luckily for our petroleum-driven lifestyles, permission to haul gasoline was granted decades ago, in more naive times. Nature went to so much trouble to get fossil fuels safely buried, but now human beings are in such a hurry to dig them up and burn them dirtily! Too often in my travels, I find denial of the filthy secret, the generator sited absent-mindedly in the lee of the house where its exhaust is permeating otherwise lovely rural living space. The best installations acknowledge the presence and need for the fuel-burning plant, and cede to it a shed of its own, sufficiently downwind to abate, at least locally, the noise and stench.

Nineteen fifties sticker-shock culture favors solutions that are cheap *in*, but expensive and dirty *out*—the economic bafflegab is "deferred costs," which is code for "let our grandchildren pay!"—and nuclear energy undoubtedly claims the championship as the worst offender. We must relearn a global, natural truth known instinctively by every other living organism: Intelligent decisions about any action can only be made after counting that action's whole cost, looking beyond the horizon in all directions, starting before planning, and ending after the equipment and any by-products are properly recycled or dumped. When all these costs are reckoned, renewable energy is generally found to be much cheaper than any other. Its sole defect is that it fails to enrich the entrenched energy profiteers . . . and so it has been shunned.

Let us work to provide clean renewable energy for ourselves, and to change the political climate in favor of renewable energy sources for everyone. What could be a better goal for those of us who wish to live independently in our homes?

As technology encroaches further on our everyday lives, we must remember that residential electricity is a 20th century idea, as new as the dependent home. Like most newfangled ideas, our ancestors treated it tentatively at first—possibly knowing that, in time, its hidden hazards might prove it to be not such a good idea after all. Caution subsides with familiarity, especially in a land where convenience rules, and now we see that we have willingly entangled ourselves in a web of wire.

The wires strung across our towns and countryside look like the kind of temporary work done stringing lights on a Christmas tree. When trying out something new at home, I drape the wires carelessly at first, in case I hate the new device, but if it works and I like what it does, I figure out a way to install it permanently so that its infrastructure and support mechanisms do not intrude on my senses of order and beauty. After a century, can we agree that electricity is a good idea, and we would like to keep it around? Very well, can we now agree to hide it in the walls and under the floor where it belongs? Then we can begin to handle it in a way that doesn't make us constantly conscious of its haphazardness. In a recent study of outages in the western U.S., it was determined that 98% of the failures were due to overhead lines, and 65% were due to collisions between exotic (non-native) trees and overhead lines. After almost a century of impermanence, planners and customers in upscale communities are finally forcing utilities to put the rat's-nest of utility lines underground where they are safe and invisible.

Apart from their offense to the eye, can powerlines be good for the plants and animals that live near them? Stray current, electromagnetic pollution, and incidental microwave radiation are apparently a problem wherever power is moved overhead. This is doubly offensive, because this health risk is the result of distribution loss, for which we pay. To many independent power producers, the grid seems an elaborate, unnecessary, fragile, and poorly designed scheme; increasingly, when power and communications companies and the government require reliable, uninterruptible power, stand-alone renewables are their source of choice.

Huge, centralized, profit-oriented power companies are politically strong and jealous of their hegemony. One standard argument they advance against widespread use of photovoltaic power is an assertion based on the prodigious energy consumption of that energy horror, the 1950s-era All-Electric Gold Medallion Hog ... sorry, I meant House, originally promoted by Ronald Reagan. Producing enough power for such a "standard" home requires so many hundred square meters of PV modules, they assert, that all of Nevada and Arizona would have to be covered with PVs to provide what their nice nukes and coal-burners do now. By curtailing waste, which we have seen typically amounts to more than half our electrical use, an average home's supply can be harvested with modules covering an area about half the size of that home's roof. After purging our homes of needlessly wasteful energy equipment, a program like the Million Solar Roofs, which enlists residential roofs and covers them with grid-interconnected modules, could supply more than enough power, using currently available technology, to satisfy all residential and office requirements; existing conventional capacity could then accommodate requirements in the manufacturing sector while it shifts from Victorian "heat, beat, and treat" techniques to earth-friendly technologies. This would be a very good idea ... unless the government can find a way to mess it up. Such an effort to mass-market renewables, and the cosmetic improvement of undergrounding utility lines in every community, could provide us with a functional, long-lasting, and appropriate relationship between people, houses, and electricity.

The ugly and often dysfunctional appurtenances of 20th century technology, looping powerlines, satellite dishes, television antennae, and smoke-belching diesel engines, do not settle well into the peaceful dignity of a New England farmhouse foursquare on its land, or the rugged beauty of California's wild north coast. I wondered, as I visited independent homesteaders across the country, how these high-tech gadgets would look if they had been developed at the same time the saltbox house was evolving in gradual, unforced response to art and necessity. I found a hint in the wonderfully ornate, wrought iron, glass-balled lightning arresters on some New England barns: Humorous and functional, these finials remind us that invisible powers such as electricity are not far removed from necromancy. We live in hasty, greedy, unfinished times, and our styles reflect this. Good planning and proper selection of technology has repeatedly shown that a solar house can be as beautiful as any other, and twice as sensible.

CHAPTER 4

MAINTAINING HOME UTILITIES

To take responsibility for our own power, we must manage and maintain energy generation equipment, as well as the infrastructure and hardware that we need to transmit, store, distribute, and measure the electricity, heat, and other forms of energy we use. We take over responsibility for energy flows normally overseen by profit-oriented utilities and suppliers. To do a good job, we must have a better load management plan than they have.

The starting point for any such plan, whether for a small remote home or for a utility monopoly serving millions, is a reasonable estimate of the power required. How much power is needed, and when is it needed? Utility dispatchers describe the process using their lingo: They predict peak and base loads—the largest and smallest amounts of energy the system will be called on to provide—and then dispatch adequate resources—turn generators on and off and negotiate purchase of power from neighboring systems—to satisfy the daily and seasonal demand patterns of their customers as influenced by weather, time of day, and even what day it is. In a household grid, where you and your family are the customers, and power consumption is limited by your system's generative and storage capacity, a realistic load management plan reconciles capacity with your requirements. This may mean that two big energy loads, such as the washing machine and the iron, cannot be used simultaneously. It may mean that at times when generation is at a low ebb—

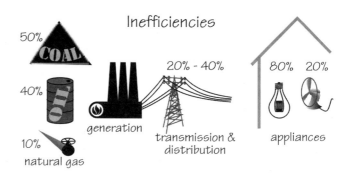

Inefficiencies

50% COAL

40% OIL

10% natural gas

generation

20% - 40%

transmission & distribution

80% 20%

appliances

Most American electricity starts in a thermoelectric generator plant, where fossil fuel is converted to rotational energy which is then converted into electricity. This process can be as efficient as 90%, or, in the case of old coal-burning plants, below 50%, meaning that the plant produces electricity amounting to less than half the original fuel's energy content. From the the generating plant, the electricity is transmitted and distributed, losing as much as 40% of its energy along the way. In the home, appliances are also more or less efficient at converting electricity into measurable work. Fans, motors, and fluorescent lights are better than incandescent lightbulbs and heating equipment. It is worth noting that every energy stream, including the coal, oil, and gas that fuel the electricity stream, experiences the same inefficiencies from well-head and mine-shaft through the power plant to their final point of use. The Exxon Valdez *is a spectacular example of transmission loss.*

for example, if your power comes from photovoltaic modules and the weather has been cloudy for a few days—certain work, such as ironing, will be deferred so that basic power needs like lighting can be met.

Utility customers expect as much power as they can use whenever they wish to use it; they are the demand side. On the supply side, energy dispatchers use weather reports, customer history, hydroelectric reservoir level and fuel price statistics to provide what consumers—the demand side—will be using. Energy marketing schemes such as time-of-day metering (customers pay a premium for energy consumed during peak-demand periods, and pay less for energy used at other times) influence power use in terms of months and years. When they must, dispatchers use preemptive power management tools to adjust loads to local or purchased power availability. These supply-side tools include brown-outs (utility voltages are lowered, putting some equipment, especially motors, at risk, and shortening their lives) and rolling outages (neighborhood- or town-sized areas of service are shut down temporarily). Any of these undermine the autonomy of your household, and therefore advance the argument for energy independence.

How much power do you need? There are three answers, two simple and one good. The simplest, most pessimistic answer is that we always need more power than we can possibly produce sustainably and locally. This answer assumes that we insist on using any amount of energy at any time, without considering load management, and therefore unless we learn to work efficiently and within our energy budget, we will need the

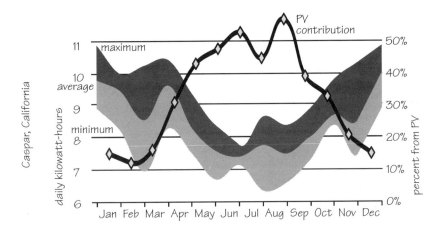

Average daily electricity demand varies with season and weather, but may also reflect other patterns, such as our propensity for abandoning frigid Caspar for sunnier places during December, February, and March, creating minimum demands that reflect lower occupancy. More northerly locations may experience doubled demands during winter months, especially for light and heat, and severe air conditioning climates will show demands peaking in August, a peak that could be matched by the PV productivity curve. Overall, the national power budget peaks in August.

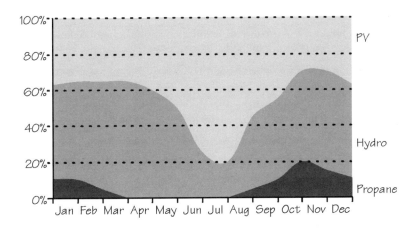

David Palumbo's homestead electric company relies on three sources. The propane generator runs most during "swing season" after summer's sun but before winter's wet, but it never provides more than 20% of their power. Year-round, PV and hydro are about equal as sources.

HOME ENERGY SYSTEM OVERVIEW

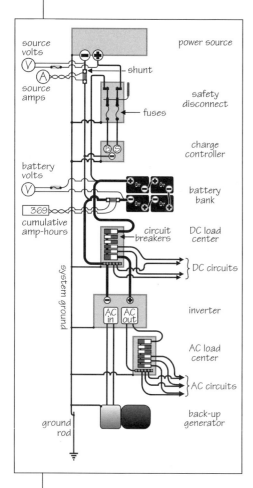

The parts of a typical independent electrical system. Energy is harvested at the source and transmitted by wires to balance-of-system control devices, including an inverter that converts direct current into household alternating current.

Renewable energy sources for a home power system can be photovoltaics, wind, water, biomass, or any of these in combination; the accompanying diagram starts after the source. Whatever our sources, there will be times when we will wish to disconnect them from the house's wiring, and the Safety Disconnect accomplishes this. The S-curved symbols represent fuses, which are the first line of overcurrent protection. When too much current flows through a fuse, it quickly heats up and "fuses" or melts, thereby breaking the electrical circuit. A fuse is a one-time device; once fused, it must be replaced.

The Charge Controller controls the amount of electricity from the source that is applied to charging the battery bank, based on the batteries' state of charge. Each DC branch circuit is connected to the positive bus bar within the DC Load Center through a circuit breaker, which is an electrical analog of a fuse. When too much current flows through a circuit breaker, it "trips" and interrupts the circuit; when it has cooled down, and the overcurrent condition has been corrected, it may be reset by switching it off, then back on. These circuit breakers protect their low-voltage circuit from overcurrent—a "short circuit"— and allow a branch to be turned off for servicing. Two wires conduct electricity between any two components in the power system, because electricity only flows in circuits; imagine a stream of electrons flowing away from the minus side and toward the plus side. For safety, a redundant third wire connects all metal frames and boxes to each other and to a ground rod driven into the earth.

The batteries and instrumentation are the focus of a DC system, and safety is the guiding principle. Note that even the meter is fused.

The inverter changes low-voltage DC power into house current, 110-volt alternating current. In the system diagramed, the inverter includes a battery charger that can accept AC electricity from the generator; when the generator is providing power, the inverter uses electricity from the generator to recharge the batteries while protecting the DC system from irregularities in the generator's AC output.

The AC Load Center provides convenient connections and circuit-breaker overcurrent protection for AC circuits in the same manner as the DC Load Center.

grid, which is designed to deliver unlimited energy without regard for the sustainability or closeness of its sources.

The second simple answer, if you buy electricity from a utility, can be found by reviewing your monthly electric bills and calculating a seasonal daily average. (I will ask you to join in that exercise in the next chapter.) Normal households use between 500 and 50,000 watt-hours per day, which is a very wide range; my house in Caspar consumes between 6 and 11 kilowatt-hours per day, half of it used by computers. On average, a third of my electricity is generated photovoltaically.

The good answer calls for mindful selection and use of energy-consuming devices, as if you were responsible for gathering the power yourself. In this chapter, you will meet people who do just that.

Planning domestic power: Felicia and Charlie Cowden's story

Charlie and Felicia Cowden are back living on-the-grid after a stint in an off-the-grid house they built on Kauai's North Shore. In the aftermath of Hurricane Iniki, which bludgeoned Kauai in September 1992, they helped their neighbors get their power back. In sunnier times, they run the Hanalei Surf Company. As you will see in their story, moving off the grid and back required learning and adaptation, and the conclusions they drew will work for anyone who uses electricity.

Felicia: The biggest stumbling block on the way to power independence is our culture's custom of wasting incredible amounts of energy. Americans waste more than they use. You can't simply buy energy independence: You have to make a brain investment, and learn some things.

We started moving toward conservation when we were living on-the-grid in Princeville by replacing incandescent lights with compact fluorescents.

Charlie: When we were deciding to buy our property and build, whether to pay the money to trench and connect to the powerlines at the highway or go solar, I didn't know anything about alternative energy sources. So I asked my contractor, should I do it? He didn't know much about it either, but he said, "It's the future, man, you gotta do it."

Felicia: We shopped around and bought parts from different sources; sometimes they weren't even meant to work together. Because we got different parts from different people, we had some problems getting people to help us put things together right.

Charlie: So I drew a picture of my charge controller and sent it off to the guys that made it and said, hey, I'm thinking about putting this stuff together like this, what do you think? And the same with the batteries and inverter. I did it the way they told me was best.

Felicia: We throttled our energy use way back when we moved to our new house. We were limited financially, and could only get six PV modules. We figured we would need more, but during the sunshine months, we're full by noon, and we haven't given anything up. We just changed our energy habits.

When we first moved in, it was like living in a science experiment. It soon became very apparent that there were things we hadn't thought of before. Living in my house has made me much more conscious of consumption, so when I visit other people, I see incandescent lights, and can't help but notice all the inefficiencies. It feels sinful! Solar power's biggest gift to the environment is showing people that it's possible to live well without being wasteful. Since moving in, I've become an energy evangelist, always trying to get more people into conservation.

There are so many things we wish we'd known! Nobody told us about phantom loads, devices that use electricity whenever they're plugged in, like the clock on a microwave. Four little clock radios can use up everything our PVs produce in a day. We use battery-powered clocks with rechargeable batteries. The real energy criminals of the small appliance world are the remote products that require chargers, like electric toothbrushes and cordless razors. They draw substantial current to charge their batteries, then when the batteries are full, they continue to trickle energy. That's a lot of burnt dead dinosaurs for a few minutes of minor convenience, and we all have to breathe the smoke. Much better to use appliances that plug in directly and only burn power when they're in use. If push comes to shove, you can always brush your teeth yourself.

When we go to bed, it's usually really quiet, but if I still hear the inverter humming, I go look at the meter. I've learned my phantom loads: If it says 2 amps it's the microwave, 4 amps is the TV and VCR, and 8 amps is the sine wave filter for the stereo.

Charlie: After Iniki, I was helping out an electrician friend installing gen-sets—generators, you know?—and when we'd get ready to start the engine, we'd tell the owner, "Go in and turn everything off, okay?" and pretty quick he'd come out and say, "Okay !" So we'd fire up the generator, and immediately see 100 amps being drawn out of the system. We'd figure there was some kind of short or something wrong because of the storm, so we'd shut down and go looking for the problem. We'd find a dustbuster drawing 30 amps over here, and a little convenience fridge drawing another 30 over there. . . .

Felicia and Charlie Cowden.

In our house we put our phantom loads in power strips, but it would be better to put switched outlets in the wall.

Felicia: A perfect example: a recessed microwave with the outlet behind it, where you can't reach to unplug its phantom load. Lots of times, people want to install their microwaves that way. If you put a wall switch on it, you can turn it off when it's not needed.

If I leave all our phantom loads—the VCR, stereo, TV, microwave, and adding machine—plugged in, even if they are turned off, they consume more electricity than my PVs produce. If I can turn their outlets off, I produce more power than I need. So you don't have to do austerity things, you just have to be careful not to waste.

We learned a lot since we built our house, most of it what the power company calls load management or demand-side management. I can't run the stereo when I'm doing a wash, because I chose a power-hog double-agitator washing machine. Now, I'd buy a front-loader.

Charlie: You've got to make small amounts of electricity do your work. I saw a front-loading washer in the store, and I thought they'd made a mistake because the energy tag said it used a tenth the power of the top-loaders.

Felicia: It's all in the torque. You've got to understand the difference between inductive loads, like motors and little transformers, and resistive loads, like lights. Inductive loads are real power hungry, especially when they start. For example, you learn not to run the vacuum cleaner when the washer's running. The garbage disposal puts a big strain on our system . . .

Charlie: . . . and the television picture gets real small.

Felicia: If I wake up in the morning and see the batteries aren't depleted, I look for high-demand activities, like washing and vacuuming. Load management depends on weather. If it's cloudy, and the batteries are depleted, I think about conservation.

Charlie: At first, demand-side management is hard for people to think about. They still want to do things as if they had infinite power, and they have equipment already . . .

Felicia: . . . their energy-hog refrigerator, which would take twenty-four modules by itself. . . .

Charlie: We tell them they can't take that equipment along, they need to start out energy-thrifty.

Felicia: I start out with Kauai Electric's energy pie, showing what the average household uses, and tell them that you can do all that only with a $300,000 solar installation, so instead let's start managing. Solar power is an excellent application in the tropics, because we don't have to heat or cool space. We install solar water heating. We substitute a Sun Frost or a Sunshine Coldmizer for a conventional fridge. Manage lighting and phantom loads. We personally reduced our power consumption by 95%. That's substantial.

Another thing I wish we'd done: Our inverter puts out a modified square wave, not the sine wave that most appliances are used to. Before buying an appliance, find out from the manufacturer if it needs a pure sine wave to work right.

Charlie: Felicia bought a stereo, and when we plugged it in, the lights went on, but no sound came out. I went, okay, no tunes, and turned on the TV, and the stereo started up! So I turned off the TV, and the stereo stopped. How's that? I checked with the stereo techs, and they told me the TV had capacitance that smoothed out the power. I called the inverter techs, and they said, "Capacitance, hunh? We can send you some of that!" and they sent us this big capacitor which I put in the system. It worked okay, only seemed to draw 50 watts, but when I went to check the inverter, it was really hot. Whatever kind of 50 watts that capacitor was using, the inverter didn't like, so I disconnected it. Finally we got a sine wave filter that uses lots of electricity, but lets us use our stereo.

People think they've got to make sacrifices, give up TV and live in the dark. That's not necessary, they just have to pay attention. Take glowplugs for example. . . .

THE HEART OF A POWER THIEF

A power cube steps the house current voltage down (transforming it from house current to a lower voltage) and then converts or rectifies it from Alternating Current (AC) to Direct Current (DC) if required by the device it is meant to power. The wasted energy turns into a surprising amount of heat and electromagnetic radiation (EMR). Existing electronic devices can easily be converted to higher efficiency by connecting them directly to 12-volt DC circuits if (and this is an important if) they use 12 volts DC, which many of them do, and you have 12-volt electricity available. The nameplate on the transformer will specify output voltage; anything in the range of 9 to 15 volts DC will probably work fine, but I specifically refuse responsibility if it turns out voltage is critical and your little device gets fried. If in doubt—especially if the device is costly—check the manual or call the technicians. Most telephone devices will *not* work: While answering machines and cordless phones may say 12 volts DC on their nameplates, there's a gotcha! in that telephones define the ground or zero voltage side differently than other DC devices, and if there is an internal linkage between the two systems, there will be a conflict. As long as the power source "floats" and the grounds of power and telephone systems are not connected, the potential difference creates no problem. If an inverter is attached anywhere in the DC power circuit, a noisy disagreement takes place over which ground potential is right, the sounds of which drown out anything one might wish to hear on the telephone. A DC answering machine solves the problem. I use a tiny inverter to provide for my electronic answering machine's eccentric needs.

Felicia: Yes. When you're buying a gas stove, it's a big clue if you have to plug it in.

Charlie: I think it's illegal to sell a stove with pilot lights in some places, because they use so much gas, but the glow plugs burn an awful lot of energy.

Felicia: We like to put inverters and batteries in an outside building. Our inverter is in the bathroom, and if you're in there and the toaster's down, you hear the inverter whining. It's putting out a worrisome amount of electromagnetic radiation.

Charlie: So when the washer is going, we scatter.

Felicia: When you don't meter your system, it's like having a car without any gauges, no odometer, no gas gauge.

Charlie: So you wouldn't know when to go to the gas station. I tell people that if you've got a PV system, you've got to have meters, too. You can't make the system work right without the meter. I went to a house where the guy had been living off-the-grid for fifteen years, and all that time the system has been growing, adding batteries, modules, and circuits without labels, mostly running off his generator. But he doesn't understand why he runs his generator so much. He's finally redoing his system, and adding an amp-hour meter. It will be a born-again system.

Felicia: Whenever we go into a house without a meter, we find that even after years and years, the people don't understand their systems. When guests come to our house or people come up to work, they are fascinated by our system, the fact that we have all the electrical devices we need. When their friends come by, they like to play with the meter, show what the system is doing. It's instructive and fun.

It helps if the system is shipshape, like it was meant to be part of the house. When people see batteries in a corner and all sorts of cables, it makes an image of alternative energy as a subculture life style. That's not necessary or even true. It works better if it's clean.

It's worthwhile to check local requirements. Here, you have to get past the inspector to get fire insurance, and you can't get financing without fire insurance.

Charlie: You've got to find out the rules—and there are so many rules. You must be careful to comply with building codes, or if there's a fire, the insurance company will find a way to deny your claim. I'd be scared to death to work on an unsafe, non-code system. If you don't use the right practices, and something goes wrong. . . .

Felicia: When we talk about energy self-sufficiency, we try to get people looking away from the money issues. Maybe you save money going off-the-

"People think they've got to make sacrifices, give up TV and live in the dark. That's not necessary, they just have to pay attention. Take glowplugs for example. . . . "

grid, but there are more important reasons. When you go to sleep at night, there's a zero electromagnetic field. And think of the tons of stuff you aren't putting into the atmosphere. It's a bonus if you save money.

People on the North Shore have a wonderful attitude about work, which taught me a lot about pride. Here, if you ask somebody what he does, he's likely to say, "I surf." My friends here are carpenters, plumbers, electricians, but they are all surfers, really. With them, I want to be me, not what I used to be, or what I do to make money. In fact now, that's how I answer that question. Somebody says, "What do you do?" and I ask, "You mean, to make money?" It makes them think, sometimes, that there might be more important things to do.

1999 Update from Charlie and Felicia:

In 1996 Felicia and Charlie moved their family to a new home they built right in the middle of the village of Hanalei. As you will read, their new home incorporated lessons they learned building an off-the-grid house. They sold their off-the-grid place to a friend.

Felicia: For me, the hardest part of selling our last place was leaving the fruit trees, which are getting big and producing a lot of fruit: lemon, lime, two oranges, mango, two avocados (one of which is bearing heavily), a macadamia nut tree, two sapodillas, atemoya, lychee, a stand of banana and papayas, and a hedge of guavas. We sold our old place to a contractor who's intrigued by the land's possibilities. Having all that fruit would have been wonderful, but the new buyer liked that, too. He's planning to add a well and several sustainable demonstration projects to complete the place's independence.

The main concern that motivated our move was the drive to the shop. I'd have to bring the baby with me, and it just wasn't practical. I might need to be at work for a couple of hours in the morning, and again later, and I couldn't stand to make the fifteen-minute trip back and forth. Dragging Matthew with me and staying for ten-hour days wasn't working either. I like not taking my car. I like the exercise better, and I'm not a car kind of girl, I guess. Now with the kids I don't have as much time to exercise in other ways, so I like walking with the stroller and riding the bike. It's a block to the beach, a block to shopping, a block to work, and a block to school when Matthew starts. I'll go six days without getting in a car. Mostly I drive only for meetings and groceries.

The first year we moved, I was very proud of the fact that I only put 2,000 miles on the car. We probably have a much smaller impact than we did living off-the-grid.

The biggest thing for me is how "second nature" conservation has become.

Charlie: If you build your house right and buy the right appliances, it's hard to go wrong. Our last house was really hot, but this one is cool. This house has a white roof, then an air space. I didn't want to put on a roof-mounted fan because the hurricanes we get here rip anything on the roof off, leaving a big hole. We placed our windows to get good cross-ventilation, which is important here.

We have two big AC load centers, all the lighting on one system, and the heavier stuff on the other. I'm waiting for a deal to come along, so I can buy a used renewable energy system, split the loads up, and go solar. Eventually, I'd like to be running this house with thirty 50-watt PVs.

The roof was an happy accident. We have a perfect south-facing roof with a 6:12 pitch. We bought the lot, and found some plans we liked,

The Cowden family.

and the only way it fit on the lot just happened to face the roof south. Felicia designed the house; I was too stuck in retail prison. She had a lot of ideas about how to design the house for kids. The way I figure it, if I put a lot of energy and thought into a project, I want it to go my way. She always comes up with good ideas. I don't think there's anything I don't like about the house.

I got the things that mattered to me, lots of extra wiring, dual load centers, so converting to solar power will be easy . . . I've already reserved a place for my metering.

It's amazing, how much material and money goes into building a place like this.

Felicia: Since we're twenty feet from the powerline, there was no compelling reason to have solar electricity.   Yet because we know what we know about solar power, we built this house for solar. And conservation, of course. Our electric bill runs between $30 and $45 a month, at 22 cents per kilowatt-hour; most people spend $150. And we're sloppier than we were in the other house.

Charlie: We use about five kilowatts per day, where in our other house we seldom used more than one. A fat quarter or a third of our energy bills is service charges, basic fees, so instead of the 17 cents they advertise, we pay up to 24 cents per kilowatt-hour.

We have a beefy solar hot water system, 120-gallon tank with 80 square feet of collectors, because Hanalei is a pretty cloudy area. What El Niño means here is it's been sunny forever. Normally we have the backup Aquastar (instantaneous water heater) shut off and bypassed, since 120 gallons of 140-degree water is a lot of hot water. Solar hot water is probably our main energy-saving feature. In Hawaii, water heating in an all-electric home is usually about 50% of the average electric bill. We got a tax credit of 35% of the total cost, which including the Aquastar was $5,000. Now I figure we get a $50 monthly solar subsidy. It probably takes three to five years to get to payback. Meanwhile, it's nice to get free hot water—less guilt, leave the water running to do the dishes, because we don't have a water shortage around here.

State tax credits have lapsed, the argument being that such subsidies favor the rich, because only they own land. It's hard to get a landlord to put solar water heating on a rental. But environmentalists want to reinstate the solar hot water credit.

Felicia: I have some good habits, and I designed the wiring for the TV and microwave with wall-switched outlets. I even designed the wiring for my gas range with a switch—it's got a light on it; stoves all come with electrical functions nowadays—but I don't usually turn it off.

Charlie: We bought an Amana refrigerator, really efficient, and a front-loading washer and dryer. Felicia, I saw some new ones the other day, and they're huge. That technology is really developing well.

Felicia: In a lot of ways we are still independent. We have our own little septic tank to think about and take care of.

Charlie: When you tell people around here your monthly electric bill is $34, they can't believe it.

Felicia: We are still aware of the environmental guilt factor. We feel good about our participation in keeping the air as clean as we can.

You focus on whatever you do, so I tell people about our low energy bills to try to encourage them to lower theirs by doing what we're doing. I have two babies in diapers, and this is a very standard house, so if we can do it, they can too. Around here, the main thing people complain about is their big electric bills, but you know, they don't do much about them. Maybe we need a good energy crisis. . . .

I've been too busy with two little kids to work much with schools. I do work with the business council, maintaining community focus. In a community like Hanalei, almost everybody cares about the environment, because that's what tourists come for. Of course we have our bozos: The rich business bozo who lives clear out in Haena says we really don't need any kind of business on the North Shore, but should be commuting like him two and a half hours every day around the island to Lihue!

Charlie: I still install systems. A major press release for us was the system we installed at the ecolodge in Peru. Felicia wrote it up for *Home Power* magazine—you ought to take a look!

I've done over forty houses working with the same electrician. We have come quite a ways, have even got the county building division to cooperate with us. We had a seminar with the inspectors, brought DC breakers and disconnects, and tried to show them what they should know to do their job. We try to show them new equipment we're trying, and mostly all they want to know is where they can get one.

Felicia: A hard issue I deal with is diapers. I'm using Huggies. I have several justifications. I used to give weeks of diaper service as a shower present, and got a month when Matthew was born. Diaper services use such harsh chemicals to maintain their sanitation standards that it can't be good for the environment. Cloth diapers don't seem as comfortable for the baby, and they don't hold things together as well. (Laughing) You just wouldn't believe the advances in diaper technology! They are even using partially recycled materials now, even if they haven't closed the recycling loop. We worry about the landfill . . .

Charlie: . . . because of the hurricane, where we used up seven years of capacity in six hours.

Felicia: So that's my main hypocrisy right now. ⤙

Storing Energy

When sources produce energy at the same time that loads require energy, the match is perfect. This is surprisingly common: When the sun shines, algae grows and pool filtering is required; or when the sun strikes the solar hot water panels and the water is heated, a photovoltaically powered pump provides timely circulation for an active hot water system. In chapter 2, I described a solar-direct fan which comes on when the sun strikes the roof and heats the air in the attic, and moves the hot air out. In these cases, a PV module directly driving a motor matches the source to the demand, and eliminates the need for other controls. Year-round hydroelectric systems may produce so much full-time power that little or no storage will be required, and the problem instead becomes finding uses for the excess, like heating water electrically, or electric baseboard heaters.

More often, source and demand do not match. If photovoltaic power is required for nighttime lighting, yet the source is collecting sunlight only during the day, a means of storing energy is required. The grid has allowed us to ignore the timeliness of our sources when using energy, and until we correct this mental habit, we use energy stored in fossil fuels.

The primary reason to store photovoltaic energy is because the sun doesn't always shine when we want power. The sunniest sites enjoy more than 330 days per year without cloud cover; other sites are lucky to have 200 clear days. The problem comes, even for those in sunny sites, when storm clouds cover the sun for several days at a time. At my home in Caspar and at the Rocky Mountain Institute in Colorado, these periods have been known to last the biblical forty days and forty nights. Unless backup power is available, storage must anticipate this worst-case circumstance. The number of days a system can run on storage alone are counted as the system's days of autonomy. It makes sense to store at least enough energy to weather the average storm and minimize the use of fossil-fueled backup.

Energy can be stored in three ways: mechanically, physically, and chemically. Mechanical storage is most familiar in a wind-up clock, where rotational energy is stored by tightening a spring which then slowly uncoils, releasing energy. Another mechanical technique for energy storage is through compression of a gas, the commonest gas being the air we breathe. Compressed air is an extremely clean, clear power source which can be used to drive hand tools and small turbines such as a dentist's drill.

In a fabulous example of physical energy storage, Pacific Gas & Electric (PG&E), Northern California's huge utility company, takes advantage of the fact that an electric motor can be designed so that its shaft will spin if electricity is applied to it, and will generate electricity if its shaft is spun. Using excess base-load electricity during off-peak hours, PG&E's immense motor generators pump water from a river at a lower elevation to a higher lake. During peak hours, when more electricity is needed, the water is allowed to flow back down, spinning the motor generators and generating electricity. This technique is also inefficient, but the inefficiency here is irrelevant, because the excess power comes from generators that cannot be turned off without great expense—a nuclear power plant is an example—and this excess is being stored, then converted into power when needed. A more down-home example of good storage technique involves direct PV-powered pumping of water from a well into a water tower only during the daylight hours, so that residents can enjoy the benefits of gravity-fed water pressure at any time of day or night. The storage tank must be big enough to supply household water during the times without sun; the pump must be large enough, and supplied with enough power, to fill the tank during sunny times.

Chemical storage of energy involves a reversible chemical reaction. Petrochemicals and their younger relation biomass are forms of chemical storage in which sunlight, nutrients, and water were patiently transmuted into an embodiment of energy: vegetation. The energy content in fossil fuels has been further concen-trated by time, temperature, and pressure so their energy occurs naturally in a dense and gooey form.

Many chemical reactions are reversible, and so can be used in chemical storage of energy, and every one of these has problems. The most common chemical storage devices, flooded lead-acid, or wet, batteries, familiar to us as the keeper of starting energy in our automobiles, have the advantage of being rechargeable. The next most common form, dry cells, found in flashlights and other portable electric devices, work in a similar way, except that only certain types can be recharged.

There is no such thing as a perfect battery; a whole-life cost analysis of most battery technologies leaves an honest evaluator depressed. Every automobile, as well as most forklifts, golf carts, cordless floor polishers, and countless other large machines, contain flooded lead-acid batteries. An ever-increasing welter of "batteries-not-included" smaller devices, from flashlights and boom-boxes to power tools and laptop computers, use dry cells. Consider the creation costs: Batteries are resource night-mares containing large amounts of embodied energy (the energy required to extract their raw materials and manufacture them). They are hazardous and heavy, therefore expensive to transport. Wet cell batteries, once in service, are dangerous to handle, giving off explosive and corrosive vapors, and requiring constant tending, and careful and often frequent re-placement. Now consider the disposal cost: In a typical small appliance, battery life span is measured in days or weeks. In a residential home renewable energy system battery life span is between a year and twenty-five years (depending on type, treatment, and many other factors), and all batteries are classified as toxic waste at the end of their lives. Lead-acid batteries are recycling heroes, being about 98% recyclable, but consumer dry cells have a dismally small potential for reclamation and recycling.

Even so, with all these unendearing traits, batteries are at the very center of a home energy system, and we had better love them. For the time being, they are the best game in town. I need hardly add that the invention of a better energy storage strategy will be greeted with delirium by consumers and independent energy producers. Imagine the path beaten to *that* inventor's door!

We measure the efficiency of any storage device in terms of how much of the charging energy can be recovered, the number of times the batteries may be cycled, and the original equipment cost. By these measures, flooded lead-acid batteries perform extremely well, and single-use alkaline cells are a terrible investment. We are always looking for better, by which we mean, more efficient, storage techniques, and we may, in the next decade get them: Flywheels and fuel cells are coming, but for now, flooded lead-acid is what most people will use.

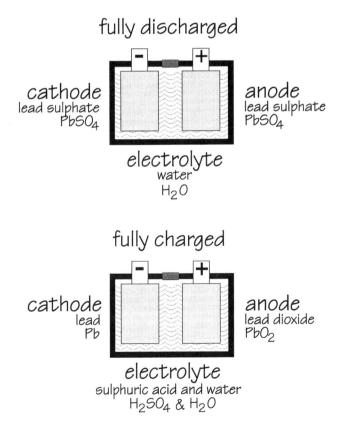

fully discharged

cathode
lead sulphate
$PbSO_4$

anode
lead sulphate
$PbSO_4$

electrolyte
water
H_2O

fully charged

cathode
lead
Pb

anode
lead dioxide
PbO_2

electrolyte
sulphuric acid and water
H_2SO_4 & H_2O

Fully spent, a battery's electrolyte has turned to water, all ions having been taken up by the cathode and anode. At the peak of readiness, a battery's plates have given up the ions to the electrolyte, which is now a corrosive mixture of water and sulphuric acid.

Balancing the Battery Bank

The best attitude to hold about batteries, I find, is to treat them the way you might a team of high-strung dogs. They respond well to love, but will find ways to get even if you neglect them. They are choosy about how, and how much, you charge them, and fail quickly when overcharged repeatedly (although, perversely, they do like to be overcharged occasionally, which battery mavens call equalizing); nor do they like to be discharged quickly or completely. They do not like extremes of any kind. They perform best at a constant temperature, are sluggish if cold, and fail prematurely in heat. They like exercise, to have electricity cycled through them, but prefer to be kept well fed, almost fully charged, which is called "floating." If left unused, batteries lose charge and fail, usually irretrievably; if left unwatered, they dry out and fail. At different times during the battery cycle, they give off hydrogen sulfide, hydrogen, and oxygen gases; the first is smelly and corrosive, the first two are explosives, and the latter encourages explosions and corrosion. Under some circumstances batteries also give off a mist of sulphuric acid which corrodes all metals, with a special preference for places where dissimilar metals touch each other.

In most books about independent power systems, batteries are described as petty tyrants and the dark side of the energy self-sufficient household. The truth is that they are safe, familiar, and remarkably well designed. Warnings are necessary because their contents have a potential for harm. People are mostly unfamiliar with their behavior even though we work with them frequently in our automobiles, where their contribution to the whole spectrum of vehicular catastrophe is negligible, and where their management is usually delegated to professionals.

A battery is a team of cells, each consisting of two metal poles or plates bathed in conductive electrolyte. The positive pole is called the anode and the negative the cathode. In a charged lead-acid cell, the cathode is lead, the anode is lead dioxide, and the electrolyte is dilute sulphuric acid. As a cell discharges through a circuit, ions of oxygen in the anode change places with sulfur in the electrolyte, and sulphur dissolves in the lead cathode. These changes produce an excess of electrons on the cathode, and a shortage on the anode, which causes the flow of electric current through the circuit. The lead-acid exchange has a specific potential, or electron pressure, stated as 1.5 volts of direct current. The state of a battery's charge may be known very accurately by its voltage when it is at rest, but the voltage of all batteries drops when their chemically stored energy is being converted into electricity and current is flowing. By the time a cell is discharged, the electrolyte has turned to water, and both cathode and anode have become lead sulfate. By reversing the current, so that electrons are forced back from cathode toward anode, these exchanges are reversed and the battery is recharged. Discharging the battery too deeply and repeatedly causes sulfate to form on the plates and fall to the bottom of the enclosure, reducing battery capacity and finally creating an internal circuit between the plates. This process takes place under all conditions, and all batteries succumb eventually, but well-bred and well-fed cells have been known to last for sixty years or more.

Batteries come in standard teams: Three

cells make up a 6-volt battery, and a 12-volt battery is made up of six cells. Each cell has its own contained electrolyte supply, so checking a battery's fluid levels will take time and care. If one cell is full, this does not guarantee that every cell will be full, but it is a fair indication. A battery is only as strong as the weakest member of the team, and a single bum cell can ruin an otherwise useful team, so affirmative preventive maintenance—keeping a happy team—is the best plan.

By keeping the batteries clean, well watered, and with all their metal surfaces greased (any grease will do) or otherwise protected from corrosive mists, the negative attributes can be kept under control. Batteries behave spectacularly badly in a fire, and are considered Class B explosives by the materials handling industry, so electrical inspectors in some jurisdictions require that they be kept in two- or four-hour fire-resistant enclosures. A well-planned battery enclosure and reasonable maintenance reduces battery risk to less than that associated, for example, with a propane stove, and far below the danger of candles, kerosene lamps, and woodstoves. Despite the greatest precautions, spills will occur, and so allowances should be made for spill containment, cleanup, and rinsing without shifting the batteries. Wise independent homeowners *always* wear goggles when handling batteries, always carefully wrap the handles of their metal tools in electrical tape, and always keep a big box of baking soda nearby.

People who live with batteries consider them the heart of their system, and develop a strange affection for the beasts. Despite their problems, batteries are by far the most efficient energy storage devices currently available, and so every independent home probably has an

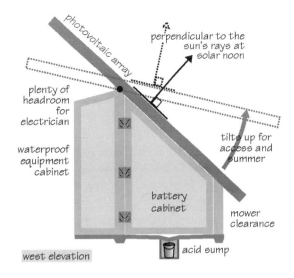

This little power house gives the batteries and PV modules a happy home. A well-ventilated battery compartment also provides excellent spillage control. The modules are above the weeds and stones hurled by a mower, and they may be aligned to optimize their output. A waterproof cabinet accommodates the charge controller and inverter.

imposing battery installation—impressive at least in weight. Because of their weight, their noxious fumes, and their aggressive liquids, batteries deserve a place of their own, removed from human living spaces but close enough to minimize voltage drop, because direct current (DC) doesn't travel well.

Plan for easy access, because batteries require servicing. It is a good idea to provide space for twice as many batteries as you think you will need; experience shows that you will underestimate the system size required, and your requirements will grow. Since battery life is logarithmically proportional to the degree of habitual discharge—typically, batteries used to 25% of their capacity before recharging last

It is a good idea to provide space for twice as many batteries as you think you will need; experience shows that you will underestimate the system size required, and your requirements will grow.

more than three times as long as batteries discharged by 50%—one theory recommends buying twice as many batteries as you figure you need. A larger battery bank is an excellent investment. Robert Hale, an experienced installer of systems on Maui, cautions that too big a battery bank may never get completely charged unless there is also a big equalizing source. If cells are never brought up to full charge, their capacity will gradually dwindle, and you will be left with twice as many spent batteries.

As with the arrangement of PVs into subarrays described in chapter 8, I favor organizing batteries into sub-banks, so that a smaller team of batteries may be thoroughly equalized by giving them the full production of the energy source. This also allows for mix-and-match battery purchasing, easier troubleshooting, and possibly the use of smaller-gauge wire connecting bus bars to the main battery controls.

Batteries do not get lighter with age, and battery acid is aggressive, so the foundation on which the batteries rest should be strong enough to be unaffected by moisture and acid; concrete is good, while wood even when protected with an impermeable, acid-resistant covering, eventually rots. When designing a battery enclosure, consider all these factors.

It seems silly to mention it, but I have made this mistake myself: The battery enclosure must be ready before the batteries arrive, because these are unpleasant heavy puppies to shift and you do not want to move them oftener than you have to.

Mush, You Batteries! Regular Battery Maintenance

Once the batteries are in place, your goal is to keep them full. Full lead-acid batteries, batteries in float, are happy batteries. The next happiest batteries are batteries that cycle gently from just below full to float. The more deeply lead-acid batteries are discharged, the shorter their life. Automotive batteries cycle as many times as the car is started, but use only a small fraction, less than 5%, of their capacity before driving charges them back up. Renewable energy systems *must* use deep-cycle batteries, which are designed with thicker plates to withstand 50% discharge several times before failing.

Batteries demand more attention than any other part of a stand-alone home energy system. The pioneering school of thought suggests that everyone must crash a set of batteries before they learn. Others, installers with dozens of systems in their portfolios, say that is old news. They say that a well-balanced system with a large enough bank of batteries, good metering, and decently trained users may skip this expensive first lesson

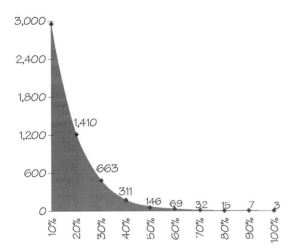

"Deep-cycle" lead-acid batteries last longer if they are never fully discharged. This graph shows the relationship between depth of discharge and number of discharge cycles for typical deep cycle batteries. I have found that the most cost-effective level of systematic discharge is about 30%.

entirely. Batteries misbehave only when treated cavalierly or ignored, and then they usually fail safely: They just stop working. In some cases, battery failure is due to the fact that batteries were consigned to a dungeon, literally out of sight and out of mind. An accessible fire- and acid-proof cabinet improves battery life-expectancy. Wise homesteaders use some of the battery bank's energy to make a light in the darkness so maintenance can be performed safely.

Most electrical system components are easy to comprehend, but what goes on inside batteries is quite complex and is poorly understood even by experts. Fortunately, to do a good maintenance job, you need not understand the subtleties, but only the broadest realities. Under normal operating conditions flooded lead-acid batteries, the kind of batter-

ies commonly used for domestic storage, convert small amounts of water into hydrogen and oxygen as they charge. If overcharged, and during equalization, their electrolyte seems to boil, although the bubbles are hydrogen and oxygen, the products of electrolysis. In either case, electrolyte levels fall over time, and must be replenished. Until a battery set's thirstiness is understood, monthly fluid checks are wise.

Batteries do their job superbly when treated correctly, but as emphasized repeatedly here, they do this job with dangerous chemicals, and so they must always be approached with proper care. Before starting, make sure you have plenty of distilled water for watering the batteries and for diluting an electrolyte spill, and an emergency box of baking soda for neutralizing any dribbled electrolyte. Wear goggles or face mask and gloves when working on batteries—you may feel ridiculous, but a splash of battery acid (the electrolyte) in your eye could cause permanent blindness—and wear old clothes, because acid eats fabric. Tools used around batteries must be nonconductive or have their handles carefully wrapped with electrical tape: Flashlights with metal cases, long screwdrivers, wrenches, and other implements with exposed metal parts are invitations to disaster, because a battery can unload a dramatic amount of current in a short time, if a conductor touches battery posts of opposing polarity simultaneously.

Cleanliness is important near batteries. Using a clean rag, either dry or lightly moistened with distilled water, clean the tops of the batteries, making sure that any dust and chemical deposits are cleared away from the area around the caps before opening them. Impurities in the electrolyte cause plate fatigue and battery failure, so start with as clean a battery as possible. Batteries exude a spume of sulphuric acid when they work, especially when they are equalized, which settles on their upper surfaces as a conductive film through which the batteries discharge themselves, so periodic cleaning increases efficiency. When you finish with the rag, burn it. You need not worry about being shocked by touching exposed metal parts with your hands, because voltages are low and you are wearing gloves and not particularly conductive yourself. Nevertheless, it is good practice to touch only the parts that need to be touched.

If cleanliness is important outside the battery, purity inside is crucial. Nothing less than distilled water will serve; purified water, filtered water, and drinking water will choke a cell quickly. One at a time, remove the caps from the cells and, using a flashlight, check the electrolyte level. There should be a split ring or fluid indicator showing the ideal level, and you may even be able to see the tops of the battery's plates. The problem here is that batteries are often put in awkward, dark places, and the electrolyte is clear, so it can be hard to gauge the level. Sometimes nudging the battery a bit will make the electrolyte slosh enough that its surface can be seen.

If electrolyte levels are found to be low, using a funnel and a measuring cup, or some other device (such as a turkey baster) designed to give good control over the volume of water delivered, gradually add distilled water to every cell in need. I am told by battery technicians that it is better to underfill slightly than to overfill, overflow, and thereby dilute the electrolyte (which also causes a corrosive mess that needs to be cleaned up). If you do make a spill, take great care not to get baking soda inside the battery, because it will neutralize the electro-

lyte. After getting exposed body parts away from the spill, replace the caps and flush the extra fluid from the top of the battery with distilled water. This doesn't make the spill more dangerous; it reduces the electrolyte's acidity and slows its attack. With the batteries sealed off from contamination, baking soda can be applied to the spill to neutralize the sulphuric acid. If any baking soda gets on the top of the batteries, it should be carefully removed.

Unwatered for too long, batteries eventually boil off so much water that they are permanently damaged, and will never again hold a full charge. Battery capacity decreases as electrolyte level drops because a smaller plate surface is available for chemical reaction. Eventually, the charging current and concentrated electrolyte attack the battery, precipitating a failure that, I can report from personal experience, is irreversible. Batteries are too expensive to be ignored; check them frequently.

While checking fluid levels, it is an easy matter to visually inspect and, if necessary, test all electrical connections around the batteries. Because dissimilar metals are enclosed near batteries that exude corrosive vapors that foster metal fatigue, these are the connections in a system that are most likely to fail. Multistranded cables surrender a strand at a time, and more and more current will be forced through the remaining strands until the cable eventually parts completely. If this degeneration is allowed to advance far enough, the heat of the current forced through the few remaining intact strands, or the sparking between the last severed wires, can damage the battery or even in the worst case cause a fire or explosion. Identify hazards at a very early stage, and replace the offending part with one better guarded against the known stresses. In the most demanding environments, where vibration from a generator or nearby motors and warm, salt-laden air adds to the battery outgassing, cables are usually soldered to their connectors, then covered with heat-shrink tubing, then sealed with grease or silicon at their terminals. And ultimately they still fail.

Taking Good Care of the Team

Each of the battery's cells, like each of the dogs in a team, is an individual, and may charge at a slightly different rate. If a weak cell runs with a strong team, it will gradually lose its ability to pull its share of the load, will weaken the performance of the others, and will finally fail. Periodic equalization, which is the intentional overcharging of the batteries, gives weaker cells a stronger, longer duration of charge during which to catch up, and helps take care of these individual differences. Equalizing need not be done more often than once a month. On a well-thought-out system, a battery equalization switch on the charge controller makes this into a simple task. Our experience suggests that typical PV arrays may not be able to provide the sustained overcurrent that makes equalization work; this provides another reason to have a nasty fossil-fuel-powered generator backup in your system. For systems with micro-hydro or wind power, if the power source is energetic enough, equalization should be possible without burning dead dinosaurs.

Battery equalization solves another problem. Over time, the electrolyte in batteries tends to stratify, so the acid is at the bottom and the water is at the top. This decreases storage capacity and also exposes the batteries to the risk of freezing. Normal use generally prevents stratification, but during times of light usage, the heavier, ion-bearing electrolyte sinks. During equalization, the batteries "boil" or outgas, and these bubbles stir up the electrolyte. Batteries freeze because the heavier sulphuric acid sinks below the water in the electrolyte cavity as the batteries self-discharge. A homogeneous electrolyte of dilute sulphuric acid freezes at quite a low temperature, but this water layer at the top of the battery freezes if allowed to drop below 30°F for an extended period, expands, cracks the battery enclosure, and ruins the battery.

Electrolyte maintenance, apart from keeping the batteries watered and keeping stratification under control with periodic equalization, is seldom necessary. Properly watered, the electrolyte lasts as long as the plates, because only its hydrogen and oxygen are lost to electrolysis. If you really want to pamper your batteries, use a good-quality hydrometer, not

Each of the battery's cells, like each of the dogs in a team, is an individual, and may charge at a slightly different rate.

the three-ball kind sold cheaply at auto supply stores. This instrument measures the electrolyte's specific gravity, which relates directly to the concentration of sulphuric acid, from which the cell's state of charge may be calculated. Every battery manufacturer specifies a different specific gravity, so ask for specifications when you buy your batteries, and refer to them. (Again, wear a face guard and gloves, and have water and baking soda close at hand whenever working with batteries. When using a hydrometer, watch for dripping acid.)

As the batteries come to full charge, the concentration of sulphuric acid in the electrolyte increases, and specific gravity rises as shown in the chart. By comparing specific gravity with other indicators of state of charge (for instance, ambient voltage and amp-hours consumed), we refine our sense of the general health of the team of batteries. Very rarely, and usually only because a battery has been watered until electrolyte pours out the top, the electrolyte will measure enough below the expected specific gravity that an infusion of additional sulphuric acid will be required. This is a nasty but quite manageable task. A good battery supplier will be able to provide guidance and the (needless to say, dangerous) chemical needed.

A final note about batteries: While much of the equipment for an independent electrical system may be purchased used, batteries should always be purchased new. By taking someone else's used batteries, you are merely solving their toxic waste problem, not getting a good deal. There is one possible exception: Used telephone company batteries are sometimes a very good deal despite their heft; these are the batteries that have been known to last six decades. Be wary, but if the batteries look clean, show good voltage, and can take and hold a charge, possibly you have found a bargain.

State of Charge

State of Charge	Electrolyte Specific Gravity
100%	~1.280
80%	~1.250
60%	~1.220
40%	~1.200

(Note: Electrolyte specific gravity varies with batteries from different manufacturers. Look to documentation that comes with the batteries for information relevant to your batteries.)

Things That Spin, Wear

Any part of your system that moves or is subject to corrosion will require maintenance. Windmills, pelton wheels, gen-sets, all have a maintenance regimen that can be ignored with complete assurance that failure will ensue. Like appliances that operate in the more benign interior environment of the household, each of these devices should come with a maintenance manual; unlike the users' manuals for household appliances that are designed to be neglected, you should read and heed these instructions. If the microwave fails, you can use the stove, but if the power system fails, you stumble around in the dark while you figure out how to pay for expensive spoiled equipment. Preventive maintenance will keep failures to a minimum.

For me and for many others who appreciate some level of energy independence, there is an honor and an honesty to the time taken on system maintenance. A system check is like a pilot's "walk-around" of an airplane before take-off. When we assert our independence and self-sufficiency, we should also be willing to don coveralls and gloves, take up the proper tools, and familiarize ourselves with the nuts, bolts, risks, and precautions that are part of generating sustainable energy. My ambivalence toward the automobile is betrayed by my failure to assume responsibility for its vile underside; I am in no way ambivalent about caring for my home power system, which presents a clean, appropriate, understandable, and rewarding challenge. Its diverse parts work so well, and are so easy to keep running, that I happily remember to give each component the small quantum of attention it requires.

Built to Be Serviced

The more complex the electrical system, the more equipment will be required to measure and control the flow of electricity. As in any system composed of dissimilar parts, thought must be given to putting the parts together, maintaining them, and eventually taking them apart. One of the most frustrating experiences of the new independent power producer is encountering a failed component that has been installed in a way to make servicing or replacement impossible without dismantling the whole system. Designers of a State of Hawaii rural electrification project in Miloli'i, for instance, thoughtlessly installed heavy metal bus bars across the battery fill plugs so

the batteries could not be watered without disassembling the whole massive battery array. Not surprisingly, a majority of the batteries boiled dry and failed quickly in sunny Miloli'i because they were never watered. One way to avoid such absurdities is to uninstall and reinstall components, adding parts as necessary, until components are well situated for servicing. Even better, install for easy maintenance from the outset. "First in, Last out" is a good strategy: Install the bulkiest and most robust equipment—batteries, disconnect switches, and the like—first, and more failure-prone equipment last. Thinking through all permutations of maintenance and the likeliest failure modalities before and during installation saves time and colorful language during maintenance.

Design for growth. Independent power users unanimously report that their demands increase over time. For some components, such as PVs and even batteries, adding a few more as finances permit or experience requires is not difficult. When faced with sizing a component adequately, if it will be expensive or awkward to expand, select the bigger one. For example, choose a battery circuit breaker big enough to handle a much greater capacity than you think you will ever need.

The simplest form of control is fusing and overcurrent protection. A battery short-circuited is an impressive thing, capable of propelling hot objects and creating an amazing amount of heat, sparks, and noxious gases in a surprisingly short period of time. Prudence requires fusing between any power-generating or power-storing device and any other device. The next most obvious precautionary measure is disconnect switching: We want to be able to sunder the connection between any two devices easily and safely at any time. Overcurrent protection and disconnects are inexpensive components when the risks to life and property without them are considered.

Balance-of-system components are the battery's accomplices, maintaining the system equilibrium by optimizing the charge applied to the batteries, and monitoring and then controlling sources to provide or store or divert current. In a simple system with a single source, the charge controller handles the whole task; in complex hybrid systems, the job requires a computer with a small raft of sensors and a custom program that accounts for the system's specific interactions. The idea is to choose equipment and manage loads so that the system remains healthy and requires as little intervention as possible.

System metering offers those of us who delight in measuring an opportunity to study the performance of each component with an elaborate metering network. For those who just want the simple facts, an amp-hour meter that keeps a running tab on the energy credits and debits at the battery bank suffices nicely.

Inverters: Different Powers

In a homestead power system, the inverter takes energy stored in the batteries as low-voltage direct current (DC), and changes it electronically into house current, also called 110-volt alternating current (AC), the kind of electricity delivered by the grid, into which we plug standard household appliances. Inverters are big cousins of those little energy criminals, the transformer/rectifiers, which do their wasteful business in the other direction, changing 110-volt AC into 12-volt DC in order to charge an electric toothbrush. As more homes

secede from the grid and the renewable energy marketplace expands, inverters are getting better. More reliable, far more efficient, and much better suited to the real demands of off-the-grid homes than devices only three or four

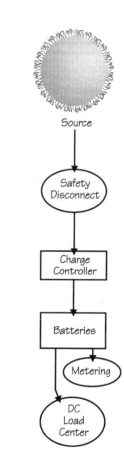

SAFETY AND BALANCE-OF-SYSTEM EQUIPMENT

Source

Safety Disconnect

Charge Controller

Batteries

Metering

DC Load Center

years old, modern inverters do a sloppy job very neatly.

As noted in chapter 2, AC house current looks like a sine wave, and most consumer appliances expect their current to look that way.

Inverters synthesize a sine wave using electronic sleight of hand that actually produces a modified square wave. Newer units come closer to matching the precise waveform of grid-provided electricity. Some amplifiers, ra-

---

A home energy system receives power from one or more sources and stores this power until it is needed. The system consists of the sources that harvest energy, batteries that store it, and other equipment that keeps the system in balance and safe.

Early low-voltage systems used simple on/off balancing devices, and safety equipment was primitive or non-existent; even so, there is no record of serious injury or damage associated with these early systems, many of which are still in service. As more is learned about maintaining systems efficiently, and as more systems are installed, balancing and safety devices improve. To date there are still no reports of fatality or serious loss due to the failure of an independent energy system.

A Charge Controller monitors the batteries and source, and keeps the batteries as full as possible while ensuring that they are not overcharged. When the batteries are full but the source is still producing electricity, some Charge Controllers can divert energy to a diversion load, which uses excess power when it is available. Fred Rassman (in chapter 1) diverts extra power to hot water-heating elements; I run fans, lights, and air-heating elements to relocate solar energy within my house. When the bright halogen spotlight over my work table comes on in summer, that signifies that the batteries are full and the sun is calling me to go out and work in the garden.

By metering the batteries, sources, and loads, we can determine the health and general status of the system, and we can track consumption. This information allows us to manage our demands and plan for maintenance intelligently.

Safety in case of malfunction and easy maintenance without risk to humans and equipment dictates most of the safety components: Safety disconnects allow all sources of energy to disconnect themselves if overcurrent or under- or over-voltage conditions exist. These high-amperage switches operate automatically or can be switched manually. In a conventional grid-powered system there is only one source—the grid—but in a renewable energy system there are often multiple energy sources, and so typically there are several disconnects. In practice, a circuit breaker should be installed in any line before it disappears into or through a wall or goes anyplace where its integrity cannot be inspected and assured.

Modern inverters now take over many of the safety and monitoring functions, although the inverter is itself not strictly a balance-of-system component. An inverter's principal function is to receive power from the DC Load Center (or from its own dedicated disconnect) and convert the DC power to AC house current, which then is distributed from the AC Load Center.

Load Centers provide a reliable and organized way to connect circuits to energy sources and protect them from themselves, each other, and problems in source and storage. Separate Load Centers for DC and AC loads usually contain circuit breakers matched to the wire size and anticipated load on each circuit.

dios, printers, and other electronic devices handle the rough edges of even the most sophisticated modified square waves poorly; their filters are designed to extract the expected sixty-cycle sine wave's hum, but they pass the square wave's buzz right through with the music or the printed page, or they just refuse to start. True sine-wave inverters are available for circuits that demand purity. Most modern inverters also have the ability to search for a load and to turn themselves on only when one is found; this makes them much more efficient, especially in a no-load situation, when they periodically emit a small intermittent pulse of power to seek loads.

The distribution system in an independent home looks like conventional house wiring, except that there may be more wires in the walls. Safety and practicality guide us: Each zone and general function in a house gets its own branch circuit, and so lighting, kitchen outlets, and the shop will each be independently wired and connected, through a circuit breaker, to the bus bar's main power line in the distribution center. In homes where two or more kinds of electricity are available, each will have a distribution center and a network of circuits or branches.

Because low-voltage electricity loses potential quickly if it runs through small-gauge wires, low-voltage circuits are wired with larger conductors to minimize the loss: A 15-amp 110-volt AC lighting circuit only requires fourteen-gauge wire, but to deliver the same amount of 12-volt DC current requires ten-gauge wire. As gauge numbers get smaller, the wire gets thicker: Fourteen-gauge is as thick as a thick pencil lead, .064 inches in diameter, weighing 12.4 pounds per thousand feet; ten-gauge, on the other hand, which is .102 inches

NO LOAD IS A GOOD LOAD

If we take energy for granted, we think nothing of leaving a light and a radio on so a room will be "friendlier" when we re-enter it. And we are impatient: When we poke a button, we want instant TV, light, or other electronic gratification. Appliance engineers have taken the cue, and designed our appliances so they are never really off. Futurist-designers proudly tell us that microprocessors, the little silicon computers that make microwave ovens and televisions "smart," will soon be found in all our appliances. Off-the-gridders ask: Must I really have a clock in my coffeepot? We make jokes about how difficult it is to program a VCR, and how few VCR owners "time-shift" programs by recording them automatically and playing them back when convenient, but every VCR with a clock frivolously guzzles electrons twenty-four hours a day, and succumbs to attacks of PMS (Perpetual Midnight Syndrome) whenever grid power hiccups.

What does it feel like to use no electricity? When the power fails, we know. Afterwards, owners of grid-powered houses scurry around like medieval mendicants resetting clocks. How many grid-dependent clocks do you have? I now have six, four of which are unnecessary. When I need to know the time, I rely on the clocks that are powered by rechargeable batteries or are plugged into the never-failing DC system.

Off-the-gridders breathe more easily when the last load is turned off and their homes go silent for the night. For them, electric power is a hard-won energy source to be used only when needed. At night, or in mid-day, when daylight is ample and no electrically assisted work is being done, the only electricity flowing in their homes is from source to storage. They are puzzled by on-the-gridders' infatuation with electricity and eagerness to bathe themselves in its constant invisible electronic smog. For them, no load is the natural load.

Inverters have become so efficient and robust that career electricians, including Wes Edwards (in chapter 7), recommend "transparent systems" to their clients. In such systems, householders have no access to low-voltage electricity, and may not even know they are using an unconventional power system. The system designer insists on high-efficiency appliances and lighting, and space conditioning and cooking are transferred to another fuel, but otherwise there are few notable differences between a "transparent" independent system and a grid-connected home. Guests come and go without noticing that there are no electric lines swooping in from the road. For convenience and simplicity, all loads use house current, and everything including phantom loads is plugged in and drawing current at all times, and the inverter never shuts down. To accommodate this consumptive arrangement, every part of the system must be oversized. Owners of truly muscular systems can persist in using electricity without conscious load management, as if they were on the grid. Yet electricians who design such systems for their clients install modestly sized systems in their own homes, and are enthusiastic about the natural rhythms that sustainable power adds to their lives. Rather than trying to pass this enthusiasm along to customers who are newborn energy independents, they have found it simpler to leave the incurable electron addiction undisturbed.

I have a strong personal preference for homes where all energy, including electrical power, is suited as precisely as possible to the most appropriate source for the task. A bright 12-volt compact fluorescent (CF) light, for example, may consume less than half the energy the 110-volt version of the same light connected to an inverter due to conversion inefficiencies at low rates of demand. Inverters are most efficient when producing lots of AC, and are most inefficient when producing a trickle. Furthermore, the single 110-volt CF may not be a big enough load to trigger an inverter in its power-saving, load-seeking mode, and so the bulb's inefficiency is compounded because something else may need to be turned on to get the inverter to take the demand seriously.

In a fast-paced consumerist society where we are required to assimilate new technologies every few years, we tend to sacrifice efficiency and ease of use to convenience and ease of learning. While we are learning, this simplicity may be helpful, like training wheels on a child's bike, but when we master the technology we should want it to serve us not only powerfully but efficiently rather than simply and conveniently. After talking to many independent homesteaders, including some who started in "transparently powered" homes and asked their electricians to give them more responsibility, I believe that we are all capable of becoming electrical sophisticates; we can understand that different tasks require different electricities; we are ready to shoulder the responsibility of efficiency over convenience. People who harvest their own electricity are proud to understand their electrical systems, its demands and special traits. As they learn, they aim to use less and less power more and more appropriately.

MISSION CRITICAL POWER

My home has two delivery systems. Besides a conventional AC distribution center which controls conventional 110- and 220-volt branches, my eight PVs charge a 12-volt 400-amp-hour DC battery bank of forty 1.25-volt nickel cadmium industrial cells. During the winter, my Trace 2512 inverter is programmed to keep the batteries at float in expectation of a power failure, but in summer my excess current is converted to AC and fed to other users on my domestic grid. A DC Load Center with branch circuits supplies power for lights and instrumentation, and the inverter powers an uninterruptible AC system for the computers. When utility power fails—which is often here on the edge of the continent—the computers are blissfully unaffected, and my work is not lost. When the outage persists for more than a few minutes, I shut down the power-hungry computer and put the house into stand-alone mode: I unplug the crucial circuits that power the refrigerator and pump from the useless grid and plug them into the uninterruptible AC supply.

During long outages, our battery bank-stored energy will keep our carefully designed home's essential energy-frugal equipment running for four or five days. So far, outages have extended beyond that duration only twice, at which point we are forced to eat all the ice cream, and carry water in buckets from the rain barrel. We cheerfully give up our big computer during outages in favor of lights, running water, refrigeration, and tunes. We welcome such outages as natural holidays.

Would a system like ours make sense for you? If you want to be absolutely sure that your power will not fail, you certainly need to provide some form of independent power supply. Outages are on the increase. We like our solution because it will work without surprises for a couple of decades. If you are prepared to wait for the power to come back on, you only need enough uninterruptible energy to make an orderly exit from any mission-critical processes that might be underway at the time of failure. If you are as intrigued as we are by energy, and want to make it a hobby, this is a good and serious project to undertake. Does it make economic sense compared to the grid? That depends on how valuable you consider your work.

In a utility-interactive system, the inverter acts as broker between the local supply and the grid. Since electricity flows both ways through the electric meter, the disconnect must be lockable so that linemen can work on the grid safely.

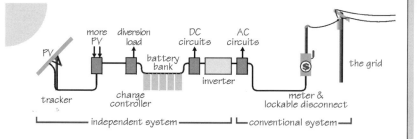

in diameter, is more than twice as massive, weighing 31.4 pounds per thousand feet.

Electrical loads are permanently wired into the system with switches and fixtures, or attached more temporarily to the system using outlets. In a household where more than one type of electricity is delivered to outlets, distinguishing between different outlet types is essential. It is best to employ outlet configurations that only accept plugs from devices that are looking for that specific kind of power. Unfortunately, there is no standardization; the commonest 12-volt outlet (a cigarette lighter socket) is not particularly safe, and no relief is in sight. Even sophisticated off-the-gridders report "frying" the occasional TV set by plugging it into the wrong kind of electrical supply. In most cases, only a fuse has blown, but that is embarrassing enough.

Light for life: Robert Sardinsky's story

The original reason for the electric grid was home lighting; at the time, no one imagined the welter of labor-saving electrical gadgets with which we have since complicated our lives. Many end-of-the-roaders left behind electricity along with civilization in a neo-Luddite frame of mind, but find themselves drawn back because they miss that sweetest of its applications, bright light at night.

Many lives have a theme, a realm of knowledge that provides a lifetime of intrigue and interest. Robert Sardinsky's theme is light. Talking with him, I become more conscious of light and of how I can improve my life by managing illumination well. Wherever I look, in my own work room, in conventional and innovative homes, off-the-grid and on-, I see that we often use light no better than we use electricity itself.

Robert Sardinsky and energy conservation got serious about the same time. After a stint as a pioneer in the northeast he came to Colorado to evangelize, first for energy consciousness, and now for light. He lives with his wife Colleen and their son above Snowmass Creek, near Aspen, Colorado, in an on-the-grid house with off-the-grid and energy-optimized systems.

I figure these four items high on my list of the necessities of civilized life: Light, Music, Refrigeration (for ice cream and beer), and Running Water. Light is the one that captivates me: Light is art . . . or it can be. It can make a home cozy, give a store pizzazz. I relate to light as an artist and a scientist.

As a way to educate people about energy, light is perfect. We know people are moved by light, and we can swap out a light in a minute, which can be very symbolic. Heating and cooling are important, but there's no subtlety to them: Either you're cold or you're hot. Light has so many levels.

We're solar-based, and our whole biology is based on light. We are clocks regulated by the sun, the seasons. Light is food, like vitamins.

In fifteen minutes—although I'd prefer an hour or two—I can teach you the basics. It's subtle, but not complicated. We were raised with incandescent light, which has more warmer colors in it and a lower color temperature than daylight, so that's the kind of electric light we are used to and think of as normal. Fluorescents and halogens have a higher color temperature, which means we see them as whiter, bluer. I'm curious: Would I be different if I'd been raised with cool white fluorescents?

It's important to ask, to what use will the light be applied? If you'll be reading or doing black-and-white symbolic work (writing, line drawing, calculating), a cooler color temperature, richer in blue, green, and violet may be preferable, but where true color is important, you'll need warmer light. If I'm lighting a salad bar, I want a source rich in greens, reds, and oranges. We can selectively choose what we want to accentuate. Here's an interesting one: Different races of people look better under different color temperatures, so you can see that cultural issues enter into consideration.

Another important measure of lighting is the Color Rendering Index (CRI). Incandescent light has a CRI of 100 by definition, but standard fluorescents range from 58 to 62. The good stuff we work with is 70 to 90 plus. You always want the best CRI you can get.

People are aware of gross lighting phenomena, like glare or dimness, but are not as sensitive to nuances of light as they are, for example, to scent or air quality. Yet, using light, I can create many effects: soothing, exciting, dramatic, oppressive. The first thing we can do is pay attention to the light where we live and

work. Notice a light out of your peripheral vision: Does it flicker? Can you hear a hum? Good-quality modern fluorescents don't hum, but the old ones do, and it's very distracting, to say the least, so fix it! Take a look at flesh tones: Do you or your co-workers look like they just got back from holiday . . . or from sick leave? Flushed, tanned? It's not so much a question of right and wrong, but knowing that you can have it the way you want it.

The worst offender I encounter is glare on video display terminals (VDTs) like computer and television screens. Look closely: Do you see multiple images, light fixtures or windows, reflected in your screen? Is the reflected image brighter than the screen image? We're drawn to light like bugs, and our minds must resolve the conflict of multiple images at different focal lengths. When we force ourselves (even unconsciously) to adapt to brightness and focal distractions, we waste our own energy and attention. A big part of my work is simple ergonomics: table heights, screen angles, so that people can come to their work directly, without distraction.

Of course, there's more to light—safety, security, and technical issues like efficiency, energy and maintenance costs, and the interaction in a building between lighting and heating, ventilation, and air conditioning (HVAC). There's so much to know and to do, there's no way one person, or one company, could do it all.

Thinking about how I got to this place, I remember my first business. I had a travelling light show when I was thirteen, fourteen, fifteen. It was before I could drive, so my parents had to pick me and my strobe lights up at three in the morning from high school dances. So you could say I've been studying light all my life.

I grew up in suburban Philadelphia, and another important source for me was Fairmount Park, which gave me my first sense of real wilderness. The Institute for Social Ecology at Goddard opened up my world; I studied with some of the most radical and innovative thinkers of our age. And I kept running across real inventors making extraordinary things: a guy in Vermont doing passive refrigeration by making a giant ice cube in an inflatable wading pool nestled in a silo; another guy working on a passive hydronic cooling system in his kitchen's northerly outside wall. I heard about the New Alchemy Institute and had to visit. My first impressions: a sailwind windmill's slow bright colors creaking in the wind, lightning bugs and black-light bug zappers, fish grabbing at the comfrey or whatever in the pools. It was awesome. I'd been at Penn State for two years, studying Art and Environmental Education, and I was ready to really do something.

At New Alchemy I helped oversee the Ark, but I was involved in everything. I studied carbon dioxide (CO_2) dynamics in bioshelters, and the process for cycling municipal waste into CO_2 and food, and the relationship between light and microclimate . . . I created the Institute's education program and self-guided tour. I lived in a yurt, built a solar shower, then a solar shack with a half-acre garden: raspberries just beside the door so you could step out to pee and pick berries at the same time. There were volunteer tomatoes, and windrows of carrots, parsnips, kale, and parsley that wintered over under the oak leaves.

The bioshelter is a special place: It brings the summer garden indoors. In winter in that climate the bioshelter is in the doldrums two months of the year, but the rest of the time . . . the flowers! the smells! I particularly remember skiing in to the bioshelter and picking fresh tomatoes on Christmas day, and watching a butterfly looking out of the Ark at the foot-and-a-half-deep snow. The hardware is fun, but it's the living stuff that turns me on, and the integration of the two is best of all.

Somewhere along in here I invented the sunbrero (a solar-powered beanie) and started wheeling and dealing PV modules because Boston was the center of much PV development at the time. I found this warehouse full of PV seconds, an eclectic mixture of stuff, and I started buying station-wagon loads way below the going rates, and distributing them around the country.

My idea about houses is, small is beautiful. The ultimate challenge for architecture is to make small spaces feel roomy, to have both intimacy and openness. Home can represent a model of what needs to happen on a larger scale. The wholeness, sense of place, and richness can be, should be, inspiring. Not everyone wants a garden, but I've never met anyone who

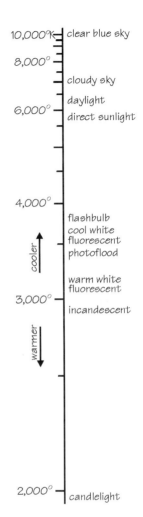

Light sources have characteristic color temperatures, measured in degrees Kelvin.

Robert Sardinsky.

isn't moved by a garden. My grand vision integrates people, plants, and architecture to create a stronger sense of *oikos*, home, a sense of family diverseness, richness, light, growth, and the senses of space and place. I get disoriented easily in cities: It all starts blurring, it's scary. Where there should be mountains, there are buildings, and there's no sun, it's usually overcast. I think we all sense this, and are drawn to the rural, want to move there. But we bring our consumptive garbage with us. We need to redefine and rethink . . . and we've got to do it quick.

I figured, we need to make models, and start close to home: our own homes. Do we know where our power comes from? We sure know bigtime when it isn't there! I found this rental house, and started making it into a demonstration of energy efficiency, doing the best I could with what I had. I didn't plan to secede from the grid, because the house shell itself is too inefficient, insulation at R-30 or R-40 at best, heat-losing glass. Naturally, the first thing I did was install all high-efficiency lighting, mostly compact fluorescent. I put up photovoltaic modules, a battery bank and inverter, and separated circuits out logically so that solar powers all but the resistive loads. It's fun to turn the lights on, and know they come directly from the sun. When the sun runs low, the system switches over to draw from the utility.

All the glass in this house works well—too well—on sunny days, but the house has too much volume and very little thermal mass, so even if we are very frugal, during the worst months space heating can cost $140. Conceptually, it would be good to increase the thermal mass, but that's not easily done in a rental. We did a blower test, and found serious infiltration, which we mostly solved with forty-eight tubes of caulk. We converted the fireplace to a state-of-the-art woodstove (a Vermont Castings Defiant Encore) and switched from electric to propane cooking. We made window covers by cutting up $32 queen-sized Wal-Mart Indian blankets, copper pipe, and leftover leather grommets from my sunbrero-making days, and they work great. I priced them, and the drapery people wanted $7 per square foot for something that wouldn't have worked as well. Now, we're puzzling on how to get the heat trapped against the high ceiling to circulate through this big, impressive, useless volume.

I've got a mess of micro-PV systems around the house for the fun of it: step lights that look like runway lights by the front door, and reading lights and a stereo in the bedroom. My truck has PVs in every window to recharge the cordless tools.

Knowing this is a rental, and that whatever gets installed might have to get uninstalled, I've been looking for low-tech solutions. I've got a qualitative problem with discount stores—our values are getting royally screwed up. Colleen and I like really fine things that last, that don't lose their beauty. There are wonderful old classic cars, for example, that look as fresh as they did when they were new. It's dangerous when we make do with temporary, cheap goods that don't last. ⌐

More Power to Us

I hear many justifications for getting off-the-grid and moving forward under our own power, the passions of pioneers who have taken responsibility for the energy in their lives. These reasons often start small and specific—"I didn't want to use nuclear energy"—but become more encompassing as these well-meaning people comprehend the interconnectedness of the kinds of energy we use—light, electricity, biomass and fossil fuel for food, heat, and transport, running water, just to begin the list. Common themes emerge from their experiences with managing energy, all leading to the discovery of more sensible and deliberate uses for valuables like time, natural resources, and community. Felicia and Charlie Cowden and Robert Sardinsky attend to matters at once more elusive and more important than many in the cultural mainstream can find time for. While they have opened the territory and shown it to be fruitful, they cannot enter it for us; this we must do ourselves.

In the next chapter, I want to introduce you

to some of my energy heroes, innovative thinkers who have helped me refine my thinking about energy. In different ways, they suggest that much of what we are taught to want may not help us understand what we really need. With this fresh range of possibilities in mind, I will ask you to join me in making an honest inventory of the energy we actually use.

PREHEATING WITH COMPOST

The King family of Riverside, California, uses ingenious appropriate technology to run their farmstead with minimal waste. This homemade passive solar domestic hot water system features a compost bin preheater that preheats water to 100–110°F regardless of the weather. A solar collector panel made of scrap wood, copper tubing, salvaged glass, and flattened beer cans, a breadbox heater, and a woodstove serve as backup for the coldest days. The Kings raise chickens, guinea fowl, quail, and rabbits for their table, using the compost from the preheater for bedding and situating the birds' pens on top of raised planters which subsequently get used to grow vegetables and flowers. They also raise pigs, goats, and bees. Oh yes—some of their power is supplied by PVs, and they heat their house with wood.

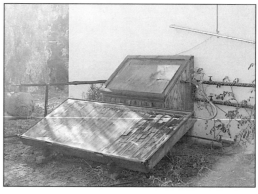

CHAPTER 5

COMPREHENDING ENERGY

THE STRUGGLE TO RECONCILE OUR DESIRES FOR WEALTH, FULL EM-
ployment, and environmental responsibility have devolved into an incessant squabble in which self-assured advocates of differing approaches
try to convert entrenched, interdependent, and often monopolistic
forces to their points of view. Those who are less than sure have learned
to shut up. Smart, busy people tune out the controversy, but their inattention may turn out to be more costly than they or our species can
afford. Predatory corporations, reflexively fearful of any kind of change,
are quite right to be afraid that rational energy and resource priorities
might change the way they are perceived by the people on whom they
rely. Deprived of their hard-won protective laws and subsidies, and required to do business responsibly, these monsters may not be able to
survive. The path to sustainability, the fabled level playing field where
everyone pays the true, whole-life cost of every kilowatt, therm, and
gallon, is proving to be fraught with potholes and landmines as these
brutes thrash about trying to protect themselves.

While governments and corporations fight their titanic battle over
the right to use the planet, a growing group of people take energy into
their own hands. By broadening our understanding of energy to include
any power that induces change, and by seeking to clarify our purposes for
altering the order of nature, we find ourselves asking fundamental questions about the forces that motivate us and the goals we strive to attain.
My personal energy heroes, several of whom you will meet here, are brave

explorers who have looked deeply into their needs and motivations, and discovered unexpected limits and surprising opportunities. Their lives and works help me formulate my own true requirements while extending the limits of my possibilities.

We have come to the point in this book when we look beneath the practical and concrete to explore the deeper meanings of "energy." After meeting my heroes and comparing their decisions about regulating desire and consumption with your own, I will ask you to conduct a reckoning of your own energy circumstances and habits. You will be asked to consider the place where you live from the perspective of energy. Then I will urge you to compile a baseline inventory of the energy you require to conduct your life. With these useful tools, we will be able to move on to finding better solutions for your unique energy puzzle.

Voluntary simplicity: Lennie Kaumzha's story

Before deciding to leave the grid's peculiarly addictive pleasures behind, we must consider how many of these, or how few, we really need in order to be happy. Could we be happy without all the newfangled electrical inventions and distractions of the twentieth century? If not, which would we care to keep? Voluntary simplicity is an idea that has ebbed and flowed periodically throughout history; it rose to a particularly high mark at the same time many were moving back to the land, in the late 1960s and 1970s. At the brink of the millennium it is rising again, stronger than ever.

For some of us, electing a simple life restored our natural perspectives and provided us with a steady foundation on which we eventually elaborated a more complex but still suitable life. For others, simplicity continues to be the basis for happiness and productivity. Lennie Kaumzha lives simply in a cabin near Brattleboro, Vermont, and tells his story of purposeful simplicity.

I have been a member of the Common Ground Restaurant, a worker-owned cooperative, for fifteen years. We emphasize vegetarian meals, and recycle many of our "waste products." I manage shifts, wait tables, and have numerous behind-the-scenes responsibilities, but my favorite job is dishwashing.

I live in Happy Valley, an enclave of eight cabins all basically without modern conveniences, one of which I rent. I supply electricity for a couple of lamps and a boom box from 12-volt marine batteries which I charge at the restaurant in the winter. I guess you could say I come to town for electricity. In warm weather, I plug into my car battery. I have running water, a gravity-fed system, most of the year; the pipes froze a few days ago, and so I haul water from the restaurant, maybe two or three months out of the year. I have what I consider a very good, simple shower system, right next to the woodstove, using a five-gallon jug mounted on the ceiling rafters, which drips into a plastic pan surrounded by sheets. I mix water heated on the woodstove with cold to the right temperature in the jug before hoisting it up. I spend about $140 a year on wood scraps, bought from a local lumberyard for heating the cabin. Refrigeration is a plastic ice chest; I freeze ice for it outside in winter, and at the restaurant in warmer weather.

Happy Valley sits nestled up against a mountain and reminds me of a hamlet out of time (except for the vehicles sitting in front). I write poetry, haiku for the most part, play fiddle, Irish for the most part, paint with acrylics, and make tin-can candle holders and lampshades with an oxyacetylene torch. I also garden and make maple syrup.

I am forty-seven years old, never married, no kids, and live currently by myself in the cabin. I don't know if I will ever have enough money to buy land and a cabin elsewhere, but I would be content to stay where I am now for the rest of my life. I live fairly simply, not because it's the politically correct thing to do, but because I

Lennie Kaumzha's cabin.

don't feel right any other way. I consider myself a peasant and feel uneasy around professional circumstances. I have been influenced by my Peace Corps stay in Gambia, West Africa, by the Native American spirit, and by Zen philosophy.

I feel that the more we live simply, using the least amount of money possible, working at jobs as little as possible (I average twenty-four hours a week at the restaurant; someone told me work is what we do in our spare time), the more we live closer to dirt and nature in general, the happier we will be. ⚊

Interdependence

Lennie Kaumzha's spartan existence strikes me as an extreme response, but I am grateful to him for providing such a clear standard of simplicity, a starting point, for the continuum of acceptable life styles. Recently, many observers have noticed that while Americans have doubled their possessions in the half century since the Second World War, happiness has decreased. To practitioners of voluntary simplicity, the cause may be obvious, but their wisdom does not communicate itself well through our marketing-oriented media. America's cultural obsession with possession and control blinds us to the idea that the way we work, spend, and play may be a form of addiction. As the planet becomes more insistent in its demonstrations that our substance abuse cannot be continued without harshly unacceptable consequences, let alone shared with other less-fortunate humans in developing countries, we must focus more clearly on our true needs and talents.

Pure water, warmth, electricity, and food are basic needs that Lennie Kaumzha enjoys in abundance, possibly more than most of us because they shine for him with simple clarity. In preparing this book, which is meant to explore the practicalities of harvesting enough energy to sustain a family and a homestead, I was thrust again and again into the tangle of interdependences and denials that comprise a culture but fail completely to propose a useful model for living well. In the first edition of this book, I somehow missed the wisdom of dear neighbors, who have been discovering and inventing ways to craft a rich life from the land and the elemental lessons it can teach if we pay attention.

Homestead energy: Stephen and Christiane's story

Stephen Heckeroth and his wife Christiane McLees live in a comfortable off-the-grid house a few miles south of Caspar. Steve has helped elevate my consciousness about buildings and energy for a quarter century. He worked with me on the design of my house, one of us holding a ladder atop the chicken coop which previously occupied the site while the other perched above to map views and energy incomes. We have both learned a lot since then, and traveled parallel energy paths. More than an architect, also a fine woodworker, hands-on builder, electric-vehicle builder, and a deeply inquisitive thinker, Steve is better than anyone I know at finding ways to express the importance of living within our energy income.

Steve: When I was young, my favorite times were our summer camping trips to the California north coast, where I learned to respect nature. My dad was a highway engineer. He remains my mentor, but we didn't always agree. He'd say, "We've got to build these freeways so people from rural areas can get into town." And I'd say, "But the freeways cause growth in the rural areas." Things like roads can be viewed as solutions, or problems. I don't think he and I have ever resolved that.

From the time I was very young I wanted to be an architect. In 1970 I was in college studying architecture at Arizona State Universtity. This is where I participated in the first Earth Day, and it changed my priorities in terms of architecture, from aesthetics and the way things look to concern about the way things work. An article by Malcolm Wells influenced my thinking tremendously. In this article he defined the necessities of life as air, water, food, and last, shelter. Of course up until that point I identified with different necessities like a car, a house, and all the consumer items that we've come to associate with success.

Another influence at that time was *Mother Earth News* and the back-to-the-land movement. The very first issue contained a pattern for a teepee, and we promptly sewed one up on our treadle sewing machine, and started searching for land.

Being clearer about the true necessities allowed me to concentrate on living rather than worrying about how I was going to make a living. That's an important distinction for me: the difference between a career and a life. My original architectural vision was to find an ocean point somewhere and build a monument that invented some great architectural style, but when I came to Northern California in 1970, after my experience with Earth Day, I looked for a place that would be able to sustain my family's life. We needed a fertile south slope with enough water for a garden and an orchard.

At that time, there were the two centers of alternative thought and lifestyle, one around Taos, New Mexico, and the other on the Mendocino coast. We quickly found a piece of land on the Mendocino coast three miles back from the ocean on a south slope above a river, five acres where I could put down roots. The first summers I lived here in our teepee, hand-dug the 20-foot well, and started planning a barn built with timbers from trees felled on our land and recycled materials.

I graduated in 1973, moved to the land permanently, and held a barn raising. All it took was a little notice at the local grocery store, and over a hundred people showed up. There were so many people here, it was hard finding work for all of them. Many of the people who showed up are still friends today. We happily moved from teepee to barn. I put up a 1925 Aeromotor windpump that same year and our first water system was completed with a tank on top of a redwood stump. The next project was a tower house with a 3,000-gallon water tank on top. The height of the tower gave us good water pressure. We also put up a wind generator, and for seven years all our energy came from the wind.

JOHN BIRCHARD

Steve Heckeroth and Christiane McLees.

In 1974 I was asked by the local college to teach classes in Self-Sufficient Homesteading and Alternative Sources of Energy. These classes forced me to articulate my vision. The students were hungry to know how to live lightly on the land . . . but I hadn't figured it out yet!

I read everything I could find on alternative energy and self-sufficiency. I asked a lot of emerging experts to do guest lectures on different topics. Jeannie Darlington, on organic gardening; Ken Kern on owner-building; Michael Hackleman and Windy Dankoff on

Rafter-raising at
Heckeroth's homestead.

wind power; and a few others on subjects like buying land and farm
animals.

Reading Schumacher's book *Small is Beautiful*, published in 1973, led
me to the idea that happiness and meaning are found in sustenance, not
consumerism. He brought me to another realization: Happiness doesn't
depend on consumerism and consumption.

I have always felt a need to refine things down to their essence. Recently
I found my ideas confirmed by The Natural Step when I attended one of
their workshops last year. Their four principles and four system conditions
closely match my own ideas. Thinking about it since, I have reduced it to
three life-sustaining principles: (1) We should use resources equitably and
efficiently. (2) We should not disrupt life cycles. (3) We should not disperse
poisons.

Let's take a closer look at the first principle, fuel efficiency: The net
energy value of the food we eat each day is about two BTUs. One BTU is
the amount of energy necessary to heat one pound of water one degree
Fahrenheit. Using those two BTUs we eat each day, we maintain our bodies
at 98.6°F. Recently a 14-year-old boy survived for a week in the snow, and
when he was found, his body temperature was still 96 degrees; he lived for

another week before dying of respiratory problems. This is an extreme example of how efficient life is. We're 80% water, so if he weighed 125 pounds, a hundred pounds of him was water, about twelve gallons. Without the life in us, that's what we are: twelve gallons of water held together with some carbon and other trace minerals.

To heat twelve gallons of water from 32°F to 98.6°F takes about 6,600 BTU per hour. And that was what he was doing there for a week. It would take a gas fired heater over a million BTUs to accomplish the same thing. He had eaten two BTUs of food energy, and probably lost quite a bit of weight, but still this demonstrates the efficiency of the life force compared to a water heater. At the same time you're using your food energy to warm your body, you can also perform work. You could ride a bike 100 miles a day—although you might like to eat another BTU of food afterwards. Let's compare this with a car: There's 150,000 BTUs in a gallon of gasoline. At fifteen miles per gallon, you'd use almost one million BTUs to drive 100 miles. So riding a bike is about a half million times more efficient than burning fossil fuels. Seems impossible? Try burning an apple to stay warm or power your car.

Let's talk about solar efficiency. The world's whole petroleum resource is estimated at a million terawatts, which happens to be equal to the amount of solar energy that reaches the earth every day. So if you multiply 365 days times the three billion years it took to create the fossil fuel resource you conclude that using direct solar energy is a trillion times more efficient than burning fossil fuel. And you also realize that all energy in this solar system comes from the sun.

When we use direct solar energy, we can calculate solar efficiency. PVs are about 10% efficient converting the sun's energy to electricity. Thermal collectors are as much as 90% efficient at converting the sun's energy to heat. Photosynthesis is only about 1% efficient at converting the sun's energy to complex carbohydrates which we eat to energize our bodies. Riding a bicycle is over three times as efficient as walking. Riding a bicycle is about 500,000 times (5×10^5) more fuel-efficient than driving a car. When you look at the overall solar efficiency of driving a car, multiply the fuel efficiency by one trillion (10×10^9) to establish that riding a bike is 5×10^{14} times more efficient. Personally, I don't want to be responsible for denying any future generations access to the resources that have taken the earth all these billions of years to create.

Of course the automobile also violates the other two principles: Roads, parking lots and driving totally disrupt life cycles. And I could rant on about the dispersion of poisons. . . .

The world's whole petroleum resource is estimated at a million terawatts, which happens to be equal to the amount of solar energy that reaches the earth every day.

What I want to impress on people is that we can't sustain life on geologic resources; they're in a different time frame than we are. We must slow ourselves down until we are able to work with the resources that come to us in our time frame.

When I really want to drive the point home about the automobile, I refer to the Baby unit. Burning a gallon of gasoline consumes about 700 cubic feet of oxygen. In the first year of life, a baby breathes about the same amount, 700 cubic feet of oxygen. Every time you burn a gallon of gasoline, you're burning enough oxygen to support a baby for its first year of life. And it's even worse than that: The baby can survive in a closed system because its exhalations are easily processed back to oxygen. The baby participates in a life cycle; the automobile does not.

I'm working now 90% of the time at home. We should make where we live be the place where we want to be.

Christiane: You've got to plan communities where people want to live and work. Go back to the village idea: You have your home or homestead and you are connected with other people through the work you do in the neighboring fields, offices, and schools. We need to move away from the arrangement that makes us get in a car and drive to get services. We need to be able to get the services we need by getting on a bike, walking, or over the internet. This would save time, so we could partake in the activities we don't have time for now: families, garden, a swim, just because we haven't given two or three hours to the car. And we'd get to appreciate our neighbors. But all this takes a revolution in planning.

Steve: That's why I'm on the county planning commission, and pushing for the use of GIS (Geographic Information System). GIS gives us an important planning tool we've never had before. It's easy for me to envision a county or city website accessible by parcel number. A complete GIS map might have a hundred layers: transportation routes, political boundaries, watersheds, flood plains, vegetation, solar access, climate, existing housing, employment, schools, shopping. And all the information about a parcel, such as building permit applications and planning decisions, are entered into the database. It would also be a welcome tool for surveyors, planners, contractors, the utilities, and others working with land use information. People looking for land today have no access to information like the history and conditions that exist on a particular site. GIS could provide that information plus every decision ever made by a government agency that affected a particular parcel. Land use decisions could then be based on accurate information instead of speculation. Realtors will love it because it's an instant sales tool; when you want to move, you'll access the database, enter information on the location of your new job, what your school and shopping needs are . . . and out will come all the available housing within walking distance. This could dramatically change the present pattern of commuting.

The only way I could see to accomplishing the transition to sustainable communities in the short term was to build a solar charged electric vehicle (EV), but that's only an interim solution. We must develop other ways to achieve mobility without pollution.

In our society, we seem to equate efficiency with convenience. And they're really different things. Somehow, we even confuse convenience with comfort. And then our level of comfort depends on whether we are being entertained.

Geographic Information System (GIS) mapping combines information about the lay of the land (contour and slope), and shading due to land forms and trees, into single map identifying the best home sites.

I look at everything now in terms of those three principles. Are we after comfort, or are we after health? Are we after convenience, or happiness?

If your goals are comfort and convenience, you're a couch potato.

If you redefine the success of shelter in terms of health and happiness, your shelter becomes a participatory environment which invites you to interact in its operation.

When I design shelter for someone, I strive to make it work so it provides those necessities I learned from Malcolm, light and energy from Sunshine, and healthy Air, Food, and Water. The size of the south-facing wall and roof of a building determines how much of the sun's energy can be collected. Glass makes solar heating possible because it's transparent to the short-wavelength energy from the sun, and opaque to the long-wavelength radiation from heated objects in the house. The challenge is to provide enough mass to store the incoming heat for whatever time period the microclimate requires. The mass acts as a flywheel to maintain comfort. Another necessity provided by windows and skylights is enough natural light to eliminate the need for artificial light during the day. The size and placement of windows is determined by the local climate. A good rule for most of the U.S. is that from 7 to 10% of the floor area should be window area on the south wall; 2 to 4% on the east and west; and less than 2% on the north.

For thermal mass I use materials with the least embodied energy that will persist in the structure for as long as it took nature to create them. Rock from the site is good, but be careful with old-growth redwood, because it should last as long as it took to grow: 2,000 years.

Since I like efficiency, I try to get building components to do more than one thing: Synergy is the root of this concept. Starting with the foundation: Rather than spending a lot of money on a foundation and then burying it forever, there are two other functions it can serve and increase the overall efficiency of the building. It can be thermal mass (which means it's got to be included within the insulating envelope), and it can serve as a room. In most parts of the country that's a basement, but in places where we have a temperate climate and don't need the basement for structural reasons (such as frost heaving) we waste that potential.

Going up into the house itself, I like the idea that rooms can change their functions. So one person might live in a single room, with a bed in one corner that can also be a couch, but can be curtained off. Throughout the day, the room changes its function—bedroom when you awake, then a kitchen, then an office, then the gym—more like the concept of living on a boat, where space is at a premium. It is at a premium, because all the embodied energy of the materials is so precious!

Every surface that encloses the space offers an opportunity to serve more than one function. Every wall has its particular synergy: The south wall is where we let in the majority of the light, collect the heat; thermal collectors mounted vertically replace the need for siding, and have twice the heat gain in the winter as in the summer, because of the sun's lower winter angle. I like the east side to open into the garden, because in this part of the country it's the most protected side; the entrance is there, too, and early morning warm up comes from that direction. The west gives a window on the end of the day, and must shelter us from the prevailing wind, and prevent the house from overheating in the late afternoon sun. The north is well insulated but still offers ventilation. If you lack suitable wall-building materials on the site, I like soil-cement or regular concrete on the inside of a good insulating envelope. Straw bale has great insulation, but offers little or no mass; rammed earth is exactly the opposite: tremendous mass with no insulation. Straw works well in dry northern climates if you also build in plenty of thermal mass. If you have a diurnal cycle with cold nights and hot days through the year, rammed earth works well, but in climates with cold winters, it only gets colder and colder.

The ideal wall might be a strawboard core, which has the same insulative value as petroleum-based foam, with a thickness of durable shot-crete (sprayed-on concrete) on the outside for durability and inside soil cement for thermal mass. Interior walls provide privacy, out-of-sight storage, services—if you put plumbing in exterior walls, you have to give up some insulation, so I like to put the plumbing in the inside walls. I like to cluster functions that require the same service, such as kitchens and bathrooms sharing a plumbing wall. A woodstove is only a backup to solar heating.

The roof serves many functions, from the obvious—keeping out the weather—to the creative—integrated PV collectors on the south-facing roof, eliminating the need for a separate structure and membrane. By using thin-film amorphous technology, which handles diffused light more efficiently, the need for a tracker is eliminated. Whatever the material, the roof must be lightweight and well insulated, and should provide ways for exhausting excess heat and letting in light. My first choice would be wood framing and straw board insulation with a 50-year-plus metal roof. I learned an important lesson in the 1970s when Ken Kern died because heavy rains caused his rammed earth dome to collapse on him.

Back in the 1970s, my focus was self-sufficiency, but the lessons I learned helped me make the leap to sustainability. Self-sufficiency exists in isolation, while sustainability requires that everyone works together for a better future.

Christiane: There are other options to the self-sufficient homestead, like work trades, barter . . . we need to work out complementary arrangements with suppliers of the things we need. We need to build community.

Steve: Serious gardening over a half acre requires a tractor. When I worked a big garden by hand, I saw that time was a limitation. Living in a peasant mode, subsistence farming in our modern environment, doesn't work. A human can produce a horse-power-hour per day, 2,500 BTUs. I tried Morgans, learned how to work them, and realized that it takes 13 acres of land to feed a horse in this climate, at least twenty times more than

Back in the 1970s, my focus was self-sufficiency, but the lessons I learned helped me make the leap to sustainability.

it takes to feed my family. This taught me to look to the sun for my energy source. My experience with EVs showed me that I could easily charge an electric tractor from the sun.

Regular gas or diesel tractors need weight for traction. The batteries provide that for an electric tractor. By using wheel motors, with the batteries right between the drive wheels, you optimize traction. You can have a PV shade canopy to charge the batteries and with the addition of an inverter you have a mobile solar power source for your electric chain saw and power tools. Plus all the farming benefits of having a tractor. Every implement that currently exists for a gas tractor works better on an electric tractor because the PTO (Power Take-Off) is powered by a separate motor and does not depend on ground speed. Electric forklifts have been in use for over a hundred years, and so the battery and motor technology is highly refined, and crosses over perfectly to the tractor. So my totally solar homestead is now a reality. ⌐

Mapping Home Energy

At home, we throw away one out of every two units of energy we buy.

⌐

Up to this point, although other forms of energy have been mentioned in passing, the emphasis has been on electricity. Now is the time to get serious about all the other energy uses in a home, and what will be required to achieve independence. How does your home life fit into the larger planetary energy budget?

We all consume fossil energy. In the last century North Americans have learned to use up phenomenal volumes. In one year, the average North American is responsible for reducing the sum of global stored energy by more than ten 18th-century North Americans used in their whole lifetimes, a five-hundredfold increase. Of course, these heritage fuels cannot be replaced in millions of years. There is much written about this subject which we do not want to duplicate here.

For the most part, we are unconscious of the energy we squander. Our vehicle addiction is beyond reason. At home, we throw away one out of every two units of energy we buy. Our appliances are inefficient, we use too many of them, and we use the wrong kind of energy in most of our activities. For example, using electricity for heat (in electric water heaters, stoves, or ovens) wastes up to 80% of the original energy. The all-electric dream home of the 1970s has turned into a costly energy-hog nightmare in the 1990s. Many of us seem to be incapable of having fun unless we are wasting energy with leaf-blowers, jet-skis, hot tubs, and a million unsustainable pursuits.

You are here invited to participate in two exercises needed to make a sensible energy budget. In the first exercise you are asked to make a drawing of the south elevation of your home, and a map of the space you inhabit extending out about 100 paces in every direction. In my work, I am constantly astonished by the degree to which homes (and most other structures) built since 1920 turn their backs on the free energy around them—sunlight, wind, and running water. To start using this energy well, you must first understand where it is, and this exercise provides an instrument for finding it. Your job is to map your home site's energy opportunities, or potential income, as well as the liabilities inherent in the site.

Although site is all-important when building a new house, I am often amazed at how much energy can be found and used in the ambient of existing houses. My students systematically find a quarter to a third of the energy they need to run their homes just "lying around" the site.

The south side of a house is the business end from an energy perspective, although all exposed surfaces, even the one exposed to the earth, play a part. Who would have thought that a roof's slope had any but aesthetic importance? But if we want to get the best performance from expensive photovoltaic modules, they should be pointed to within ten degrees of perpendicular to the sun. One of the most anguished moments for "tree huggers"—city-bred back-to-the-landers —is when they understand that not all trees should be kept. In a net heating environment, where heating degree-days exceed cooling degree-days, trees that shade the house or energy harvesting equipment are ready for conversion into firewood and building materials, and we gratefully replace them with lower-growing, more thoughtfully chosen flora.

One of the most interesting and curious discoveries for me, as I became accustomed to my land, was the patterns of weather. Hard to miss that the cold, sharp winds dominating our clear weather come from the northwest off the Gulf of Alaska. Storms, which had seemed a random and chaotic phenomenon, attack my house with greatest fury from the southwest here on the northern California coast. In lucky poverty, I chose not to build a driveway to my house . . . and now I am glad that cars must stay a hundred meters away. I recommend this separation to anyone who cares for a sweet-smelling, peaceful place to live. I am quite sure the map I would have drawn during the first year I lived on my land would have missed significant influences, and I am still learning. When I finally acquired a computerized weather monitor and started graphing weather, I discovered that a significant early morning wind blows from the east, the

A well-organized storm hurled itself at Caspar on February 5, 1998, dropping as much as a quarter of an inch of rain an hour. About the time the photo was taken, the power failed.

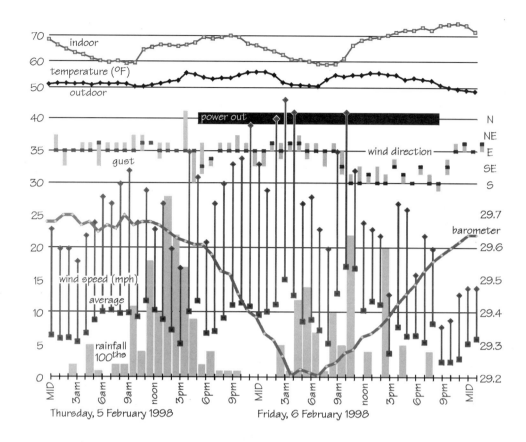

Thursday, 5 February 1998   Friday, 6 February 1998

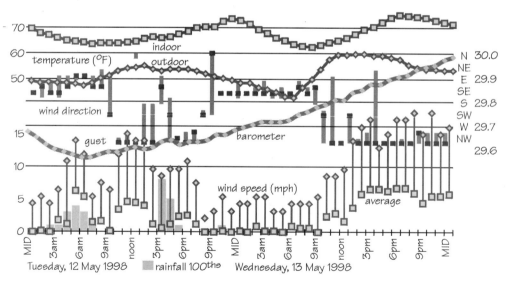

Tuesday, 12 May 1998   rainfall 100ths   Wednesday, 13 May 1998

The upper graph shows a pair of winter days in Caspar, California, and the lower graph shows two changeable May days when we had rain, calm, and a brisk nor'westerly in a 48-hour period.

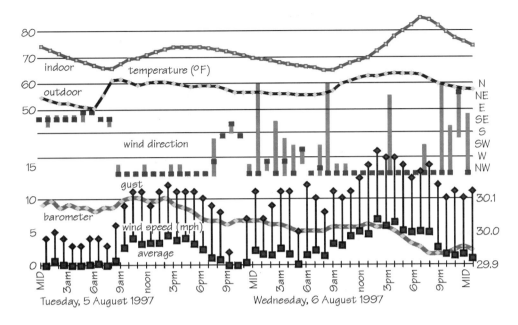

Caspar's mid-summer days are foggy or sunny. When sunny, like the second day shown here, indoor temperatures often rise into the eighties if doors and windows are kept closed.

"quiet" side of the house, about a hundred mornings a year. During the winter of 1997–98, which was exceedingly wet thanks to El Niño, vernal pools and swamps appeared that we had never seen before. A wise neighbor told me when I moved here thirty years ago, "You won't even *start* to know your land until you've lived here seven years." Thirty-two years later, I begin to think it will take more like seven times seven. How long will it take to learn the land if the climate is truly changing dramatically?

It will take most people at least a year to complete this mapping process well enough to use it seriously. Elsewhere in this book, wise homesteaders advise that you not build on a property until you have lived a circle of seasons on it. And you will need to be superlatively attentive to nuances during that year. Often, when trying to do this exercise with old-timers, I fail to get good answers about where winter storms originate. "The sky," the more perspicacious hazard. In summer I have been told by old-timers that there are no winter storms in places known for harsh winters. When I insist that the blizzards for which the region is famous must come from somewhere, I often get a blank stare, and the grudging admission, "I don't rightly remember." Civilization disconnects us from the influences of nature, and it takes a conscious effort to reconnect. This is your invitation.

In the back of the book, you will find a form you can use in mapping your own site. I suggest you work not in the book, but on photocopies of this form, possibly enlarged ones, and hereby grant you unlimited right to make and distribute copies of the forms for personal use.

Urban energy treasures: Berta Nelson's story

Berta is one of America's energy treasures. She lives right in the middle of a small town in Connecticut, yet she has taken firm control of her energy destiny. To accomplish this takes effort and consciousness, and Berta has plenty of both. She makes the point that careful attention to the energy we use, and attention to finding better, less wasteful ways, can make sense.

My main issue all my life has been electricity. I was born and raised in Seattle, where my dad was an engineer. He loved rivers. He was always telling us about great construction projects. One I remember particularly was the Grand Coulee Dam. He worked for the state Department of Fish and Game, and we used to drive all over the state looking at his fish ladders. I grew up thinking in terms of how many gallons it took to make a kilowatt, and about those fish ladders, so I've been really shocked to learn that they don't work. Oh, Dad, if you only knew! It just proves that we have to be more holistic in our thinking.

I was ten or twelve and very excited when I finally got to Grand Coulee. I was so disappointed! The technology was so brutal and archaic. All that cement! Huge machines! I remember thinking, there's got to be a better, gentler way to get electricity out of a river. I imagined some sort of cable that could be laid down the river so the crawdads and gravel could live right beside it, if we only had something sensitive enough to pick up the flow. I saw it as a problem then, and I still do, that engineers want to solve problems with bigger, bigger, bigger versions of the same old, same old.

During the late 1970s, when computer chips started coming into common use, I figured we'd have less gross machines, and we could finally graduate to less electricity. I didn't think of this as a soft path, just something logical and inevitable. Early on, during the Carter era, when everyone was talking about the possibilities offered by the sun, I was very interested in solar stuff and PVs. Making electricity directly from sunlight seemed really cool, so it was too bad that only NASA could mesh on that. When Carter left, I went to sleep like everyone else, I guess. There were a few years there when I just stopped thinking and kind of gave up.

I've lived in Connecticut for 25 years, and moved to Norwich in 1984, three doors down the street from here. I liked the neighborhood; you'd never know it, but we're three blocks from town, right over the hill, so you can walk to the Post Office and the bank. I hate having to drive everywhere. I knew the little old man who lived here, and something of the house's history.

It's a small house, about 1,100 square feet, built in the 1840s by a millworker who took a great deal of pride in his workmanship. The sills and underneath structure are quite heavy and in good shape. He used traditional windows with recycled glass or seconds, so they were wavy. He was building as cheaply as possible, doing his best to save money. I like that.

The last family to live here were the McCarthys. Mary McCarthy raised seven children, planted that big Norway maple, and had a big garden out back. I'm the seventh myself, and this is just like the little house I was raised in. The house's biggest defect is its perilously steep stairs, but I can handle that: I fall down, I get up again; I don't care.

One day a "for sale" sign suddenly appeared on the place, and I thought, this is my chance to own in this neighborhood. Couple, three

Berta Nelson.

days later, the sign was gone: My landlord got it. Fortunately we were friends, so he let me look at it, and I watched him work it over, doing stuff I could never do. He took out the horsehair felting in the walls, replaced the spool-and-spindle wiring. He ripped out all the inside walls and threw the plaster down the well, which made me mad . . . but now it's all gone! I could see he wasn't doing the best job, but was just looking for the fast turnover. This was just before the real estate boom of the late 1980s. He bought it for thirty-six, sold it to me for sixty-five, and one year later I was offered ninety. Now it's probably back to sixty-five, even with the improvements. I didn't mind paying for his work.

I chose this house for a lot of reasons, but as I said, my main issue all my life has been electricity, and it seemed to me this small house would be perfect. Right after buying it, we took a hot-air balloon ride over it, and we could barely see the house in amongst the trees: That was cool!

About the time I bought the house, I did a self-realization course and realized that I like being Ms. Solar, and set about finding out what was new with that industry. I was driving a school bus, so I had plenty of time to call up everybody in the phonebook with "solar" in their name. Everyone referred me to someone else, so pretty quick I had quite a list of local sources. It must have been about 1988 or 1989 that I found a Real Goods *Sourcebook* at the Yale Coop, and I grabbed it. I still have it, too! Wow! all this stuff available to regular people. Of course this was twenty years later, but that's really a big change in such a short time. I could understand that PV would let you do your own electricity, but not too much. The real question was, would it take lots of tinkering? I didn't care.

At the time, I was teaching gardening on the side, and had invented my pyramid greenhouse, so I was in touch with what the sun could do. I was thrilled to get back into electricity.

I had a big garden, and was doing a lot of canning, and so I got this ancient beast of a second hand refrigerator, *El Monstero*, we called it: noisy but I needed the big freezer. Suddenly my electric bill was through the roof. I knew it couldn't be just the fridge. I read Joel Davidson's book, *The Solar Electric Home*, where he said to audit all your electrical uses. I was thinking that solar was perfect, because we hardly used electricity anyway . . . was I surprised! I thought we had single 40-watt bulbs, but we were burning double 60s, things like that. When I compared the audit with the electric bill, it came out pretty much right. They really have you by the nostrils! Three hundred kilowatt-hours a month, how are we using all that electricity?

I'd already heard about Carol Levin and Richard Gottlieb, but I saw an ad for one of their classes, and signed up for a one-day class. It was early spring I remember, May or June, a blustery day. We spent all morning doing theory stuff, and that was all right, but we didn't get things wired up until . . . well, first we set up under a tree, then there was a little shower, then pretty soon the sun was setting—such a comedy!—but we still made enough electricity to light a couple of compact fluorescent bulbs . . . well, maybe one and a half, but I was so impressed! Out of a class of seven or eight, I was the only one that was really excited; I was sold! I wanted some.

See, I was afraid PVs would be fragile, with all that glass, and too expensive, but at Richard's workshop we were slinging modules around, turning them upside down. They're not fragile. I'm also sold on them because they don't make noise, and they're not dangerous. PVs are the most benign source of electricity I know about.

I got into this because I wanted to know, can a person who's used to having electricity and who lives on a street in a city . . . could a person like me put in some PVs and a few batteries, and really live without being a burden to the earth? Or would it be a constant hassle? See, people don't have to mess with their electricity now, they just take it for granted. It seemed to me that was an experiment worth five thousand dollars. Hopefully, it would fit in as seamlessly as the old way.

But the first thing I had to do was get my electrical consumption under control. I changed all the lightbulbs. A neighboring utility was doing a deal on compact fluorescents for $4 each instead of $20, but Norwich is a public-power town, so the utility sent my order back saying "nope, talk to your own power company." I got my friends out of town to send the order in for me, and when I got the bulbs, I put them everywhere. Luckily, they all fit. I liked using less power, but in the summer, less heat, too! I love that.

I wanted a super-efficient Sun Frost refrigerator in the worst way, and was going to have one. I like the design; if I had time, that's the refrigerator I'd build. I knew that building such things cost more, and I wanted to support the Sun Frost guys. If you like something, you've got to support it or else it might fail and you'd feel terrible. I talked my Mom into a little early inheritance money—she always liked new technology—and ordered the Sun Frost. When it came and we moved *El Monstero* out, we found that it had burned a big black hole in the kitchen floor!

That took care of reducing my big electrical loads. We've got a washer but no dryer. Sometimes when there are days and days of rain like this

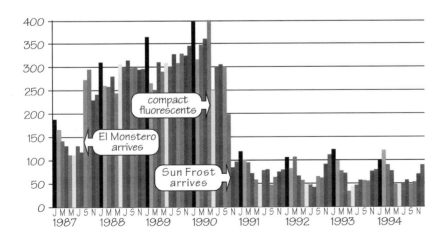

Berta's consumption of electricity shot up when she bought El Monstero, *dipped noticeably when she replaced the incandescent bulbs with CFs, and dropped to a reasonable level when she unplugged* El Monstero *and installed her Sun Frost. During* El Monstero's *reign, it was impossible to see seasonal patterns in electrical usage.*

August, the solar dryer—some folks call it a clothesline—is a problem, but in the winter the clothes freeze dry. Everything else in the house that uses heat runs on gas. We cut our electricity use from 300 to about 80 kilowatt-hours a month. I was pleased with that and figured I was ready for PV.

My original idea was to wire up just the kitchen and laundry room and convert the fridge, washer, and kitchen lights. Richard came down and looked things over and said "I think you'll get more energy than you expect; let's do the whole house." I was skeptical, but said okay, and it works out fine. See, I was impatient. I wanted to see if this stuff really worked, and I was willing to support the industry. So why wait?

I'd measured the distance from the power closet to the best solar exposure out in the yard away from the Norway maple's shadow. She's our air conditioner and saves more electricity than any of the rest of us. We had the perfect setup, with a 220-volt line for the dryer we didn't have running straight to the basement. Richard took charge, put in a new electrical distribution center in the power closet and the batteries, inverter, and instrumentation beside the washing machine. As long as I knew where the circuit breakers were, I didn't care. Nothing he did seemed too hard. If I'd had help sizing the wire, I could have done it all myself, except that it would have taken me a year, like finishing the sunroom has, to do what Richard did in three days. I wanted the electrical part done right, and I liked the way Richard and Carol do their work.

When the system was new, I was tracking cloudy days, daily amps generated. We've produced more than a megawatt in the three years the system's been in place! I suppose I should take more time to maintain the system. When it's sunny out, I say "it's sunny, so let's go off-grid" and throw the switch. There's a little buzz in the boom box from the inverter, but that's the only problem. Once in a while I check the batteries, but mostly I'm too busy to mess with anything. It's so easy, it's transparent.

Here, we're on a main with the hospital, so the grid never fails. Right after we installed the system in 1991, we went away for three days and there was a big outage, but when we came back, we could find no evidence that we'd been without power, so the system obviously saw us through. I think a lot about others. I have a friend who lives in Hampden, near New Haven. Every time the wind blows, she loses her electricity. She's electricity-dependent, with pumps and all. I've lived like that. I was talking to her last winter while a big storm was coming, and she was filling her bathtub so she wouldn't run out of water. I said, "Wouldn't it be great to have your own power if you mean to stay there?" There're lots of people in Connecticut

I'd give this whole experiment a B+: quiet, safe, able to deliver what I need.

—

like that. When I lived in Monroe, we'd get a little breeze and lose power. For five thousand, it's worth it.

I have a problem with batteries even though I have a love affair with my PVs. We need some better kind of absorption and carry-over. I'd give this whole experiment a B+: quiet, safe, able to deliver what I need. It's easy flipping back and forth to the grid, which I don't even have to remember because it reminds me because the lights flicker. If we need something big, like a saw, we switch back to the grid. In the wintertime, we need extra electricity to run the fan for the heating system.

That's the next project, now that the PV is finished. I've gone about as far as I can with electricity, so what else can I do? In the winter we pay between $150 and $250 a month for heat. That's too much, and I want to cut it down. Again, we started out by reducing our loads. The landlord took out all but two of the north windows, and I put in new double-paned windows all around, vinyl sash, low-E. I did the best I could to tighten up the whole house, although I can do an even better job blocking the wind through the baseboards. The winter of 1993 was brutal and there's no way to know how our bill would have compared if we hadn't made the improvements. This winter will be the test.

There seems to be a big barrier between seeing that energy conservation is a good thing to do, and actually doing it. Maybe people think that what they do makes no difference, but I think that's wrong. If we do it, house by house, hundreds by hundreds, we can make a huge difference. I even got into a fight at church about plastic cups, my friends saying it's a lot of work to bring cups from home and wash 25

cups a week 52 weeks a year . . . but now they see it my way.

There are two other barriers. Electricity is absolutely invisible. People are energy illiterates. They say, "Do you use your solar panels for heat?" and I say no, for electricity, and they say "What do you use that for?" Well . . . people have no idea what they use electricity for, and yet they still use enormous amounts. To conserve, you have to make a consciousness shift and even a lifestyle change. I don't burn as much electricity now, but I surely wouldn't want to give up the stuff I do use.

The third barrier is "what's the payback?" Well, for me, knowing that I'm not supporting James Bay, not damaging the air or resources, knowing that our household contributes as little as we can to the destruction caused by generating electricity. The way we produce electricity wastes land and oil; spoils air, rivers, shorelines; reduces biodiversity. What, besides the car, is our main use of oil? Making electricity, that's what. Getting out of that loop is enough payback for me!

My other payback: When I'm seventy, I'll already have electricity. Nobody knows what electricity will cost, or what my income will be. Buying PV seems like the most conservative investment I could make, as well as the smartest personally and most socially responsible.

I'm fascinated by a new technology involving low-temperature phase change, invented by a guy down in New Jersey who's working with freon—I know, that's a problem—in half-inch-thick four-by-eight-foot panels which gather heat and store it as hot water. If we can produce heat for almost nothing like he says, and I know that's a big if, then we should be able to produce all the hot water and heating we need for only the cost of the equipment, just like the PV electricity which looks free to me. At least it's worth investigating.

We had a contractor build a sunroom with heat-mirror glass on the roof, buffered to the southwest by our Norway maple. I had him build the shell only, so I could finish the inside myself. I'm disappointed with the work the contractor, who's a friend, did. I told him early on that I meant to do a radiant floor, which should have been a slab poured over gravel to make a good earth connection. He's just a sunroom installer and I guess most people just want a glass room and don't care if it really works. He stonewalled until we couldn't pour proper cement, and used the wrong kind of decking for a decent thermal mass. Right now I'm putting in a hydronic floor, with Wirsbo tubing under gyp-crete, working like a fiend to get it level. Amazing, how long it can take when you do everything yourself.

I didn't like the glass sunroom roof at first, but now I can see my maple. I can't tell you how many people have said, "You'll have to cut that tree." But I love the way she lovingly drapes herself around the house . . . and I can see her so well through the sunroom roof. A lot of things happen when she loses her leaves in the fall. Her branches lift and she pulls into herself. In the summer, she fills back out into a wonderful canopy that keeps the house cool. She reminds me of what I want to do: find the most benign way to live on the earth, lightly, in cooperation and concert with everything, without giving up what I really need. ⌁

Reckoning Home Energy

In the second exercise, you will be asked to record a baseline of the energy you consumed in the last four quarters. By noting how many kilowatt-hours of electricity, therms of heating fuel, cords of firewood, gallons of gasoline and water you consumed, and how many gallons of sewage and barrels of garbage you produced, you quantify the impact of your energy behaviors from the immediate past.

The point of this exercise is to promote awareness of energy consumption. In the back of the book, on the Home Energy Audit form provided, you will note a column marked "something happens here" . . . and we sincerely hope something does. We simply cannot continue wasting energy at the present rate. In fact, many futurists believe we will find ways in the next decade to reduce our per capita energy consumption by a large amount . . . or perish. We believe that awareness is the first step to resisting this undesirable result, and invite you to work with us.

Using this form as your starting point, you can get a clearer idea of how your whole energy regimen shapes up. Those who submit to the harsh self-evaluation of this work tell us that the information and techniques contained in this book, applied to the inefficiencies and out-of-control costs revealed by this audit, have brought about immediate savings. Most are easily able to reduce energy expenses by 50%, using this liberated surplus to finance further measures for energy efficiency.

(Again I recommend that you work not in the book, but on photocopies of this form, possibly enlarged ones, and hereby grant you unlimited right to make and distribute such copies for personal use.)

Our goal is to take a year of old data, and use it to plot a course for the next year. As we saw with the chart of Berta's kilowatts, a longer period often shows interesting and unsuspected trends. In most places in North America, weather breaks into four roughly equal seasonal periods; in Caspar, they happen to coincide well enough with bookkeeping's quarters, January-February-March and so forth to October-November-December (JFM-AMJ-JAS-OND), but your weather year might be different. How you quarter your year is not especially important as long as you consistently use meaningful periods for your climate. As soon as you start penciling in data, the seasonal differences should become apparent.

Note that I am asking you to look at two measures of nearly every consumption, dollars and some appropriate unit of volume. As we have seen, the real work is to reduce consumption, but reduced cost gives you the budget you need to implement changes. Most of us live close enough to the edge that our economies must be real, and must pay for themselves in a reasonable period of time. The electricity industry is rife with enthusiasm for "energy restructuring," which means many of us will get to choose our electricity provider. Whether this is a good idea in the long term is anybody's guess—it will depend on the games played by those with the power. (Paul Gipe in chapter 3 and Leigh Seddon in chapter 14 offer their differing opinions.) But in the short term, suppliers and governmental agencies can be counted on to confuse the issues with bait-and-switch tactics, misrepresentation of renewable energy shares, and promises of reduced and guaranteed energy costs. In many communities sewage charges are based directly on water use, and you may be paying unfair sewage charges for water used in your garden. The best solution for this particular conundrum may be to get an *agricultural* water account for your farming efforts.

I ask you to total all costs for the year, but you may also be interested to calculate seasonal sub-totals. This exercise would work better on the

Many futurists believe we will find ways in the next decade to reduce our per capita energy consumption by a large amount . . . or perish.

computerized spreadsheet available at the book's website:

www.chelseagreen.com/IndHome

The really important part of this form is the "something happens here" column. I hope you have gleaned some ideas from this book to help you make the magic happen. One friend did this, and at the end of the process exclaimed, "I can't believe it! I'm working to support my car!" Others have reported that simply becoming aware of the unreasonable costs, and applying simple common sense to reduce waste, saved as much as 20% per month, enough to afford some serious efficiency measures, which of course reduced costs even more—a cycle both virtuous and enjoyable.

The smaller grids at the bottom of the audit should help you get a closer focus on efficiency projects that will pay for themselves. Starting with the home's north wall, which is usually the coldest, a policy of keeping rooms closed and unheated or at lower temperature on that side of the house in winter, plus a program of replacing old single-pane windows (starting with those that can be felt to be the greatest sources of cold) should have a significant effect on the cost of heating fuel, electricity (if used for local heat), and firewood. Earlier chapters provide suggestions for reducing costs of refrigeration and hot water.

Getting Specific

In the following chapters, readers getting ready to build new, energy-independent homesteads will find advice on suiting a home to the site and the inhabiting family. Readers who have already completed this adventure may find these chapters elicit more than a few winces at what might have been, interspersed, I hope, with Aha!s. As I visited homes in various climes and building styles, I was glad for my home building past, which heightened my appreciation of the thought and care builders focus on important details I missed. Readers who acquired someone else's ancestral home and are working to make the best of it may join me in sifting the next chapters for ways to refine my home's relationship with its land and my family. Before casting ourselves adrift on an ocean of dead dinosaur ooze, let's look closer to home for the energy we need.

CHAPTER 6

CHOOSING A SITE
FOR ENERGY

AN INDEPENDENT HOME IS BUILT LIKE ANY OTHER, ONLY MORE MIND-fully. The builders of an independent home must understand all the relationships that master builders of previous centuries understood: the relationships between the home and its inhabitants, its surrounding biosphere, the materials from which it is made, and all its energy sources. Consumerism and dependence made homebuilders sloppy. Desiring to live within our current renewable energy budget offers us challenges and opportunities that enrich the planning of an independent home. We are invited to become partners with the land, and let the landscape itself show us where it would be best to build. Where the housing industry bruises the land and flattens its particularities, independent home-builders gratefully accept guidance from the land.

Until the early 1980s, the pattern of home life in the United States was absolutely determined by two systems laid down upon the land by humans: a pervasive network of roads, and the only slightly less far-flung network of the utility grid. Our lives are so intertwined with mobility and electrical power that between 1910 and 1990 the presence or absence of the grid nearly always determined where and how we built our shelters. In the United States in 1980, there were about one hundred million houses; fewer than ten thousand, or .01%, mostly low-income families and remote hunting cabins, were not connected to powerlines. In the last two decades, the number of independent homes has increased by a factor of thirty to a little better than a quarter of a percent, or about one of every four hundred homes. Still, the tide of settlement is turning away from the tract-home model and the deterministic grid. As people become aware of the brutal life of cities and yearn for the remembered natural connections of country living, more and more of them gladly disconnect from life support, and live independently. (See page 3 for a graphic representation of this trend.)

Declaring Freedom

Most of us find it impossible to imagine a house without electricity. Now that we can easily and reliably generate all the electricity we need, the grid no longer defines the perimeters of settlement. We can select our home site for its merits alone, sure in the knowledge that we can harvest electricity anywhere we choose to settle.

Electricity is very likely to cost more and more as easy energy gets scarcer, and as visible and invisible subsidies are rationalized and monetized. Prices asked by utilities for line extensions have also risen abruptly, so the cost of stringing powerlines across a mere quarter mile of emptiness between a home site and the nearest grid connection makes a convincing argument for independence. As you will see in subsequent chapters, there are other reasons for disconnecting from the grid, or for connecting as lightly as possible, but cost continues to be a most compelling justification for creating a self-sufficient home.

If we have the inclination or the need to build independence into a home, we should pause first and review a list of the possible independences we may wish to attain. Homesteaders who have built independently tell about the freedoms they cherish with admirable vigor and intelligence. As I visit more and more homes, the list of freedoms keeps growing. Here are some basic freedoms worth considering, organized into the three phases of homemaking: building, occupying, and maintaining.

Freedom While Building

Faced with the demands of "curb appeal" and the real estate market, choosing a happy situation for a home has become a lost art and building a house can be frightfully wasteful. Developers flatten the land and parcel it into convenient lots without regard for natural contours, existing vegetation, neighboring views, or the vagaries of the weather. Builders of conventional custom houses typically waste almost a third of the materi-

Now that we can easily and reliably generate all the electricity we need, the grid no longer defines the perimeters of settlement.

als purchased. These two mistakes have a major impact on the cost and quality of our residences. An independent home will be sensitive to the environment from the very first footstep on the site. In building, we try to preserve the spirit of the following freedoms:

— Freedom from the insults of local microclimate. It is important to build the home in harmony with local weather. Shaded north slopes are clammy in winter—treasured in a hot climate, but uninhabitable in the northern forest. Unshaded exposure to direct sun without access to tempering breezes can make summer hellish in hot climates. A home's main entry should be on its most protected side, so that opening the front door during a windstorm will not rearrange all the light objects in the house, or rip the door off its hinges.

— Freedom from the hypocrisy of making war on the land before settling. We need not destroy the landscape and its life forms in order to occupy it. Where developers terraform, leveling mountains and filling swamps, purists attempt to move as little earth as possible. If earth must be moved, precious topsoil should be carefully set aside and lovingly redistributed after construction calms down.

— Freedom from reliance on rare, endangered, or strategic building materials. Local or recycled building materials are invariably cheaper and better suited to local conditions. Imported materials benefit the importers, distant owners, and possibly the builders, but are seldom helpful to the residents.

— Freedom from pollutants and extractive or exploitative materials and techniques. Homes made this way are healthier for their inhabitants, and demonstrate that we can live happily without oppressing others.

Freedom to Live

Once the home is built and we move in, another set of freedoms becomes important. When we live independently, we dedicate ourselves to treading gently on the land, knowing that such a life will give us freedom from many old patterns and dependencies:

— Freedom from dependencies on governments, far-flung monopolies, and extractive and polluting technologies. Local energy sources are in tune with their environment; to the extent that these are truly free even from hidden pollution and deferred costs, they bring a gratifying economy to our lives.

— Freedom from the necessity of daily travel. By integrating life and work, we may find that both are enhanced. Few who manage to work at home are sad to give up the daily commute, the inhaled exhaust, and the frustrations of traffic and parking. If we limit travel to necessities, share with neighbors, and adopt vehicles that minimally impact the biosphere, we may be able to preserve our treasured mobility.

— Freedom from estrangement with Nature. We know the restorative and inspirational value of taking nature into our lives. Why not live always with beauty?

— Freedom from food and clothing dependencies. Living independently, many find that their tastes in clothing and food change dramatically, needs becoming more manageable within a simpler context. To the extent that we can produce our food and clothing on the homestead, we liberate ourselves from dependence on external banking and marketing systems, which devour time and reduce our ability to concentrate while enforcing on us a culture of scarcity and perpetual obsolescence.

— Freedom from propaganda and regimentation. Only by taking responsibility for our own entertainment and education, by recreating neighborhood and community, can we free our minds.

— Freedom from our own garbage. By intensively reducing, reusing, and recycling our waste stream, we bring consumption under control. This, more than any other action, runs counter to the prevailing disposable culture, and emphasizes our eagerness to depend on ourselves and the place we have chosen as our own. In this freedom, civilization is slowly coming to meet us. Overflowing landfills and a new understanding about pollution in our air, oceans, and rivers has changed our national garbage behavior significantly over the last decade.

Freedom to Last and Restore

We may wish to build for the ages, but this is unlikely: All things fall, and are built again, and the building of them fills us with joy. We must accept the inevitability of decay, and from the start we must plan for maintenance of the home during its useful life, and the eventual restoration of the site to its natural state. Planning for the long term gives us access to further freedoms:

— Freedom from wastefulness. If we have built with materials and techniques selected to be durable, maintainable, and finally, capable of being dismantled and salvaged, the materials in a home are truly invested in it rather than consumed by it.

— Freedom from knowing we will be remembered for our selfishness. By treading lightly on the land while building and living, we leave a haven that may be enjoyed by our descendents or restored to its natural beauty after we are gone. The notion of "ancestral home" has practically vanished from the American experience, although we and our children increasingly yearn to go there. Any action that outlives us bequeaths its own burden of care.

Building in Our Minds

The elements of an independent house are assembled first in our minds, and then with our hands. To the extent that we anticipate what we must do and what impediments will stand in the way, our progress will be orderly. Therefore time spent in careful selection and planning repays itself abundantly. When we rush to completion before visualizing the process step-by-step through to the end, we condemn ourselves to live within a monument to our mistakes and impatience.

There are three dimensions to suiting a family to a home. Most of us naively think that the house design process starts with graph paper and a straightedge, but those with experience in finding fit homes for families suggest we save straight lines for later, abandon our conventional definitions, and honestly confess what we really do at home. This requires a careful study of the things you and your family like to do, which architects like to call "the program." Next, having decided on the important qualities and activities you seek, you start prospecting for a suitable site. Finally, having refined your family program and used that to find the perfect site, you can begin to think about walls, windows, floors, and doors . . . and the budget that will be required to secure and prepare the site and build.

Of course, your "perfect site" may be terribly elusive, and the search frustrating. Every candidate may violate one or more of the freedoms on which you insist, or prove to require compromises of life style and finance which are not manageable. Many homesteaders find themselves repeating the three-step design cycle—program, site, design—more than once before they can settle into something even remotely resembling the home of their dreams. The challenge stubbornly embraced by independent homesteaders is to keep seeking, studying, and planning until all the elements coalesce, a process which may take years of dedication.

When I explain these preliminary steps to a new group, I have learned that groans and complaints are inevitable. "Buying a house was never this hard *before*!" they moan. Yes, how remarkable that is! The most important purchasing decision in our lives, the largest single expenditure, is one for which we are unprepared by education and custom. Little shared wisdom exists about how to find a happy home, but a great deal is easily available about the buying and selling of real estate as a commodity. I assure you that if you follow these steps, even, at first, tentatively and superficially, you will learn much of value about the differences between your real needs and the requirements you may think you have. Invest this time at the beginning of your search, and you are likely to discover that your inquiry and quest become something which, done right, need not ever be done again. Compare that with the upheaval and expense experienced by the average family who moves twice a decade, and I think you will see the importance of this preparation.

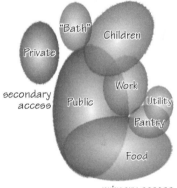

This bubble diagram shows a typical family's program elements.

Defining the Program

Every family has a unique set of activities and values it holds dear; sensitive architects spend quite a bit of time teasing this "program" out of prospective homebuilders. So, how do you use your house? Most people sleep (33% of their time), work (33%), eat (10%), maintain (10%), and entertain themselves (14%). A family's uniqueness is in the way these activities overlap and interrelate. Start out with a collaborative and inclusive list of every activity you can think of, from sleeping to emptying the compost, being sure to include new ones you would like to do in your new place. Completing this list could take weeks. When you are fairly sure you and your family have listed almost everything, put the list in order of importance to you, and ask other family members to do the same. Negotiate the differences, and map your program with circles and arrows to show the way you would like to steer your lives. Try drawing bubbles for each important area of activity, scaled according to how important they are and how much space they are likely to require. When you get a grouping that feels good, save it. By drawing and refining a bubble diagram of your family's program, you have a tool to use in the next exercise, and a much more powerful understanding of what will really take place in your home.

The Right Site

Now that you have an idea how you want to live, you can set about finding the best site. Peasant wisdom decrees that there are three important considerations to make before we even begin to think about creating shelter for ourselves: *location, location, location.* Our contemporary architecture of dominion arrogantly pretends that a seemly house can be built anywhere: in the darkest gulch, on the polar ice cap, at the edges of the solar system. In theory, such a house can be built anywhere, but it does not become a home until it is lived in graciously. Centuries of experience suggest that a home can be no better than its site, nor can it be better than its accommodation to that site. With your program before you, you need to deduce a list of the considerations your family wants in a home site. A checklist of site requirements might look like this:

CHECKLIST OF SITE REQUIREMENTS
❑ Freedom, quiet, and privacy
❑ Breathable air
❑ Good water
❑ Attractive surroundings and suitable interior space
❑ Utilities: electricity, phone, cable TV, sewage
❑ Good neighborhood, good neighbors
❑ Close to school and shopping
❑ Available transportation
❑ Community services: police, fire, ambulance, maintained roads
❑ Retains or appreciates in value

Site selection is never simple, and always involves compromise and uncertainty. Since a sufficiently "perfect" site may take years to find, you must find ways to weigh a variety of elements and locations with a satisfying degree of logic and objectivity, or your quest becomes frustrating and meaningless.

What do you require of a home? Enumerate the important elements that will figure in your family's life in their home, being sure that anyone who will have to live with the decision has a chance to contribute to the list. Some requirements are pass-or-fail—someone with asthma should never be made to live in a pasture of goldenrod. List these criteria first. List preferences as well, distinguishing them from requirements and placing them in a hierarchy of other preferences. Perhaps you prefer to walk to work, or for your children to be able to walk home from school. Some elements, preferences as well as requirements, may turn out upon analysis to be problems that go away when money is thrown at them. (Electricity far from the nearest powerline is a fine example.) If the amount of money required is beyond the sum available, or overshadows the rest of the project, then a site must be disqualified. Otherwise, a remediable problem bears a cost as well as a weight of importance. Devising a logical method for evaluation allows you to quantify the compromises associated with a site, compare sites with each other, and examine the requirements that cause the most trouble in finding a suitable site. The object of this exercise is to enumerate completely and prioritize fairly so that the compromise that emerges is best for all concerned. The evaluation of a pair of sites might look like the sidebar on pages 156–157.

Real Estate Traps

Independent homeowners love to tell stories about finding their land. There is often a magical recognition: a secret ring of trees, an unexpected nap, a dream, an awakening to the realization that this place is home. The land speaks, but its language is not inscrutable. Hilltops are windy, cold pools in north-running valleys in the early morning, and floodwaters stack up high against the outside bank above the sharp bend in the river. Physics rules nature as surely as it rules the built environment. Nature's imperatives are consistent: thrive using current solar income; create no waste that cannot be used by some other system for fuel; encourage diversity. Nature played by these rules for billions of years before humans walked the planet, and humans lived happily within these rules until a century or two ago. Since we changed our rules, we have attained much

Since a sufficiently "perfect" site may take years to find, you must find ways to weigh a variety of elements and locations with a satisfying degree of logic and objectivity, or your quest becomes frustrating and meaningless.

of value—leisure, freedom, a facility with facts and ideas—along with many habits and preferences that are not sustainable. Even the notion that land can be owned, which is central to buying land, is a recently invented fiction we can work with only if we remember that all our acts have consequences. Acting within nature's reasonable rules may well seem impossible in an era conditioned by rampant consumerism, but we are fools to think we have a choice.

Real estate salespeople love to sell mountaintops. "Isn't the view *great?*" they ask. Do

EVALUATING COMPARABLES

This chart represents the actual comparison for a young family of four considering buying or building their first home. The family used this comparison a decade ago, while deciding between an independent home on a mountaintop, twelve miles from town, and a house in town, but the costs have been updated. This site evaluation exercise, like any effort to quantify intuitions or impressions, makes some broad and subjective generalizations in the interests of practical comparison. The family in the example conceded that a good, solid, five-thousand-dollar backyard fence will be required for privacy and quiet in town. Assuming that the two properties are equal in monetary value, the cost attributed to "Attractive / suitable" represents the extra cost of the family building their own home at the rural site, and so, since they will be building the house themselves, it should suit them very well. The absurdly high cost of driving to and from the rural home includes the expense, at thirty cents a mile (up 20% in five years), of commuting to a neighboring town four days a week, carting the children to school thirty-six weeks each year for twelve years, and two shopping trips per week the rest of the time, calculated over thirty years; the smaller cost for the urban site assumes that the children only cadge two trips a week to school in the family vehicle, carpooling or walking the rest of the time, and that the family sticks to a twice-a-week shopping regimen. In reality, the trips to and from the remote site may cost even more, because rural roads eat cars, and few of us are as good as we should be at abiding by a twice-weekly shopping schedule. Rural costs for water, utilities, and community services represent estimates of infrastructure costs—well and pump, solar modules, storage and power-conditioning equipment, heavy road work, and additional rural insurance cost —over the same thirty years. Given the importance of water, and the promise of sweet well water in the rural location, many consider the investment in well, pump, and filters much better than drinking over-treated city water.

In this example, the family estimates that it costs half as much, and will only be a third less satisfying, to live in town, assuming a direct relationship between money and suitability. Their fence serves as an important clue

they ever think that if one can *see* forever, one can *be seen* forever, and a house on that spot may defile a view treasured by others? Do they consider that hilltops are windy, and wind strips heat right out of a house? Do not buy a mountaintop; buy a valley. You thus control your viewshed, and the home's inhabitants will not be indentured to the perpetual labor of compensating for height and exposure by carrying water and firewood uphill. The harsh fact is that the most habitable lands are owned, and available properties are troubled. Fortunately,

SITE COMPARISON MATRIX

	Weight	Site 1: Rural			Site 2: Urban		
		Points	Total	Cost	Points	Total	Cost
1. Freedom, quiet, and privacy	25	9	225		3	75	5,000
2. Attractive/suitable	20	10	200	100,000	6	120	75,000
3. Breathable air	15	10	150		4	60	
4. Good water	15	8	120	5,000	4	60	
5. Good neighborhood, neighbors	10	5	50		8	80	
6. Available transportation	5	0	0		10	50	
7. Close to schools and shopping	5	0	0	26,496	9	45	3,900
8. Retains or appreciates in value	3	7	21		6	18	
9. Utilities	2	0	0	20,000	10	20	
10. Community services	0	0	0	15,000	10	0	
	100		766	166,496		528	83,900

about the family's real preference, since fences never substitute for space. And yet, of the five most important elements, the town dwelling comes first only once, and total rural suitability points are twice the urban total. This family, already leaning toward life in the country, decided a decade ago that the rural life would be much more gratifying and that money, after all, is only money. The site comparison process was worthwhile for them because it gave them a clear and arguably rational justification for their decision. Their experience since has shown their decision to be right for them.

the troubles experienced by conventional developers may be opportunities for anyone willing to look at the land honestly, but without prejudice. The agent may accept the practice of hurling heroic amounts of energy at an unsuitable site in order to erect a house, but this will not suit those who mean to build an ancestral home. Realtors think in terms of the quick sale, repeated every few years if the fit is poor enough, and so expensive, unsuccessful houses serve them well. Builders of ancestral homes think in centuries, and plan to be remembered gratefully by their heirs. If we select a site for the value of its unspoiled woods, pure water, and generous environment—the qualities that inform nature—and then build to preserve and enhance these qualities, the home will be an adornment, not an eyesore.

The Site Rose

Once a potential site is found, its uniqueness, which architects call the *genius loci* or "genius of the place," must be mapped and described

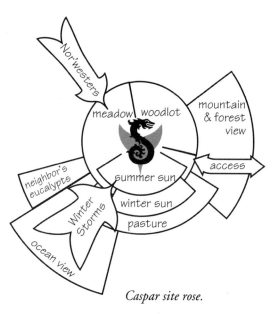

Caspar site rose.

before a satisfactory house can be built. Although this information may take years to gather, it is important to make a preliminary estimate before a commitment is made to purchase the property, so you can be sure that you will be able to build a suitable home.

Geographic and seasonal factors are critical considerations in choosing a site and situating a home. The sweep of winter storms, the direction and frequency of spring and autumn gales, the source of the prevailing breeze that tempers summer's heat, can all be ignored in favor of other preferences, but at considerable cost, because an ill-sited (or ill-suited) dwelling is at the mercy of the earth's forces.

Many modern houses are built on flat, agriculturally valuable land because this makes production-line building easier, and maximizes the developer's short-term profit. These homes, of course, are horrors. Few independent homes are on flat land; their owners explain that they looked to the little hills, not to the cornfields, for their home sites. In the hills, favorable dispositions of sun, slope, wind, and water make for a home that is more interesting to live in. Because homes serve their residents, not their builders, commodity housing seldom suits householders well. Could this be one reason that the average tenancy in a subdivision house is less than four years?

One difference between a house and a home is the harmony (or disharmony) of shelter and site. For centuries, oriental geomancers have studied the feng shui of building sites, the concordance between a shelter and the energy flowing in its surroundings. Westerners take similar readings, and call the results their site study.

A site rose is a graphical way of depicting the currents and flows that a site brings to bear on a house. (You did something very much like this in the last chapter when you drew your site map. As you glean new information, you should go back to review and update your site drawing as necessary.) A solar home's principal determinant is its relationship to the sun. Houses in any net-heating regime are usually best built with their longest dimension running east-to-west, their longest side facing the sun. In net-cooling climates, the shortest side should face the sun to limit unwanted solar gain. Conventional American homes slavishly align their longest dimension with the street, thereby enhancing what realtors call "curb appeal." (I am always reminded of the way cats turn sideways and fluff themselves up to look bigger to rivals.) Since streets usually reflect nothing more sensitive than a developer's whim or a cat-skinner's convenience, it will be only by accident or dint of great heart on the part of residents that dwellings like this ever become homes. (I hold that love can overcome almost everything, including bad feng shui.) The forces indicated by the site rose are inevitable. Atop my bluff in Caspar, December through March brings powerful storms from the southwest; all year long, dry, cold northwesterlies cleanse the air two days out of five. Here as everywhere in the northern temperate zones, the sun rises to the east, arcs high to the south at noon, then angles back to the west. These are the forces to which a house must respond if it is to become a home.

In Caspar, winter's defect is horizontal wind-driven rain and biting cold, and summer's is fog. Across the northern tier and in Alaska, winter is the problem, with wind, interminable biting cold, and its cohort sleet, ice, and snow; summer's only flaw is brevity. Even if you have never before given a thought to

Asian builders observe the ancient practices of feng shui (wind and water) in the proper siting and interior design of their homes. This ba-gua *can be applied to any home by orienting it with the entryway at the bottom, although agreement with the compass directions is considered by geomancers to be preferable.*

mapping your home's relationship to its ruling forces and features, you could sketch the most basic features of your place's site rose. Some influences exist everywhere, including the sun's light, but others are subtle and may take years and special attention to discern.

Regional Design

As you accumulate testimony for and against home sites you evaluate, you will also encounter a range of more- and less-successful homes in the neighborhoods surrounding your candidates. In America, there are few examples of old, successful homes, and you may need to search them out. By this time, with your family's program and a selection of sites and their evaluations, you are right to be looking for inspiration and example in the communities where you may soon be settling.

There is wisdom in traditional design even deeper than the satisfying coherence it lends to neighborhoods. Distinctively "regional" homes and other vernacular forms reflect the experience of generations with the

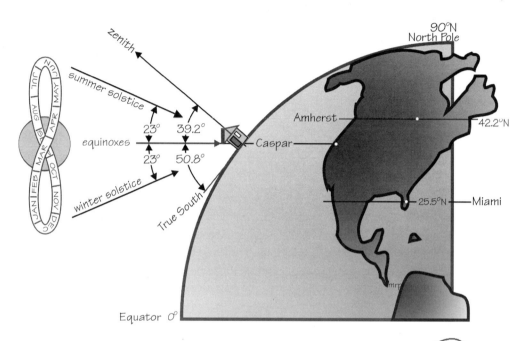

To get the most out of a building's roof, align it with the sun's best angle, which is determined by latitude and the season. An analemma—the figure-eight shape—represents the sun's motion during the year. On the equinoxes in Caspar, California, the best roof angle is 50.8°E or, in roofer's terms, about 6.75 in 12. In Amherst, Wisconsin, the angle would be 47.8° (10.9 in 12); in Miami, Florida, 64.5° (5.75 in 12). The general formula for the roofer's measure is 12 times the tangent of latitude in 12.

The earth's wobble, or precession, is 23°E which means the sun's direction changes 46°E from north to south and back every year. For the optimal alignment at the equinoxes, the roof's angle from the horizontal is equal to 90°E minus the latitude. Steepen the angle by as much as 15°E to gather more solar energy in the winter.

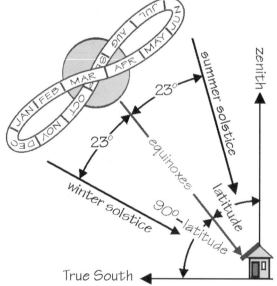

weather and the style of life imposed by the land in that place. I have already noted that architects refer to this as the *genius loci*—the specialness of the place—in recognition of the "mind of the land." It is perilous and expensive to ignore the land. In the twentieth century, coerced by mass-production values and styles, we have disengaged structure from surround. We are then forced to compensate by massively consuming fossil fuels to enforce our own preferred conditions upon our environment. Cheap resources, and our heedless rush to consume them, made it possible to export some strange notions of suitable houses to unlikely locales: sprawling split-level ranch-style houses in Vermont, and gingerbread-encrusted New England saltboxes beside Kealekekua Bay. The unfitness of these architectures to their environments will have a direct effect on their longevity and the success of the families who inhabit them. To make an out-of-place house work requires heating, ventilation, and air conditioning (HVAC) equipment reckoned in the tons. Such a structure makes arrogant claims about human dominion, and nature contrives to make such claims costly. It is better to build in accord with the environment of a region. We do better when we bend to nature's powers and adopt the forms that time has proved best.

Improving the land: David Palumbo's story

David and Mary Val Palumbo live with their three children in a comfortable off-the-grid house near Hyde Park, Vermont. Before building, they spent years preparing themselves to build, and years living with their land. Since building, David has consciously polished his stewardship skills, and the land glows in response.

I like doing things myself, and I want to have firsthand knowledge of what it takes to make things work. Technology excites me, not the research and development part, but taking proven technology and helping people apply it properly is what I like to do.

I was lucky when I started, because I got some good advice from the pioneers. Peter Talmage showed me by his example how to scale my business so I could make a comfortable living without being greedy, and always take time to be sure my customers are happy.

I got a lot of help on my own house, which is also my biggest and best project, from Milford Cushman, a local architect. I've learned that, whenever I need technical advice, it's best to ask, and the pioneers and experts are very generous. But let me start a little closer to the beginning of the story.

After college—studying business, no science or engineering—Mary Val

I often have to hold people back when they move away from the city. . . . You don't want to buy land without studying it carefully.

and I worked for Outward Bound in northern Minnesota, and I learned to love the woods. After a stretch in the wine business in Santa Cruz, I returned to work with my family's real estate business in Massachusetts. It took seven years there to build up my share and get Mary Val through her professional training. When we cashed out, the stake looked small for what we wanted to do in Massachusetts in 1984, so we came up here where land was much cheaper. We knew we were looking for a good-sized piece, with woods. I guess we just assumed we'd be connected to utilities, but when it became clear that the properties that were right for us didn't have power, I set about getting myself educated. We were glad not to add to the need for more nuclear power, and we're individualists who pride ourselves on adventure, so relying on ourselves for our own power fit right in.

We looked at a lot of properties. I often have to hold people back when they move away from the city; few people have the necessary discipline. You don't want to buy land without studying it carefully: You want to make sure you understand about rights-of-way for utilities, about neighbors, about how the seasons and weather extremes treat your property, before you commit. We went slowly, and ended up with a better site than we knew.

We wanted a place with privacy, but close to commerce and Mary Val's work. We wanted to be at the end of our own road, so we could choose our neighbors. We wanted woods, at least fifteen acres, we thought. After we got over the surprise of going off-the-grid, we knew we needed a good balance of energy sources.

We were very lucky finding this land, because we had made an offer on a smaller site, for more money, but this came up and that fell through. This land needed some bushwhacking before we could appreciate it. We knew from checking the soil and drainage that there were two good house sites. We lived in a tent on the best site while we built a small, efficient, quick house, twenty-four feet square, facing exactly south on the second-best site, so we could start living on the land. We did a lot right with it, and it's still a very workable place. The PVs are an easy broom's reach for clearing the snow. We learned what we needed to know to start planning this house.

This is a cloudy place, and PV is not enough by itself this far north. Our original concept was to use as much PV as we could and make up the difference with a propane generator. There are two houses and a big shop on our homestead grid, so we manage it like a small utility. We use some big power tools, so we knew we would need a good-sized generator for backup.

Our favorite time of the year is July and August. There are a few days when it gets so hot, you want to spend time in the pond. That's how we got the idea for hydro, prospecting for pond sites. We found three streams on the land, and one of them runs year-round. Getting the hydro system working was more hit-and-miss than it had to be. I know how to do it now with much less fuss, and I'm looking forward to doing this one over, but it works well. We only use the propane generator when we run the bigger tools, like the planer.

We didn't know how many children we would have when we built this place, so we made it big. All the framing and trim lumber came from the land—we used as much indigenous material as we could. We used a lot of wire, because we wired for 12-, 24-, and 110-volt.

During the winter, it takes some time to manage all the systems, the electricity, the heater, the children. The kids go to school and childcare, and Mary Val commutes to her work as a geriatric nurse practitioner, specializing in continence care using biofeedback and exercise. It's a new field, and she also teaches at the University of Vermont four days a week. During the week I run the house and do my work helping people put their technology together properly.

The big Essex wood gasifier runs all year long. It burns at about 1,800-2,000 degrees Fahrenheit; when it's cold out, I fill the firebox two or three times a day, but when it's warm, I fill it once every two or three days, enough to keep us in hot water and the chill off. We thought a lot about the way the systems would interact as we planned and built the house, so there's a wood chute in the garage for getting the wood to the basement and the gasifier. And my workshop, which is above the garage, has a trap door and stairs so we can move equipment up and down easily. The root cellar turned out well,

Adjusting the flow to the micro-hydro.

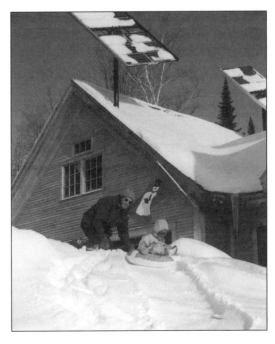

The PVs are snow-covered, but the Palumbo hydro is operating at peak.

because I can work the variables; many people forget that you have to control humidity as well as temperature, to keep things from rotting. Domestic water is another problem. We have plenty of pressure, but if everyone starts a shower at the same time, the gravity-feed line just can't supply enough, so I've put a pressure tank in for just such times, and now we can hardly tell there's anyone else on the system.

I like the flexibility of 24-volt storage, and most of our appliances and loads like it too—including, I guess, about 95% of the DC loads: lighting, the Sun Frost refrigerator and freezer. The little task lights, draft-inducer fan, hydronic pumps are 12-volt. The washer, dishwasher, dryer, microwave, TV, and VCR are all standard 110-volt AC. I have my batteries laid out as two 12-volt banks. I anticipated that balancing the loads, and keeping both banks at the same state of charge, would be a problem, and I set the 12-volt draft-inducer, the fan that makes the gasifier go, so I can switch it to draw from whichever bank is stronger. I have three sources of power, which is a little complex. Especially in cold weather, generators give off-the-wall power when they start up, both frequency and peak voltage. I learned the hard way to be sure there is no load until the generator's RPMs smooth out.

I studied the way Native Americans, especially the ones who lived around here, treated the land. Flatlanders get emotional about cutting trees, but the regenerative capacity of the northern woods is staggering. You don't need to replant, and you can't keep the woods back. There are two hundred and fifty wild apple trees on our land, planted, I suppose, by the original farmer who cleared it. The forest is crowding them out, but they are important for the wildlife, for bear, deer, and grouse. We've

worked with the Department of Agriculture to release those trees from competition to the south, and now we're investigating some edge and patch cutting. Patch cutting (clear-cutting a small area, no more than a half-acre, defined by the forester) can make flatlanders really howl, but it turns out the edge of the regenerating forest, where the poplars start, is a crucial part of the woods habitat, where most of the animals thrive. The young poplars, it turns out, are necessary for grouse reproduction. We're doing what the original inhabitants did, bringing sunlight into the forest. ⚊

Meeting Our New Neighbors

When we come to a new place, it is wise to seek out its oldest inhabitants, to learn to honor them, in order to appreciate and secure the future of the *genius loci*, the genius of the place. I mean "appreciate" in both its senses: In addition to enjoyment, we should be consciously striving to add value to everything we touch.

The oldest inhabitants are usually members of the native plant and animal community, who can testify in enormous depth about how best to live with the land. Where human encroachment has reduced the nonhuman populace to mice, insects, and ornamental foliage, relevant information and experience is lost. In the parts of the world where humans have overpopulated and driven other species away or to extinction, this loss is tragic and often irreparable. Conscious people everywhere are beginning to see (though in many places, too late) that native flora and fauna are inestimably, inconceivably valuable, and that they must be preserved and restored at any cost. When we favor indigenous plantings over exotics we begin to appreciate the importance of an ecology that is attuned to its specific place, and that maintains itself without expensive and labor-intensive intervention.

The appearance, habits, and homes of our native, nonhuman co-inhabitants yield crucial clues about our chosen place. Citified humans may bring a sort of nature blindness with them, an inability to see a butterfly, to hear a bird. A wealth of literature has been written by people who are recovering from this handicap, and are drunk with the joy of finding that the minuscule and majestic survive despite humankind's pretended dominion (to which I happily bring this small offering). For those whose previous experience of nature has been derived from books, movies, television, and the denatured world of modern landscaping (a narrow pseudo-ecology contrived for its visual effect), riotous nature comes as a delight . . . and a shock.

The redwood trees that dominate the forest on my eastern horizon have never colonized the windswept blufftop where I built my house. The earliest photographs of Caspar, taken just as whites began thrashing through the forests, show a wonderfully thick stand of shore pine right to the rim of the ocean bluff. The shore pine are largely gone, replaced by unsuitable exotics, cypress and eucalyptus, which are not happy. (For more on natives and exotics, see chapter 11.) On the stillest days, you can read about winter gales from the way these hardy natives are sculpted by the wind. During the dry days of summer, vernal pools and winter watercourses can be discerned by the plants that remain, if you only learn the local flora's habits.

A seven-year drought lasting from 1986 until 1993 forced Californians to face the fact that we live in a seasonal desert, and awakened our interest in xerophytes, plants that tolerate dryness, such as the native plants that had been grubbed out and replaced with thirsty exotics decades ago. Climatic change, perhaps the result of global warming, may cause climate to intensify. Indigenous nature has been through all this before, and can be counted on to adapt. Permaculture, a garden design strategy based on the idea that appropriate plantings should naturalize and take care of themselves perennially, provides us with important tools for influencing the crucial relationship between weather and the carrying capacity of the land.

As out-of-place as eucalypts and cypress, domestic dogs and cats have devoured ground-nesting birds and hounded the large game away from settlements. Hooved invaders—cattle, sheep, pigs, and goats—have had their way with meadows and wetlands where some of the tenderest, rarest plants have quietly subsided into extinction without ever being known.

When we favor indigenous plantings over exotics we begin to appreciate the importance of an ecology that is attuned to its specific place, and that maintains itself without expensive and labor-intensive intervention.

Meanwhile, the herders, citing a need to protect their valuable livestock and forage, have hunted wild creatures ever deeper into the woods.

People everywhere are beginning to long for the confirmation and context that only a healthy nature can grant. In Colorado, a jogger is attacked by a mountain lion, and the county animal control officers offer to abate the nuisance; the neighbors, jogger included, resist vigorously, defending the magnificent cat's right of place, saying, this is why we moved out here. The shamefaced jogger admits, "Now I pay attention and look around, and I don't bring my Walkman when I run at dusk." But desire and talent may not always go together, and nature is sometimes healthier and more insistent than city folks might expect. After building his dream retreat in another rural corner of Colorado, a man called the state game warden, complaining "I want you to kill all the songbirds on my property, or give me permission to do it myself. I built this summer cabin for peace and quiet, and I want to sleep late, but the damn birds wake me up every morning!" Another irate woman caller reported that deer were eating everything in her garden. The warden suggested deer-proofing her plantings, but the woman had a better idea: "Why can't you move them back where they came from? My house was here before they were."

Hummingbirds recommend migration, and follow their own advice.

What the Woodrat Recommends

Along with the redwood, the most intelligent of my surviving native neighbors is the woodrat, who recommends the following building practices for Caspar:

1. Choose the south-facing slope above the flood plain and out of the way of rivulets that run downhill in big rains. Do not select a site without surveying it in the wettest part of winter, the windiest spring day, a foggy summer day, and the driest time in autumn. A spot below the brow of the hill and out of the northwesterlies is preferable; the riparian coppice along the creek is chancy to cross in winter even in the driest years, and most years knee-deep (to a human, to say nothing of a woodrat) from first storm until May.

2. Build with local wood, in layers. The woodrat piles on twigs and sticks extra thick each fall; the heat generated by their internal composting keeps the carefully drained nest, lined with feathers and other found softnesses, cozy on even the coldest night. The

entry to the home is through a vestibule similar to the mudroom found in New England homes, a sort of air-lock which keeps the weather outside.

3. Learn to live with the outdoor wet, carrying on life in the rain as in sunshine.

Larger species, fox, coyote, puma, and bear, are so rare that we may only speculate how they live deep in the small remaining stands of ancestral forest. If brown bears still survive there, do they hibernate? In winter there are whole weeks at a time when I would if I could.

Hummingbirds recommend migration, and follow their own advice.

Original Dwellings

The white exploiters have managed, in all but the least desirable places, to dispossess the original human inhabitants of the land, but where indigenous peoples have lived, we may study their shelters, which in most cases have been patiently refined through the ages to fit the locale. The Yuki and Pomo who occupied Caspar before the whites came maintained two homes and traveled from near Caspar in summer to the inland valleys near present-day Willits and Ukiah for winter. Their comfortable, low, wintertime sleeping houses were hemispheres made from abundant, sustainable local materials: rock, mud, willow, and redwood bark. Daylight hours were spent outdoors except in the worst storms. Communal kitchens, bark and leaf shade shelters without walls, separated from the sleeping houses, were focal points for the community's daytime activities. All across the American continent, the vernacular dwellings of the original human inhabitants, and the structures of the first settlers, remind us of values and adaptations to climate and local materials we would be wise to copy.

These masterpieces of adaptation to local materials and climate offer models of careful siting and orientation—of sensitivity to common forces, summer and winter, and to the basic human needs for family, privacy, and shelter. Everywhere I traveled, homesteaders showed me artifacts of earlier occupation found at the best home sites: arrow points, stone tools, and the debris of stone-napping. In Caspar, we found points and a small shell midden. When ancient conceptions about tribal housing are translated into modern shelters for extended families and intentional groupings, we often see better adaptations between the people and the land than can be provided by the European-style single-family resi-

Indigenous dwellings, supremely well-suited to their locales. From Atlas of the North American Indian *by Carl Waldman, illustrated by Molly Braun. Copyright 1985 by Carl Waldman. Reprinted by permission of Facts on File, Inc.*

Chickee

Lean-to

Plank house

Longhouse

Pueblo

dence or the conventional arrangement of a community into tracts of houses.

When we listen to the land, and attend to the prehistoric patterns of living in partnership with it, we change the way we live. As population loading increases and more and more land is ceded to this continent's emergent top predator, the automobile, the land that remains requires us to be ever more attentive and reverent. Living in conscious harmony requires us to work actively with the land. Subdivision dwellers are spared this effort, because they live in places where nature has seemingly been tamed (until a hurricane comes along). Healthy land exacts a tribute and dedication from its stewards that stretches the abilities of the conventional isolated nuclear family. Just as it takes a whole community to raise a child, it takes a small tribe to keep a homestead.

After seeking the advice of the original inhabitants, we should turn to living masters who have been with the land all their lives. The wonderful insights of these people can bring us closer to the realities of our home place (when they are willing to share). They are wise about the weather: A cloudless azure January sky in Vermont is a "breeder sky" foretelling a coming storm; clouds and breeze coursing in from the southwest promise a storm in Caspar. Knowing these signs allows us precious hours to prepare. The old ones are also wise about shelter. Attentive old-timers may still remember, and will happily tell us, how houses worked before indoor plumbing and energy dependence changed our perspective; hearing their stories, we may re-evaluate our assumptions about what makes a good and suitable home. In Caspar, where the storms come off the ocean, old houses have few windows because old-style wooden sash windows leak water and cold air; majestic as it is, the great western ocean is identified with southwestern storms and nor'westers, the sources of domestic discomfort. We might wish to preserve an ocean view, but those who build without harkening to the voices of old-timers do so at their peril. Anything facing the ocean must be built like a boat.

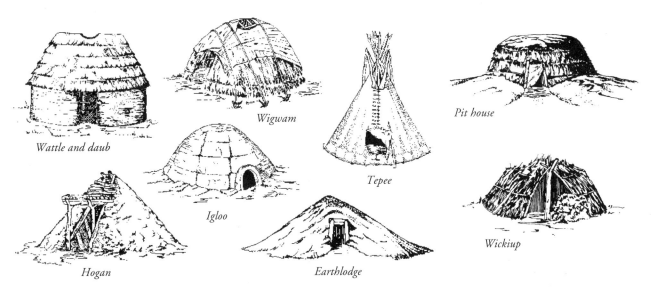

Wattle and daub

Wigwam

Tepee

Pit house

Igloo

Hogan

Earthlodge

Wickiup

The Laws of Physics: Laws We Can Live With

If, after consulting the original inhabitants, plants, animals, and humans, we need further guidance, the laws of physics can be called upon to help. Abiding by the same laws, all our various instructors tend to suggest the same solutions. For example, where cold is a major factor, thermodynamics dictates that the best shape will contain the most volume while exposing the least surface: a sphere—such as beaver lodge and Inuit igloo. Spheres are hard to build using wood and glass, and humans require floor-space with headroom, which is a little different from simple volume, so a rough rectilinear approximation of the sphere, the cube, fits us better. In cold climates, successful human-made houses tend to be cubical. Another example: Animals that do not migrate either take to houses in times of adverse weather (hibernating or aestivating in caves or tree trunks) or are superbly hirsute. My cat's winter coat makes her look twice her actual mass; arctic foxes have even better insulation. Cold-weather creatures are usually smaller and more compact than their temperate relations. So it should be with cold-weather houses, which need abundant insulation and adequate but not excessive space, because the luxury of heating superfluous space is perversely costly.

By anticipating the physical effects of prevailing winds, seasonal moisture, and the warmth of the sun, we improve our theoretical model of a house's mechanics. The classic New England saltbox is wider east-to-west, which maximizes solar exposure, and lower on the north side, often sheltering beneath a hill or woodlot for even better wind-shadowing, out of the way of prevailing northerlies. At the heart of a saltbox, a massive fireplace keeps warmth streaming outward from the protected center. If southern windows allow sun to warm the fireplace's mass, the heat re-radiates in the evening, reducing the amount of wood required to counter evening's cold: a simple application of the greenhouse effect.

Heat most often comes to us as light in the form of infrared radiation, which travels in straight lines from its source; once light strikes something—a surface or an air molecule—it warms the object struck and the surrounding air. Colder air is less chaotic, and thus denser, or heavier, so it sinks, displacing warmer air, which rises. This circulation is called convection. In a saltbox, the daytime living spaces are downstairs, where active occupancy makes cold tolerable, and bedrooms are above, warmed by the rising warm air; doors at the foot of stairways keep the heat downstairs during the day. Of course these ideas get turned upside down in warm climates: There, a physicist's home will be more spread out, containing more space, narrower and protected by trees on the too-sunny south side, which also has few windows. In the Sun Belt, thermal mass is a good idea only if it can be "charged" with cold during cool nights.

Other physical phenomena inform regionally apt shelters. Wherever summer sun causes too much summer heat, overhangs and deciduous plantings help us tune the home's exposure to the sun, blocking incoming radiation in summer, but admitting more light in winter. In the sticky tropics, where seasonal variation is minimal, the earth's surface is often moist from daily rain and inhabited by insects, because breezes do not penetrate the thick foliage; above the canopy, breezes at the treetop level promise the only relief from the humid heat, and it is sensible to build living spaces on stilts, aligned with and open to the slightest breezes.

After thirty years studying the constantly changing genius of the land I live on, the lessons remain interesting and surprising. Creating a living space, particularly along the margins of wildness, presents a challenge to nature. Even when we provide ourselves with a comprehensive set of findings covering our personal, familial, geographical, climatic, biological, and anthropological requirements in a home, we make some mistakes. The biggest mistake of all, however, would be not to try. Fearlessly, then, since you are now better informed than most architects and builders, turn to chapter 7 and design a home.

The Marx home in Taos, New Mexico, was designed and built by Ken Anderson, whose story is in chapter 8. This 1,350-square-foot passive solar house is post-and-beam with straw bale infill. Interior walls are made of adobe brick and cob. A poured adobe floor provides thermal mass. Greywater is recycled and the residents use a composting toilet.

The cistern at Chez Soleil, a self-sufficient solar home designed for the hot and humid climate of the Texas hill country. The house utilizes a solar electric system, high thermal mass, and several passive cooling strategies (including a solar chimney, radiant barriers, low-E windows, overhangs, and vents). Chez Soleil's roof harvests rainwater, which is pumped uphill to the cistern. Water is heated by solar and processed by a constructed wetlands septic system.

PAUL BREAUX

LLOYD GODING

PAUL LACINSKI

Lloyd Goding built his passive solar, earth-sheltered house in 1979 in the rolling farmland of southern Wisconsin. The house incorporates berming on the north side, 350 square feet of glass on the south side, a greenhouse, a workshop, and a sauna. The post-and-beam framework and loft floor joists are made from recycled and resawn timbers originally used in the Chicago stockyards.

This treehouse makes use of a grand old mango tree in Kipahulu, Hawaii. The outdoor kitchen and bathroom are built around the tree's base, and the bedroom is aloft to catch the breezes.

Left: The ceramics studio of Laura Fitch in Port Townsend, Washington, has straw bale walls covered with cement stucco and ceramic motifs.

This 1,340-square-foot home in southern California was designed and built by Robert Mehl and Kit Boise-Cossart to fit gracefully into its surroundings, and to operate with minimal impact on the earth. Many components are recycled: walls were constructed of broken chunks of salvaged concrete (much of it trash from the Northridge earthquake); interior floors were made from salvaged slate shingles; and only salvaged wood was used. The grid-connected photovoltaic array produces 8 kilowatt-hours per day.

Clark Sanders built this magnificent "ancestral" stone house in East Meredith, New York, by working 80-hour stretches as a veterinarian, then taking time off for construction. The stone walls are insulated on the interior with urethane foam, which was then plastered. The roof is covered with ¾"-thick slates.

Right: The Norris Tower House in California rises six levels, from its earth-sheltered walk-out cellar to the hot tub on the top deck. Solar gain stored in the massive concrete foundation heats the main floor above through passive "baseboard-heater" vents. Solar panels on the basement level thermosiphon to supply domestic hot water, while those on the upper levels heat the hot tub. This home was designed by Stephen Heckeroth (inset), shown working in the solar-powered studio of his home in Albion, California.

This renovated 1850s barn in Montpelier, Vermont, features a concrete radiant floor poured on top of straw bale insulation. The straw came from a nearby oat farm. The concrete floor is entirely hand-colored. Most of the plumbing and lighting fixtures are salvaged.

CAROLYN L. BATES

LILIANA BELTRAN

JEFF SIMPSON

Casa del Sol is located east of Lubbock, Texas. The centerpiece of this house is a combined active/passive solar system that includes south-facing glass, a large solar collector, and a rock storage-bin with controllers and blowers. The system provides space heating and heat for domestic water, a swimming pool, and a whirlpool.

Jeff Simpson sells Solectria electric cars in a five-state region around his home in Paola, Kansas. He organizes the Kansas Electric Car Rally, which in 1999 involved forty-two cars built by thirty-four teams of high school students. Jeff put 30,000 miles on this electric Porsche, fully fueled by his grid-tied 4-kilowatt photovoltaic array (average production 15 kilowatt-hours per day), until his family outgrew the car; now he's converting a larger vehicle into an electric hybrid. The house is passive solar, with a thermal mass floor, south-facing windows, and a superinsulated shell. Jeff figures he saves $700 a year in reduced heating and cooling costs.

Right: The energy-saving features of this home in Spicewood, Texas, include straw bale construction, a thermal chimney, solar orientation of windows and double-pane low-E glazing, and finished concrete floors. Doors, bathtub, tile, and furniture are all recycled. All but 10% of the landscape will be xeroscaping, requiring virtually no water. A rainwater catchment system is also planned.

DAVID SHAW

Juliet Cuming and David Shaw built Earth Sweet Home in southern Vermont to showcase sustainable, bioregional, biodegradable, and recyclable construction techniques. The two-story house is made primarily of straw, wood, and stone. The builders made a conscious effort to reduce the use of plastics, metals, and other energy-intensive materials, and to avoid synthetic products that create indoor air pollution. For instance, the bale walls are held together with oak stakes and twine rather than metal pins, and covered on the exterior by lime/straw plaster. Most of the construction materials came from within thirty miles of the site. Electricity is supplied by photovoltaics on trackers and a wind turbine. Water is heated by solar collectors, a masonry stove, and a tankless propane backup heater. Sewage is managed using a constructed wetland septic system. Note the load-bearing "living" roof on the battery shed.

PETER PFEIFFER

The Clark residence, near Austin, Texas, is a 2,200-square-foot passive solar home designed by Barley+Pfeiffer Architects. Careful site orientation, a massive thermal foundation, and a radiant roof barrier made of recycled steel from automobiles keep the house warm in winter and cool in summer. The exterior siding is made of cement and recycled wood chips, and the stone is locally quarried. This house is water independent, relying on a rooftop rainwater catchment system and large cistern for all its water needs. Greywater is used for landscaping.

LEIGH SEDDON, SOLARWORKS

The Larsen residence in Vermont was designed by Bill Maclay, who appears with Dimetrodon in chapter 9. A grid-connected photovoltaic system provides electricity; hot water is from a solar collector; solar hot water also heats the radiant floor. The energy systems for the house were designed by Leigh Seddon, whose story is in chapter 14.

This wonderful earth-sheltered house with a two-story attached greenhouse and a living roof was designed by visionary Malcolm Wells, the Gentle Architect. It rests lightly on the land in eastern Ohio.

Above and above right: Buildings made with tires rammed full of earth can break out of the linear mold and assume fanciful and practical shapes. This is Nautilus House, designed and built by Michael Reynolds and his Solar Survival Architects near Taos, New Mexico, whose story is in chapter 13. It features passive solar heat, hybrid photovoltaic/wind power, an internal greenhouse, rainwater catchment, greywater recycling, and a solar-assisted septic tank.

Kevin Jeffrey and his family built their home on Prince Edward Island in maritime Canada to demonstrate that resource efficiency and affordability can be integrated into a heritage exterior design. The house is tightly constructed using three layers of insulation. Its long dimension faces due south and all living spaces are arranged on the south side. Electricity use is minimized by a judicious selection of appliances and abundant natural daylighting. Power is provided by a hybrid photovoltaic/wind system. On the electric bill, the monthly service charge exceeds the cost of power for the home and an office combined!

Bob Herman, a self-described "expatriated Connecticut Yankee," built this solar-heated hot tub at his home in the Rocky Mountains. "This is actually a 'soaking tub,'" he says, "devoid of the pumps and jets and surround-sound stereo that hucksters peddle for five grand. I built my tub for about $100 and spare parts from my junk pile. The tub works on the thermosiphon principle; there are no pumps or other moving parts. One 40-square-foot collector provides more than enough BTUs to keep the 100 gallons of water hot." The Hermans live off-the-grid with photovoltaic-supplied electricity.

CHAPTER 7

DESIGN FOR
INDEPENDENCE

IN CREATING A NEW INDEPENDENT HOME, WE BUILD ON CENTURIES
of tradition, and strive to winnow useful ideas from empty conventions.
Sensing a sea-change in the use of building resources, we struggle to
reconcile nostalgia for gracious living and prudent investment with misgivings about wastefulness and conspicuous consumption. When we
plan a new house, we have an opportunity and responsibility to build an
authentic and enduring setting for ourselves and our offspring. We integrate good ideas from homes, books, and our imaginations, into our
design: something fresh and relevant to the life we predict for ourselves
and our families. Diligently following the money, insistent on balancing
costs with benefits, we encounter pressure from planning authorities,
lumber suppliers, carpenters, and family members to make decisions
speedily. Proven techniques, easy availability, and conventional solutions
tempt us, even as we see them crippling our vision. Faced with apparent
impossibility, we continue to negotiate and compromise, reshaping family priorities, deepening our conversation with the genius of the site, and
adjusting allocations of precious materials, time, and energy . . . until we
finally settle on the only possible plan.

Slopes, streams, sensitive habitat, the flow of weather, and stores of
indigenous building materials suggest the best site and building method.
We may be able to visualize the outside of a house long before we are able
to imagine ourselves and our families living for years within, but that is
precisely the challenge we accept. Compared to the certainties of land,

climate, and neighborhood, time and change are elusive dimensions. We live in the present, and can discern evidence of the past in the community, natural surroundings, and the houses we have lived in, but we design our homes for an unpredictable future. By anticipating the cycles through which a home must shelter us, starting with the most commonplace, we build for endurance.

The most important cycle, the diurnal progress of day and night, is seldom considered by designers of commodity homes. We are solar creatures, so our homes are the sundials our shadows move across. In the morning the sun rises in the east, and its warmth and brightness predispose us for a wakeful day. We arise from our beds and perform morning ablutions. As the sun moves up across the sky and down again to set in the west, members of the family move from sleeping area to food preparation

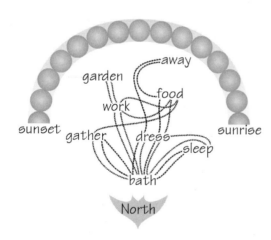

Once main activities have been mapped, an interesting way to find interconnections is to map the path family members follow during a typical weekday or weekend day. This is a map of a very simple, two-person household in which one member goes out to work and the other works at home.

and morning chores, to work, play, and family spaces, to food, to bath. Friends visit. We store, utilize, and admire our possessions. At sunset, we complete our various tasks in the fading light and return to the glowing center where we reconnect with family, prepare and eat our evening meal, and get ready for restorative sleep. In addition to the diurnal cycle, many other climatic and familial cycles shape activities within a home. When the design for a home is patiently evolved over a period of years, and is sensitive to the qualities of our specific site and the patterns of our actual family, this design is likely to serve us very well. (Much more about cycles can be found in chapter 13.)

Recovering from an Edifice Complex

Many lessons can be drawn from the recent history of the American home. In the last fifty years, houses have doubled in size while family sizes have dwindled; from this we infer that modern houses are twice as big as necessary, and thus more expensive to build and keep comfortable. Commodity housing makes large, even grandiose, gestures toward habitability without managing to create homeyness: two-story entries, unused rooms, wall-to-wall carpet, long hallways, enormous picture windows, all contributing to "curb appeal" or the attractiveness of home-as-commodity when viewed from the street. These gestures frequently aggravate the adversarial relationship between the house and natural events like the sun's rise and fall, its seasonal declination, and the march of the seasons; occupants must be comforted instead by the hum and roar of the big HVAC (Heating, Ventilation, and Air Conditioning) unit.

Starting in 1978, in reaction to the rising cost of comfort, houses have been built like space capsules, tightly sealed to conserve heat, isolated from the harsh outside environment . . . but certainly not made any smaller! Considering only energy, tightness is a good trend, but can easily be overdone. Most houses contain a witch's brew of volatile compounds, including original materials manufactured using synthetic glues and persistent chemicals, to which we add the smells and effusions of everyday activities such as cooking and bathing. Excessive tightness, resulting in too little exchange of air with the outdoors, can create an unpleasant and even toxic environment.

By seeking ways to connect ourselves and our living spaces to the pageant of weather and seasons, by involving ourselves in flux rather than denying its existence at great expense, we can design nature and flexibility into our home spaces. A properly oriented independent home invites interaction between indoors and outdoors. Independent home residents derive most of our comfort directly from the sun, opening doors and windows, controlling sun and shade, and we enjoy a participatory style of occupancy as well as a much smaller monthly energy bill.

As you design, remember that siting, space, and services in most conventional houses are arranged for wholesale development and speedy construction. From the beginning, the developer's strategy is to eliminate eccentricities and singularities in order to produce the greatest number of salable generic houses with minimal building costs. These are mistakes you need not repeat: Parcel boundaries and roads maximize density without regard for residential privacy or the dictates of the land. Sites are typically bulldozed into submissive flatness. Showy exterior features produce awkward room shapes and door, window, and closet placements. Kitchen and bathroom plumbing may share a wall, not because there is inherent sense to this arrangement based on the dynamics of the family that will live here, but because it is cheaper to build with all plumbing components consolidated in a single wall. This strategy improves resource efficiency, both in terms of the plumbing and the cost of hot water, but only when the arrangement suits the family.

As in the negotiation between site, family requirements, and the ability to pay, compromises between ease of construction and livability must be made mindfully. Reckoned over the life of our home, how much more initial cost can we justify for plumbing, windows, and other expensive aspects of the house that suit us perfectly? Can heady design and an acceptable additional installation cost eliminate recurrent operating costs? How do we recovering suburbanites learn to apply the criteria we

Independent home residents derive most of our comfort directly from the sun, opening doors and windows, controlling sun and shade, and we enjoy a participatory style of occupancy as well as a much smaller monthly energy bill.

use to evaluate the merits of most consumer commodities—suitability, ease of use, and effectiveness of function—when designing a prospective home?

As a direct consequence of generations of carelessness and ignorance, those who continue to live with commodity housing have become a restless culture. When they are at last overwhelmed by the unsuitability of a house, they move: In 1998, the average American family moved once every five years. In 1992 we lived places longer, seven years. Clearly something is going terribly wrong with American housing.

In contrast, independent home builders are stable; of more than fifty householders interviewed in 1993 for the first edition of this book, only two have moved. As might be expected, some of their pioneering experiments worked well, and some were wrongheaded. Almost every independent home builder freely offers advice about how to do better next time. And because these houses are ours, built with our own hands, we experience no qualms about ripping out wrongness and trying again.

Right Ingredients

There is something atavistically satisfying about building with materials taken from the home site. Southwestern adobe pueblos and stone farmhouses in New England and the Midwest share a homey solidarity with their land. The timbers in my house, milled by my hand from trees storm-felled in the forest a short walk to the east, make me happy in a way that is hard to explain. Obviously oversized for the loads they bear, they make the building inspector crazy. "These aren't properly graded," he complains. Economics and inspectors aside,

materials from the site build the best homes.

Scavenged glass, copper, framing lumber, and fixtures are not exactly indigenous materials, but can find useful new lives in independent homes. As I built my house, I mined wood by dismantling the old Caspar Cookhouse, a barnlike structure slumping with disuse in a nearby cow pasture. The weathered, straight-grained, first-growth redwood posts holding up its porch roof, carved by a masterful Caspar woodsman on some long-ago rainy day, are as sound today as they were when felled over a century ago. Now they support the stairway and my own rougher-hewn timbers. Architects, contractors, and building inspectors subscribe to a wasteful cultural prejudice against using old materials in new houses. They tell us it takes longer to use salvaged materials: Nails must be pulled, unusual dimensions adjusted to, plans changed to accommodate found objects. So much value is invested in current buildings, so much embodied energy, that careful retrofitting is strongly justified, or recycling is mandatory. When we buy new materials, we should be sure that they represent true and enduring value.

Unlike commodity housing, independent homes grow out of site, the requirements of the family, and the fortunate availability of many forms of energy, including indigenous, scavenged, and surplus materials. As we prepare our design, we should also be perpetually open to accidental offerings and testimonials from unexpected sources.

Comings and Goings

Our attitude toward a home is prefigured as we approach the house and enter its space. The threshold that makes a transition from ground to floor may seem difficult to imagine when you are still in the process of planning, but because it is so decisive in establishing our experience of a house, the entryway deserves special attention. In my own home, I mistakenly left the entry functionally unfinished for years, until a beloved guest slipped on the milk crate serving as the step up from the mudroom and broke her ankle. Before that, I had hardly noticed the awkwardness, nor had family members complained; we were glad to have a roof. Habit is compounded by insensitivity to irritating inconveniences until we suffer a mishap or a visitor calls our attention to an ignored problem. Our family were all amazed by how much more pleasant it became to enter through finished space, stepping on real steps.

Passing through the entry, people are generally moving fast, and are often burdened, preoccupied, in the throes of transition. The area around the entryway is a natural collecting point for packages, tools, and other gear that are in transit, are used only outdoors, or have no proper place of their own in the home. Furthermore, when we enter from outside in a storm or during mud season, we bring with us companions we would rather not bring all the way in: gusts, rain, mud, and detritus. Something must be done to control this chaos; one good solution is a mud room or vestibule abundantly provided with storage, counters, and a place to sit while unbooting. Many homesteaders, especially those living in muddy, snowy, or dusty places, shed their shoes at the door. In Hawaii, following Asian tradition, a small set of shoe-sized cubbies on the lanai keeps footwear organized.

The placement of a home's entry should be a happy compromise between the site's micro-

AN ENGINEER'S WAY TO PUT A HOLE IN A HOUSE

Flush-hung doors and windows are troublesome. The housing industry considers the milling and hanging of doors and operable windows to be too demanding for most journeyman carpenters, and responds with mass-produced pre-hung doors and windows. Ken Kern, a crusading owner-builder and iconoclast, suggested that doors and windows could be built along the lines of ships' hatches, overlapping their jambs on all sides. Storms having defeated conventionally made openings in some chronically weather-beaten surfaces of my house, I now agree with him: Give me a door made from stout plywood sheets bolted down to a rubber gasket on all four edges to keep out the driving rain, firmly battened closed like a hatch.

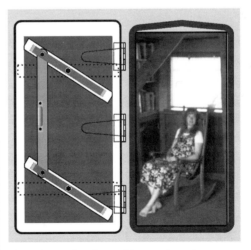

A rigid, insulated panel larger than the opening, and hinged so it can be snugged tight against a gasket to protect against weather.

climate, opportunities for access, and the house's floor plan. Since an entry is a hole in the house, it should be cut on the calmest side during the worst weather, to protect the home's hard-won interior climate from the brasher environment outside. In Caspar the east side is best, as it is seldom assaulted by storms or prevailing gusts. If you insist on placing the entry in the teeth of the gale, you will want an airlock-style vestibule, or risk rearranging papers and other light objects on surfaces near the entrance whenever the door is opened. Even the hinges on the door are worthy of your attention during the planning phase: Entry doors traditionally open inward in greeting, but to right or left? Weather, traffic patterns, and even the predominant right- or left-handedness of the residents should enter into your calculation. I decided to reverse tradition and open my front door outward, thereby preserving precious indoor space. A triumph of the millworker's art, a fifteen-light door, enables us to avoid flattening unwary visitors. Swinging out, the door takes unwelcome visitors aback and places new guests at a slight disadvantage, but friends learn to come right in.

Putting the Car before the House

Our culture has brought its infatuation with automobiles all the way home by putting the car before the house. Houses are oriented for the convenience of street and driveway rather than for the best relationship to sun and land. By connecting garage and living space, we further distance ourselves from nature, as if we were inhabiting a hostile planet where survival is possible only if we never leave the protection of house, car, office building, and shopping mall. In many independent homes, this unholy connection is severed. The vehicles (and all their drips and smells) are parked at a conspicuous distance from the dwelling's environs. At Frank Dolan's home, which doubles as the parrot hospital, cars are parked a quarter mile away, because they upset the birds. Years ago, I elected to build my home one hundred meters from the street, because I wanted to preserve the greenery and minimize hardscape near the house, so I could eat, work, and sleep well away from the noise and smell of autos. I expected my guests to be inconvenienced, and laid a lighted walkway to guide them home, but I have been gratified to hear them comment that the short walk gives them a chance to compose themselves and enjoy the garden and ocean view as they approach. An unexpected and beneficial effect is that we plan our shopping with a bit more care, and travel with fewer bags, because we consider the distance from house to car as we pack or purchase. We, too, appreciate the insulation between our home and the bustle a hundred meters away. Ungainly, necessary items come in by wheelbarrow, and in summer the house may be (but seldom is) approached across the meadow by car or truck for major deliveries. Vehicles and the home are uneasy partners, and a little separation works well for families.

Window Framings

During the day, the best light and views come to us through windows directly from the sun, but this illumination can be overabundant. In chapter 4 Robert Sardinsky told us that lighting is functional and easily ignored, but makes more difference in the habitability and friendliness of a place than any other single element.

Light is subtle, something not many of us manage intuitively; we are more likely to grit our teeth and squint in the dark than change the relationship between our work and the light. Artificial light is easy but costly to waste. Therefore task-oriented choice and placement of windows and lighting can make an important contribution to efficiency as well as livability. Planning for daylight and, under some conditions, glare, as well as the heating and cooling implications of windows over a range of seasons and weather conditions, is so challenging that we can hardly blame builders who minimize windows and specify predictable artificial lighting. Standard architectural practice places great emphasis on the appearance of windows from outside the house, but for those who live within, light and views are much more important. Architects insist that a variety of kinds and shapes of windows "look funny" but we should be concerned, I believe, with how windows look and work from the *inside*, not how they are perceived by neighbors and itinerant architectural critics. Each of the odd assortment of windows in my house is placed for reasons that are most apparent from within.

When I built my house I was so preoccupied with south-side windows for view and passive solar heating that I missed an important source of morning pleasure: nature's clarion wake-up call, the morning sun. In Caspar as in most places with cold nights, daybreak contributes needed heat, and introduces great optimism and cheer. People sometimes employ heavy drapes in their bedrooms to keep out morning sunlight, preferring to awaken themselves with alarming clocks. As an exercise in subtlety, I invite you right now to go and set off your alarm clock; even the way we say it, with the same words we use to describe detonating a bomb, warns us what to expect. Now tell me, how can we possibly arise well-disposed toward the day after such a racket? For too many, the alarm clock is an up-to-date version of the demonic mill whistle that shrieked their ancestors out of their company beds in company houses in company towns. By contrast, Thomas Jefferson, a brilliant solar-powered-home designer, imported a fine (and silent) clock for his bedroom at Monticello, and awoke every morning when it was light enough for him to *see the clock's hands.*

In too many houses, window placement is an unconscious act. There are two important issues to consider: what happens inside the house beside the window, and what is outside. When planning a house, careful consultation with floor plan, living patterns, and site rose shows where light will be required inside, where we will sit and stand to take in the grand views, and where windows will be unwelcome. In commodity

Thomas Jefferson, a brilliant solar-powered-home designer, imported a fine (and silent) clock for his bedroom at Monticello, and awoke every morning when it was light enough for him to see the clock's hands.

housing, only one window is properly set: above the kitchen sink. Oversized windows are costly to maintain, because even the highest-tech glass wastes twice as much thermal energy as a standard wall, and a single-pane window is fifteen to twenty times more wasteful. Why put in big, expensive windows and then cover them with expensive thick drapes to keep heat and privacy in, giving up the wall space to the view, then the view to the drapes? Conscious placement, in which we visualize daylighting of indoor activities, furniture placement, traffic patterns, and out to the end of each window's view corridor, the whole sweep from indoors out to the horizon, is a difficult but worthy exercise. A window placed precisely may be quite small—a living picture hung among others on the wall—and do all its job effectively.

Conventional houses seldom place skylights for nighttime views, but independent homes often have moonwindows and star viewports. My daughter's bed, at the prow of the third floor, has two skylights and an ocean window. On a moonlit night all the glass makes it seem as though you are flying slowly over an astonishingly bright meadow toward the sea.

I think this should be part of every building code: No one can steal your sun, and all houses must have correct solar orientation.

—

Before You Build: Wes and Linda's Advice

Linda and Wes Edwards moved from the suburbs near San Francisco to an isolated ridgetop near California's Lost Coast with their daughters two decades ago, where they live in one of the most gracious and well-conceived independent homes I have visited. I asked Wes and Linda to summarize their planning strategies. (You will find Linda and Wes's story in chapter 12.)

Wes: Okay, it's time to take some notes now! Here's my design strategy for a self-sufficient home. Start with a small house. That minimizes the impact on the site, and takes fewer materials. . . .

Linda: When I clean this "small house," it seems big to me.

Wes: And when the grandkids come, it would be nice if it were bigger.

Buy a plan for your house, and get an architect's help to make the changes you need. This can save you time and costly mistakes, maybe irreversible mistakes. I definitely made mistakes, so I know. I'm an electrician, not an architect, so there were things that I didn't understand that an architect sees every day. It can be done right the first time. We spent two years seriously planning this house, gathering information, studying plan books, visiting other people's houses, watching the site through the seasons. From groundbreaking to occupancy only took a year.

As you build and change things, keep the house plans current, or keep sketches or photographs of the changes, so you know, for instance, what's in the wall when you cover it up.

I think this should be part of every building code: No one can steal your sun, and all houses must have correct solar orientation. You have to think of the whole south side of a house as a solar collector.

Align the roof slope to your latitude, which should be the best angle for solar panels, either hot water or photovoltaic. They may be a bit harder to work on up on the roof, but it's easier to attach them firmly, and they look like an intentional part of the design.

We tuned the house's orientation to our solar aperture, about ten degrees east of south. We run out of afternoon sun because of the trees and the ridge. We've had to top our closest trees to keep them out of the house's sun. And we built in thermal mass: seventy yards of concrete in the full basement foundation, and dark brickwork in the north greenhouse wall. I couldn't believe how much concrete we put in, or how much it cost, but you just have to do that part properly. We tuned the overhangs for summer shading—you can also use seasonal vegetation for shading, if summer heat is the problem it is here.

Try to eliminate pollution from inside the home. Vent the propane refrigerator and the water heater, because they burn and expel fossil fuels, for example. Limit formaldehyde-bearing materials, use non-toxic paints and sealants, and avoid CFCs. Even then, you need good ventilation.

Use local materials if you can, because they will be better and cheaper. We milled the sub-structure and rafters from a fir tree that was standing right where the house is. We found a standing dead fir that we thought might still have good wood in it; the mobile dimension mill produced six thousand board feet of beautiful lumber. The deck came from a redwood sinker dragged out of the creekbed; it's only medium-grade wood, but it's perfect for decking, because it knows a lot about wet. I believe in sustained yield—you ought to, if you live in a wood house—so we tried not to use endangered materials, and used recycled wood where we could. It takes longer, but preserves the forest.

Use the highest-quality windows, the best you can buy. I had a problem with double-glazed windows in the greenhouse, and my glass man informed me that's common in very hot applications. They turned white, then their seals failed. Be sure to get good advice. Use good doors and window framing, because the operating parts of your house should be the best possible.

Build air-tight, and insulate as intensively as you can, because it's easier to heat.

Linda: We used six-inch, kraft-backed fiberglass and six-mil vapor barrier in our six-inch outer walls.

Wes: Limit windows on the north side. You see, we have none.

Around here, the best way to heat, after solar, is with wood and an air-tight, catalytic woodstove. I can clear enough from the land to keep going and hardly scratch the surface. I figure if it rots it produces the same CO_2 as when I burn it. Woodstoves with small fireboxes and the high temperature required by a catalytic converter can be a problem, so research your stove carefully.

Plumb for solar hot water even if you can't

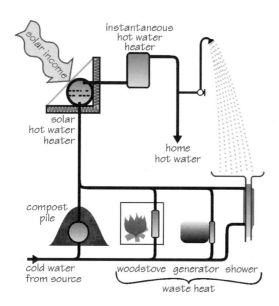

The cost of hot water can be reduced if the incoming water is preheated. By using the sun, the heat from a compost pile, and heat from used hot water, a woodstove, or a generator, it may be possible to eliminate the need for a conventional water heater entirely.

afford the panels, and even if you just stub out the plumbing in the attic. Insulate all the runs, the hot side in particular.

Active solar hot water is more efficient than passive or thermosiphon. There are two problems to be dealt with: A, panels freeze, and B, controllers fail. Antifreeze in the collector and a heat exchanger tank takes care of A, and a system where the pump is matched to the sun with a PV module takes care of B. We've got our solar hot water backed up twice, with a coil in the woodstove, and an instantaneous water heater.

Separate your greywater from your blackwater, and send it to artichokes or other crops that tolerate soap and grease. You can reclaim

the heat in the greywater by running coils through your greenhouse beds. Low-flush toilets are a must; you want to keep your septic system happy. If it's not drowned with too much water, and stays uncontaminated from soap, it makes a better class of compost.

I prefer a 24-volt system. Compared to a 12-volt system, it cuts the amperage you're moving in half. Seek professional help when you're contemplating a full-blown system . . . the same as getting an architect's help, you'll save on mistakes and false starts. Talk to people who have systems equal in complexity to the system you want.

Linda: You will be amazed at what you learn from others.

Wes: Provide for future expansion. With renewable energy, you can have it all, except heating, unless you have BIG hydro.

I no longer believe you need to double-wire your house with parallel AC and DC wiring. Direct current-to-alternating-current inverters have gotten so reliable and efficient, and the quality of compact fluorescent light is so good—expensive, but once installed, people love getting 75 watts of light for 20 watts of power. Wiring for conventional 120 AC with 12-2 Romex is much cheaper than wiring for both AC and DC. In this house, the smallest 12-volt wire is eight-gauge, for lighting; that's too much copper in the wall. Instead, wire special circuits for 12-volt DC devices like pumps that are much more efficient than AC. You'll save a lot of energy if you plan carefully for zone lighting, and suit the fixture and source to the activity. And put in lots of extra outlets; you still never have enough.

Electromagnetic Radiation (EMR) is something we should consider, even if we don't understand what it really means. We know there must be something to it, because the CIA was bombarding the Russian Consulate with it in the sixties. Wires in the wall don't worry me, but I want equipment at arm's length. Inverters should be outside the building, with the batteries.

Put in extra conduit runs between the basement, crawlspace, utility room, and attic. You can never plan for the future because of new technologies and new interests. For me it was a radiotelephone, a new rain gauge. Photovoltaic modules are expensive, so you usually won't start with enough, but plan and wire for a huge array. Spend money on the wire. Our system, which has thirty-four panels and a micro-hydro plant now, recovers quickly and our batteries are fully charged by noon, so we can use the surplus running the dishwasher and the washer in the afternoon for free.

Trackers make sense for pumping water in summer, when you get 50% more power by tracking. In winter, trackers only add 20%. You use more energy in the winter, because you're inside more and the days are shorter, so it's less costly to add 20% more panels.

Before buying appliances, check for water and energy use. European appliances are best; American manufacturers are just starting to get the idea that efficiency is a good idea.

One last piece of advice: Grow your own food. Organically.

Harvesting potatoes.

ELIZABETH HENDERSON, from *Sharing the Harvest*

CHAPTER 8

BUILDING THE HOME ENERGY MACHINE

WHILE SEEKING TO BUILD FROM THE OUTSIDE IN, WITH FULL AWARE-
ness of the particularities of region, neighborhood, site, and tradition, we
should also undertake the planning required to build from the inside out.
What functions, precisely, do we require of our home? Each of us will
answer differently, and a family's patterns will inevitably change as the
individuals change, as children leave and builders age. Anyone who
builds an innovative, appropriate shelter hopes to be free from precon-
ception and perspicacious about small, important elements. If we intend
to abandon the pretense and wisdom of conventional architecture and
still build a successful home, we must carry with us only truly essential
luggage.

Happiness is a good measure for necessity. For example, it will not
make us particularly happy to know that we are expensively hoarding
eighty gallons of hot water in our domestic hot water tank. And what is
the widest temperature range within which we can live healthily and
happily? The answer to this crucial question, factored together with the
meteorological realities of our chosen site, determines for most of us the
largest expense in our energy budget: the cost of heating and cooling our
space. If you hope to build a successful independent home, you must
identify and answer these and many other energy questions completely
and flexibly before building starts. (For instance, see pages 40–41 for

charts of climate factors in a variety of U.S. cities.)

In the past, houses have been planned assuming indefinite continuation of present circumstances: straight-line thinking. Our whole human era is called the Holocene by archaeologists: an unusual time in geologic history when nothing much changed. Nature abhors straight lines as much as it abhors vacuums, and this forecasting technique is not viable if change is imminent. Selection of heating and lighting strategies, for example, has invariably been based on current costs. Regulators, noting scarcities, pollution, and growing recognition of the health costs of fossil fuel energy dependency, anticipate steeply rising energy costs in the near future. In earlier chapters we recommended selecting for cost-effectiveness based on comparisons of energy consumption over an appliance's whole life plus its original equipment cost.

Consider glass, the most important energy appliance in most homes. Glass easily outlasts most houses, but will high-tech glass last as well? A glass-and-sealant bottle encased in metal, containing inert gas and a plastic film, is undeniably more expensive and possibly less durable than a single pane of glass, and much more efficient than a single pane as a wall, but will it pay for itself over an acceptably long life span, one that may be longer than our own lifetimes? If so, then high-tech glass is the correct choice for an ancestral home even if we shall not be around at payback time. If windows demonstrate a high rate of failure after two, five, ten, or even twenty years, the deal may not be as attractive.

Since the primary energy consumers in a conventional house are space heating and cooling, refrigeration, hot water, and lighting, conscientious examination of these largest energy uses will reveal ways to lighten our load and simplify our lives. Some reductions are available to us right now, and can be put in service immediately. Other solutions will be found and incorporated into our house plans as we begin the process of building. Many of the best ways to lighten the eventual energy budget will involve site and structure, and therefore should be developed before a final decision is made about the site.

Naturally, a home's energy budget depends on region. In a tropical climate like Hawaii's, a well-designed home on a benign site will require neither heating nor cooling. In northern Alaska, and along the northern tier of states, a 10,000-degree-day heating regime or worse, the best-sited and best-constructed home will present a heavy burden in wintertime lighting and heating. Even if energy costs are tolerable at present, there are powerful reasons to seek to reduce energy consumption from every energy-consuming device and design feature.

Paying for Too Much Climate Control

Formal architecture creates interior space that is visually connected with the outdoors yet completely isolated and regulated atmospherically. As soon as architects got the high-powered HVAC (heating, ventilation, and air conditioning) technology they needed to defy a site's natural climate, they defined a narrow comfort envelope within which, it was decreed, the human organism functions optimally: temperature between 68° and 72°F (20° and 23°C), with between 60 and 70% relative humidity. Elaborate devices, strategic materials, and awesome energy are poured into achieving this narrow goal.

Energy Domain	Coventional Source	Replacement
Heating	electricity	solar exposure super-insulation widened comfort envelope layered clothing wood: catalytic chimney excess hydro-electricity
Air conditioning	electricity	widened comfort envelope improved ventilation super-insulation architectural measures evaporative cooling
Water heating	electricity propane/natural gas	solar solar assisted wood
Refrigeration	electricity propane/natural gas kerosene ice	super-efficient spring house shady-side outdoor ice
Cooking	electricity propane/natural gas wood	raw foods solar cookers propane/natural gas
Lighting	electricity kerosene sunlight	natural light high-efficiency electric task electric
Appliances	electric	hand-powered super-efficient solar-charged
Landscaping	fossil fuel	hand tools indigenous planting
Transportation	fossil fuel	human-powered solar electric shared (public)

For nearly every conventional domestic energy requirement, alternatives exist that reduce the need to consume fuel.

Although other cultures adapt to temperature extremes quite routinely, one might almost expect a modern Bill of Rights to read: "the right to life, liberty, temperatures between 68 and 72, humidity between 60 and 70%, and the pursuit of happiness." We submit to curious climatic abuses because of this comfort fetish—I have been too hot in Boston's winter, and clammy cold in Fort Worth's summer—and to a variety of ills, from Legionnaires' Disease to Sick Building Syndrome, which result from our willingness to defeat the environment with biologically disruptive technologies. Like willful children, as soon as we found a way to defy nature, we did so, without ever contemplating the repercussions. When the OPEC wake-up call came in 1973, and energy awareness gradually took hold as a result of the oil embargo, the supposedly sacrosanct comfort ranges were redefined, and a cardiganed president begged us to comply.

Our reliance on brute force, high-tech HVAC solutions, results from our abandonment of the homestead for many hours of every working day. When we stay at home, it is easy to open and close windows, feed the fire, and otherwise regulate our home, but on most days, when we return home exhausted from a hard day's work, we expect to be greeted by a welcoming, comfortable, well-regulated environment. There is a sharp irony to the fact that many of us must leave our homes to earn enough to support the machines that take us back and forth to distant work and keep our homes habitable in our absence. Who, precisely, are we working for? The energy we buy in order to work—fuel for the commute, and the cost of maintaining a comfortable home space even when no one is home—is a hidden and rapidly increasing tax on our productivity, a levy that many independent homesteaders refuse to pay.

As my family planned our house, some of my heroes, vigorous old ones who create beauty even as they have aged beyond my youthful imagining, reminded me that a life connected with nature's grittiness is longer and more vivid than one lived within the comfort envelope. Now, a couple of decades later and with snow in my own beard, I participate in the spirit of the season by wearing several layers, or as little as possible, as the climate suggests, in-

stead of expensively denying the elements. To put it quite simply, I enjoy, as an independent power producer, living in space that buffers me within, rather than isolating me from the weather. My house invites me to participate in the seasonal responses of my local ecosystem, and I gladly respond.

Mass and Glass

Prior to stretching your tolerance for temperature extremes, designing for passive heating and cooling may allow you to enclose space that regulates itself quite comfortably. Three key factors apply: incident solar radiation (which yields a coined word, insolation), thermal mass, and ventilation. Using glass, we can invite sunlight in to heat our space to a livable temperature. Sufficient thermal mass (material that holds heat well, such as masonry, earth, or water) within our home's envelope attains a comfortable temperature, then acts as a buffer to keep the adjoining space comfortable. When the air in the envelope gets hot or stuffy, windows and doors allow the prevailing breeze to enter, cool, and refresh. Such self-regulation, with no ongoing expense, is a grand solution, but attaining it is not simple. If too high a proportion of a building's walls are sun-facing glass, the indoor temperature varies wildly no matter how much mass is enclosed (or it uses prodigious amounts of energy running its oversized HVAC equipment). If a home incorporates too much mass, the temperature will never become comfortable.

The ideal glass-to-mass ratio is delicate, site- and design-specific, and therefore hard to plan for, but instead must be found through careful analysis or experimentation. Computer programs can model simple structures, but in a

186

As described in chapter 12, the Edwards house is designed to use the sun for heat in winter, and to exclude it in summer. A massive concrete foundation serves as the thermal flywheel. When the greenhouse is warm and the house cool, the windows (2) are opened, and make-up air circulates in from the basement through a channel at the foot of the greenhouse stairs (1). In summer, the excess hot air escapes through vents (3) near the top of the greenhouse.

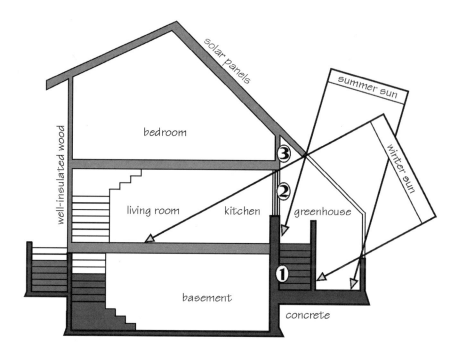

breakthrough house, one unlike any house built before, planning is complicated. Since glass and mass are costly to build into a project, and difficult to adjust after they are in place, getting the proportions right is important. Most architects sidestep the issue by specifying powerful HVAC units. One possible path to a solution is to start with a small, prototypical space such as a garden shed, and tune its glass, ventilation, and mass until the principles are grasped, then enlarge along the same proportions, taking into account the physical laws that increase area with the square but volume with the cube of a dimension. This approach may be unworkable for architects, who are paid to present a completed totality, but fits the alternative builder's gradualist mode. Another method is to isolate exterior glass and mass from the rest of the house, as Wes and Linda Edwards have done, by using a sunspace as their furnace, and interposing operable windows between sunspace and living space, so that the active thermal components may be opened to the house when they can help, and closed when they cannot.

Wood and glass houses, though quickest to build, have almost no thermal mass and are therefore at the mercy of ambient outside temperatures. Concrete, rock, masonry, tile, earth, water, and other dense materials provide thermal mass if properly insulated. Solar gain—the heat stored in the mass—must also be protected from infiltration of cooler outdoor air. Mass placed where it intercepts sunlight and is directly heated is said to have direct gain. If the mass is dark in color, absorption of incident sunlight is improved. Indirect gain is less efficient: Sunlight shines on less dense materials, such as wood, cork or vinyl tile, cloth, materials that hold heat poorly, and the heat is carried by convection currents in the air which then heats available thermal mass indirectly. In most net-heating regimes (heating degree-days exceed cooling degree-days) homes that dedicate the southern side to direct gain thermal mass are most successful; this quarter is the right place for a greenhouse. Ventilated thermal mass can help with summer cooling if overhangs protect the mass from too much solar gain, and the mass can discharge excess heat to a cool air flow when doors and windows are opened to the cooler night air.

Pounding tires: Ken Anderson's story

Earthships—houses with walls made from earth tamped into old tires—and other thick-walled houses in the American Southwest are thermally massive structures which perform well in their climate. Often adding another passive feature, earth-connection, by digging the house into the ground, earthships, kivas, and adobe houses remain surprisingly comfortable despite the Southwest's blistering summers and frigid winters. Ken Anderson is an architect who started building earthships in 1992. He and his partner, Pamela Freund, built their own earthship at Star, one of three earthship communities near Taos, in northern New Mexico. His enthusiasm and delight in this approach to homebuilding kept us both smiling as he talked.

I came out here from New Jersey, Philadelphia, and New Haven, and there's a good chance I'll stay here until I die. My goal, when I graduated, was to get as far away, philosophically and geographically, as I could. I had no idea how to build the buildings I was drawing. When I got here, Mike Reynolds put me to work on an earthship, and I stayed with it from scratch, a year and a half, from the first hellish tires. I really learned a lot on that job about unconventional construction. When you're involved in building, you need to understand the whole system.

The idea of the first earthship book was to make it possible to build an earthship by yourself. Some of the best earthships are owner-built from the first book. "Who needs to draw it? Let's just go out and build it." That worked for some, but in general we learned from people out there that

Ken Anderson and Pamela Freund in their earthship kitchen.

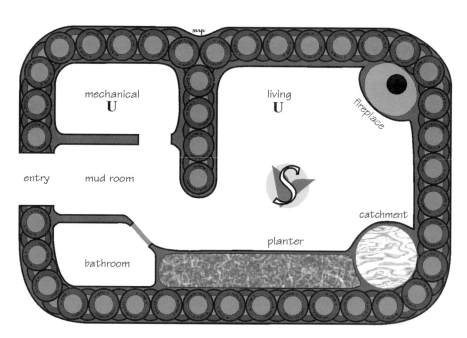

Earthships have U-shaped walls made from stacked tires pounded full of earth. Most earthships have a built-in fireplace, catchment cistern, and indoor planter for water recycling.

there was too much bureaucracy, so we made a set of generic plans that give the inspectors, building departments, and engineers everything they need to see.

Every earthship we've worked on has a building permit. We have made a conscious effort to establish the viability of this building method, and to work with building departments so they would know what we were trying to do.

It takes five or six hundred tires to make a small house; if you start with the house excavated into a hillside, you can cut down on the number of tires. There are about three wheelbarrows of earth in each tire, and it takes two people about fifteen minutes to pound a tire. Maybe you can do thirty tires a day, and at the end of a few days, you see the house forming up, and you get your second wind. As the tires fill with tamped earth, they swell and interlock with the row below, so the walls are very strong. The walls are usually U-shaped (and we call each bay a "U") which also adds strength.

Along the top of a tire wall, you put in a cement and steel or a wood bond-beam, to hold the wall together, and for attaching the roof. The sloping front wall involves some pretty special construction techniques, and the glass isn't cheap, so generally you need a good carpenter to do that part.

The traditional way of adobe roofbuilding, using vigas and latillas, works very well, but we're working with trusses, too, which use much less wood. You can buy vigas at the lumberyard for three or three-and-a-half dollars a lineal foot, which is expensive, or harvest them in the national forest, mostly standing dead, with a viga permit which costs about three dollars for each viga, then strip them, which is labor-intensive. We use some fairly high-tech sealants, six-mil plastic sheeting to seal over the tires and

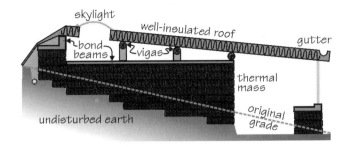

This cross-section shows how the earthship burrows into a gentle slope. Part of the original earth is incorporated in the thermally massive walls. Concrete or wooden bond beams hold the tire wall in place. Holes between tires are filled with crushed aluminum cans and the whole wall is plastered with adobe.

berm, and then a modified bitumen roof coated with acrylic over the foam roof insulation.

Mike's idea about the earthship is that it should be self-sufficient. Electricity comes from PV modules on the roof and golf-cart batteries. The electrical center makes the unusual connections, and gives the electrician a standard breaker box that he knows how to work with. We do the same thing with the water system; we want to give the tradespeople systems they understand, with standard fittings and connections for them to tie into. All the lights are 12-volt DC, compact fluorescent, and all plugs and appliances are 110-volt AC. Rain- and snow-water are collected on the roof, and flow to a catchment inside the "ship." We figure that you can live with eight inches of rain a year if you recycle the greywater into the indoor planters and use a solar or composting toilet. The water system consists of a simple arrangement of filters and a pump powered by the electrical system. These two systems cost between sixty-five hundred and eight thousand dollars, depending on what you need, although we've done much more complicated systems.

Used tires are rammed so full of earth they interlock.

Pam and I built our home at Star, an earthship community about forty-five minutes from Taos by four-wheel-drive. It's near the Rio Grande gorge, and the views are great. You buy a membership in Star based on the size of house you want to build, for two to ten dollars a square foot of livable space, depending on whether you choose the high-density or low-density area. High-density means a minimum of fifty feet between houses, and low- means four hundred feet. As a member you become part owner of the entire piece of land. We built for about thirty dollars a square foot, so for under fifty thousand, and two years of work, we have a nice home that completely takes care of us.

1999 Update from Ken Anderson:

Pam and I have our own company now, and continue to focus on building with alternative materials and incorporating sustainable technologies. The experience of hands-on building and living in our own off-the-grid home has been a great asset. We're trying to combine alternative materials with new approaches being tried throughout the country. We've incorporated some of this experience with "standard" construction techniques we learned back east, combining techniques for building cheaply and quickly with environmental sensitivity so we can design a home that is affordable, easily understood by most contractors, and still self-sufficient and environmentally friendly.

Recently, we've been doing a lot of straw bale construction. We like the thermal mass of our house, but straw is so much faster and easier to build and finish out. We work a lot with owner-builders and this technique is especially suited for them. A great solution has been to combine the insulation of straw bale walls with the thermal mass of interior mass walls. We use adobe brick, because it's appropriate in our area, for creating thermal mass within an insulated envelope of straw bale. For interior partition walls we often use a pressed adobe brick, giving us a thinner wall which still provides good thermal mass. A friend has a brick-pressing machine: You shovel dirt in the top and adobe bricks shoot out one end. Pressed adobe bricks are a little more crumbly than dried adobe bricks, so we always plaster over them.

Some sites have a lot of caliché, light colored, almost white adobe, and some clients like the lighter color, so we can use dirt from the site to plaster the walls. For other projects we use darker adobe. Our plasterer, who lives over at Taos Pueblo, hauls in a couple of truckloads of beautiful, almost black adobe from a source he knows. We still like to use vigas or heavy timbers when the project can afford it. When the budget is tighter, we do projects with joists or TJIs (truss-joist I-beams, constructed of oriented-strand board), but we like vigas because of the way the finished space feels.

We're entirely off-the-grid at Star. The nearest powerline must be ten miles away. We're happiest with the power system. We installed a Trace 2512, ten Kyocera 51s, ten golf-cart batteries. We've got an SSA power center, which shows us our charging voltage and the battery voltage. Three hundred fifty-five days of the year we've got more power than we know what to do with. We've got a Sun Frost RF16 refrigerator, which has worked well. We've never had to turn it off due to low batteries. Worst case, when the power's low, we can't watch TV . . . and that's only happened a half-dozen times. I don't see much sign that the batteries need replacement even though they're beyond their expected life span. We've taken good care of them. I have noticed that they don't hold a charge as well as they did—they'll charge to 14 volts by day, then drop to 12.5 volts by the time we go to bed. We've been careful to water them—about once a month. We don't have a generator . . . and we've really never needed one.

Friends out here have a standard fridge (which they got a great deal on) and they added four extra modules to compensate. This system has worked okay but they're usually a little shorter on power than we are. Another house has a propane refrigerator and seems to always be fully charged, although the owner has had some problems with the fridge. I heard of one house that seems to be low on power quite a bit, but they're built real close to the mountain, and don't get morning sun in the wintertime. Most people are topping out their batteries almost every day.

For heat, the combination of good siting, windows, and abundant thermal mass works very well. We've been burning construction scraps in our kiva fireplace for two years, but we're getting close to the end of that now. We burn maybe ten fires a year, half of them for fun. We built an extra-large fireplace for cooking. Last night we put our barbecue in the fireplace and cooked up some shish-kebab. The living room's nice and warm when folks gather around the fire. On cold winter nights we sometimes think about a little woodstove in our bedroom at the far end of the house, since by ten o'clock it does get a little cold.

Of course we have a composting toilet, an Envirolet. It works well enough in everyday use and the cleanup is usually okay. There are days, maybe once or twice a year, when cleanup is not as much fun as it ought to be, because the stuff in the tray is too wet. Cleanup is best if we stop using the toilet for three or four days beforehand. Outside we've got two contained compost bins into which we always dump the drawer. We mix this with our kitchen scraps and then let it cook for six months before it goes

We're entirely off-the-grid at Star. The nearest powerline must be ten miles away.

on the plants. Since we've got a lot of dogs and animals, containment is important.

We have a 3,000-gallon cistern, and I don't think we've been below 2,000 gallons more than twice. But of course, we're careful with water . . . even if we do take twenty-minute showers when the cistern's full. We just recently got a washing machine. There's only been one year we had to haul water—1996 was a drought year around the whole county. Since the sheep rancher next door was hauling water, we had him bring us out 1,000 gallons twice. The cistern was nearly full when the rains came, and we used at least 500 gallons for construction, so we could have survived on 1,000 gallons. We've talked about drilling a community well, but everyone agreed it wasn't necessary. The whole time we've been out here, we've run out of water once, so spending $100,000 to dig a six-hundred-foot well and put in a solar array to pump it doesn't make sense. We all agreed to conserve and live on catchment, and we'll stick with that.

For potable water, the last couple of years we've been using the Seagull filter. I think we may be the only people at Star not hauling drinking water. We've got a little Shurflo pump and pressure tank; we pressurize at 30 to 50 pounds, and we get a gallon a minute out of the potable water filter, which

Ken and Pam's house overlooks an arroyo in the high desert near Taos.

is under the bathroom sink; from there we have a half-inch line to the kitchen sink, so there's potable water there without giving up space under that sink.

We have a heat-pipe solar water heater with a backup propane tank water heater which we got free from our propane company. The two work well together. We've never had a problem with insufficient hot water. We also use propane in our big Imperial commercial stove, the only thing in the house that cost more than the Sun Frost! Once a year the propane company pumps a hundred gallons or so into our 250-gallon tank, so we're not using much.

We have a lot of success with our greywater planters inside, which we've customized. The original subsurface watering system didn't work well for the plants we wanted to grow, so we've adapted the system so it surface-waters. Right now we've got lots of tomatoes, broccoli, squash, zucchini, peppers, jalapeños, a couple of papayas. The best part is having fresh herbs all the time. We have about every herb we need—parsley, rosemary, basil, oregano, dill, chives, sage, cilantro, lemon grass. . . . For cooks, this a dream. We rarely if ever use dried herbs. We've got a thirty-foot-by-three-foot planter.

Our philosophy was to design the mechanical systems around the house, not vice-versa. We wanted to live in a home, not in a machine. Since we expected to live with the mechanical systems every day for many years, we designed the house and systems around our lives, rather than designing around the installation, which only took a couple of days. We may have spent an extra $200, but the living is easier. The trick is to design for comfort without sacrificing efficiency, and I think we succeeded.

At this point we've been working on the house, which is about two thousand square feet, for five years and we're about 85% finished. We still need a color coat outside and some plaster inside. We've got a concrete slab floor which we're planning to finish with Saltillo tile. Pam and I did most of the construction ourselves. We hired help to pound some of the tires and frame up the greenhouse wall. We've also gotten a lot of help from friends and neighbors, either as trade or as part of work parties. (A couple cases of beer can go a long way toward getting free labor!) At this point we've got about $65,000 in the house, about $33 per square foot. We could probably use another $15,000 to completely finish the house.

If I were building our house over again, the one thing I'd do differently is install vertical glass with an overhang. (Looking out through our windows right now, I see we could wash them more often—that's on the list.) Overheating is definitely a problem and the solutions I've seen aren't

We wanted to live in a home, not in a machine.

satisfying: ugly netting on the outside, ugly sheets hanging inside, or expensive window shades ($100 per window or more). The owners of a house we built last year really wanted the sloping glass, but we agreed to put the glass at a 75-degree slope (as opposed to the standard 60-degree) with a two-foot overhang, and it seems to be working just right. The difference in the summertime is amazing. We've also designed and installed an overhang retrofit for earthships, which considerably relieves the overheating problem.

Maintaining a successful home office at Star is impossible due to road conditions, so we've found ourselves an office in town with a small apartment attached. We're now splitting our time between here and town. Most members who have to get to town for work or school have made similar arrangements. The drive takes between forty-five minutes and an hour and a half depending on the condition of the road. We have to drive about ten miles of dirt road once we leave the pavement. The condition of the road is a deterrent—it makes us think twice about driving out. A couple of cars have rolled on the country stretch when someone hit washboard going too fast. Of course guests get stuck in the mud and we have to pull them out, and delivery trucks get stuck. The Star community gets together to work on the road a couple times a year, putting the snow fence up and taking it down. Without the fence, drifting snow made the road impassable. Each year the members living here put in a couple hundred dollars apiece to haul in some fill for low spots and grade the road. The road is hard on cars: Part of our yearly budget is a couple thousand dollars in car repairs.

Star's kind of a weird situation. There are some definite issues about property ownership which are very, very fuzzy. There are also some problems with the county, which are now being played out in the courts. And the road is still a terrible road. But these issues have become a rallying point for the folks who live here. The fact is there is a very strong sense of community here. There are now seventeen homes started with ten being lived in at least part time. There are about a dozen full-time residents.

The ownership issue is a very sticky point right now. It's kind of a legal disaster. We all started building on the understanding that we were part owners of the land. But the owner's association was never properly established, and now we're involved in legal battles to straighten all that out. We also know the county planning office isn't happy with the earthship communities as they relate to subdivision law, but there isn't much opposition to the houses themselves. We all have building permits, and we've passed our plumbing and electrical inspections. The worst they would probably do is make us move out until we've got a certificate of occupancy, but if they start doing that, I'd venture to guess that half the folks around Taos would be homeless. The fact is that most of the legal hassles with the county do not affect the individual residents who already have built. And although there is legal uncertainty about the deeded ownership of the land and problems with the way Star is being managed, the actuality of day-to-day living has been good. The members at Star share a great respect for the common land and everyone seems very committed to the community.

We're holding our fifth annual meeting this year. The members at Star work well together, and we've lived up to our end of the original agreement, essentially living as guinea pigs for the last five years. Since we've reached the point when control of the land supposedly transfers to us, we'd like to see the original agreement honored. We'll just have to wait and see how that all works out. But on a good day at Star, problems like that seem a million miles away. ⚊

Simple, Passive, and Massive

In a home using thermal mass to maintain comfortable climate, it may take a year or two for interior temperatures to come to equilibrium. After that, a certain amount of supplemental heating or cooling still may be required. In the Taos area of northern New Mexico, where massive, small-windowed adobe houses are the vernacular form of building, ventilation must be provided or temperatures indoors will exceed those outdoors where prevailing breezes blow. In winter, small infusions of heat are needed to keep the dwelling's temperature from dropping into the chilly high 50s during stormy periods when outdoor temperatures drop to -10°F. In contrast, wood-frame houses without heating and cooling around Taos are intolerable most of the year, and those with fossil-fueled heating and cooling have intolerably high heating and cooling bills.

Structures that heat themselves, then radiate the heat long after the fire dies—massive stone or brick fireplaces, concrete and masonry furnaces—are a wise addition if the combustion techniques used to heat the mass are clean and rationally fueled. Siberian stoves, for example, need only be fired with a modest bundle of wood once a day even in a harsh climate; the sinuous smoke path captures all the heat, and the high-temperature firebox assures that combustion is as efficient as possible. After the fire dies, this substantial pile of masonry continues to radiate heat for hours. Savvy homesteaders time this cycle right, and bake pizza in the cooling firebox.

High Ceilings, High Heating Bills

Heat rises. Without active ventilation, the air just below the ceiling may be ten degrees warmer than the air around our knees. People in motion set up air currents, but these stir the air's stratification for a limited distance. Low ceilings keep precious heat closer to the entities who need it. We have come to accept that anything less than a foot above our heads is too low, although I have been in experimental homes with very low ceilings and found myself adapting quickly. How many of us are taller than six and a half feet? Eight-foot ceilings, we are told, feel dignified, and are quick to build because paneling customarily comes eight feet tall, but in a temperate environment, eight-foot ceilings have nothing other than industrial convenience to recommend them. Building codes generally permit ceilings as low as seven and a half feet. Considering that it costs approximately 10% more to heat six more inches of ceiling height, I hope we can agree that seven feet six inches is ample.

Doors Indoors?

Doors represent a compromise between conflicting needs: We want be able to move from room to room, to foster or prevent air exchange, and to interpose temporary walls for privacy. Conventional hinged doors waste floor and wall space within and behind their arc. Pocket doors waste little floor space, because they live in the walls when open. Archways and carefully planned sight-lines can eliminate the need for most doors in a private dwelling. Privacy can be created by juxtaposing walls, screens, and archways so that rooms have the sense of privacy desired.

Insulation Is Everything

Thermal mass heating, and any other strategy that is meant to separate indoor and outdoor temperatures, justifies the best insulation. The wider the differential between indoor comfort and outdoor ambient temperatures, hot or cold, the thicker the insulation should be. Over the life of a house, the added cost of the thickest insulation, the highest R-value high-tech glass, the most thorough measures for reducing infiltration, is much less than the fuel required to compensate for the omission of these measures. Good siting, excellent design, and superb insulation make it possible to build a passively heated house in almost every habitable climate. In other words, most energy expended on space heating and cooling is a direct result of bad planning, negligent design,

shoddy insulation, and failure to recognize lifetime costs.

Low-emissivity, high-tech glass was not available in California when I put windows in my house, nor was it in my budget based on short-sighted, pre-1973 energy cost-benefit calculations. I was wrong. Two decades later, the cost-effectiveness curves have crossed: Energy costs are stubbornly climbing, and the cost of high-tech glass is lower than ever due to its enthusiastic acceptance by the building community. Replacing my single-pane glass with low-emissivity windows will reduce my reliance on the woodstove by more than half.

Adequate insulation, ventilation, and thermal mass make all but the most unforgiving climates livable without using extra energy. In places where ambient temperatures remain far above or below the comfort baseline, day and night, for long periods, even the best designs need help. In the desert example given above, where daytime temperatures climb to 100°F, insulation (including shaded southern glass areas and exterior shutters on western win-

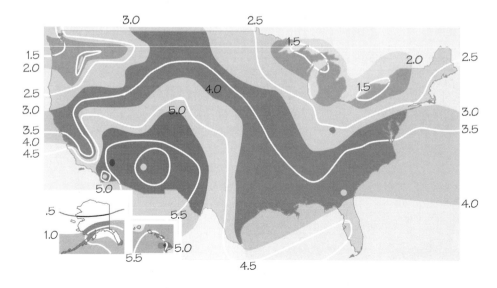

Solar zones of the United States, based on average hours of "full sun" per day, corrected so that an hour in Fairbanks is the same as an hour in Phoenix. The infamous "blue zone" covers most of the U.S.'s northern tier, and dips into California and Tennessee.

dows) and thermal mass should keep the house comfortable. Everything depends on nighttime; if outside temperatures drop to 70°F, nocturnal breezes may enter through opened doors and windows to cool indoor thermal masses. When nighttime temperatures remain high, due to large thermal masses like cliffs in the vicinity or lack of wind currents, air conditioning is the only remedy, although we might be right to question if such sites are habitable at all.

In high degree-day cooling locations, intelligent use of shutters, overhangs, and seasonal shading are crucial, as is sensitive siting, to take advantage of shade and to capture any breeze. Indigenous structures built in such unforgiving sites employ ingenious passive measures for surviving the withering heat. Often, not far below the blistering surface, the earth maintains a constant and relatively low temperature. Houses built partially or completely underground can connect to this cool earthen mass and draw comfort from it. Cooling towers, which use convection to circulate cool air from underground or from cool sheltered gardens, or which draw air across underground pools, provide precious air movement and cooling. Wind scoops can catch cooler breezes a dozen feet above the ground and direct them into the building. These measures cost nothing to operate after they are constructed. All else failing, the sun itself may be harnessed, using sufficient photovoltaic modules, to run cooling equipment, but this is shockingly expensive.

Insulation and proper siting are also the best—and practically the only—passive resources in high degree-day *heating* locations. In arctic, mountainous, and far northern lands the sun is too feeble, and too often obscured, to offer much opportunity for collecting the sun's heat. Earth-bermed houses and houses with full basements again connect with the relatively warm planetary mass. Streams and micro-hydro systems may freeze solid, and the only possible source of excess energy may be the wind. The indigenous response is to use the snow itself as an insulator, to dress warmly, and to hope for spring.

In less extreme cold locations, such as the Colorado Rockies, where degree-day heating requirements are nevertheless high, solar-charged thermal masses harvest heat effectively because the sun still shines brightly, despite persistently cold winds and snow. Solar thermal panels, if well protected from freezing, collect heat which can be stored in active radiant floors and used for domestic hot water. Running water provides hydroelectricity, which is the only renewable electricity source that can be economically used to supply heat. In these regions, a modest area of woodland provides abundant winter fuel.

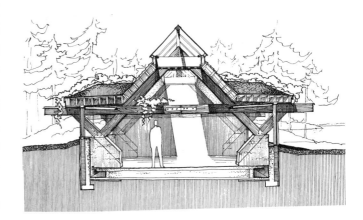

Building underground can reduce heating and cooling costs. Drawing from The Earth-Sheltered House *by Malcolm Wells.*

Radiant Floors: A Case in Point

For a civilization addicted to forced-air heating, radiant heating provides certain lovely advantages. The perennial battle for control of the thermostat* ends, because the heat comes from the whole floor. Babies and children love to play in rooms with radiant floors. From an energy perspective, forced air heating and cooling are wasteful, because air is a terribly inefficient heat-transfer medium. The challenge of harvesting the sun's heat and using it to keep a home warm overnight drove physicists and chemists crazy in the 1960s and especially during the petroleum-crisis years in the 1970s. High-tech schemes using electricity to circulate warm air through rock bins and dimorphic salts sounded great when explained in popularized science magazines, and attracted the attention of government boondogglers, but proved to be too complicated to work well. Scientists tried Rube Goldberg designs bristling with sensors, fans, pumps, and other active components that seldom worked together properly for long: pebble beds mildewed, trickle trays of black plastic leaked, and eutectic salts phase-changed well in the lab but separated out and stopped working in the field. In space heating and cooling, simple, passive, and massive are best.

As early as the 1930s and 1940s architects, led by Frank Lloyd Wright, revived the Roman idea of the hypocaust: The best way to deliver heat to a room is through the floor. Rather than routing hot water from the boiler through unsightly radiators, they sent it through copper pipes buried in the concrete slab floor, thus heating the floor directly. By the early 1980s, when capturing solar energy for heat graduated from experimental to practical, radiant floors were making a similar transition. Within thirty years of service, a pattern of massive and spectacularly costly radiant floor failures was emerging. Concrete is not good for copper tubing; the lime in cement reacts with the copper, corroding pinhole leaks. The worst failures flood rooms, but most leaks invisibly pump expensive hot water into the ground beneath the slab. Ants love it! In Scandinavia, where nasty winter weather makes active hydronic floors especially desirable, chemists developed cross-linked polypropylene plastic tubing with thermal qualities almost as good as copper but which should last for more than fifty years.

In the early 1990s all the parts came together. The demand for solar domestic hot water allowed plumbers to learn how to work with a new set of components. The sun's heat could be harvested with the same familiar, efficient, and cost-effective solar panels. The plastic tubing buried in the slab had been in the field long enough to demonstrate its superiority to copper. Compared with air, water is an extremely efficient heat-transfer medium, but hot water is also surprisingly corrosive over the course of years, and so handling it requires special materials and technology. Necessary pumps, sensors, and valves were improved during the shakedown of solar hot water, and became affordable due to mass production. Innovative new components appeared: Because a slow flow is best for thermal transfer, new pumps were designed in which the impeller—the device that makes the water flow—is the only active part in contact with hot water. The impeller is driven by magnets spun by a small motor using a minuscule amount of electricity. The only remaining puzzle: What kind of battery do you use to store heat?

The solution to the puzzle of thermal storage turns out to be quite simple: Include adequate thermal mass within the insulated envelope of the building. One good method is to pour a thick cement floor over insulation and hydronic tubing, and then pump the finished slab full of heat, which radiates out at a comfortable rate. Because the pressures and temperatures involved in a solar-powered system are moderate, the whole apparatus is simple and cheap. The capacity of the "heat battery" is directly related to the mass of the insulated slab; in Wisconsin, builders have installed 60-inch slabs with two and three levels of hydronic tubing so that excess solar energy harvested in September can still be radiating out of the floor in late February. Deep-slab systems can even accept energy in summer and give it back in winter, although generally hydronic systems are shut off or redirected (for pool heating in a high-end house) in summer.

A solar hydronic floor's double-headed power source basks in the direct sunlight: Heat is harvested by hot-water panels, and electricity by a little photovoltaic module. The hot water is pumped from the top of the hot-water panel down through a well-insulated pipe to a few hundred feet of plastic tubing entombed in the massive concrete slab floor, the "thermal battery." Most of the heat is transferred to the slab before the fluid drains into an insulated storage tank. Sun shining on the little PV module runs the pump that circulates cool fluid back up through pipes in the wall to the bottom of the hot-water panel basking in the sun.

In a drainback system, fluid is present in the solar panel only when there is enough sun to drive the pump. If the panel and its support pipes drain themselves dry when the sun goes down, freezing temperatures are no threat. Pressure throughout the system is low, and so even when leaks occur they are not very serious. Over the years, hot water, heat, and

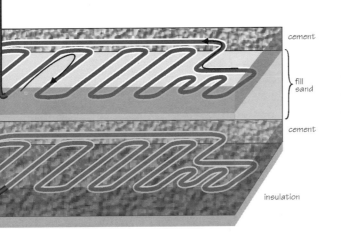

PV
runs
pump

solar
collector
heats
water

117°F

67°F

top
bottom

cement

fill
sand

cement

insulation

Hydronic floor-heating system. In the diagram, the pump and insulated water tank are on the left.

pressure will attack plumbing parts, and so systems should be designed and installed so that maintenance is easy. The components of these systems should last a hundred years.

One noteworthy feature in this system's minimalist design is the absence of controls. Controls are the parts that most commonly fail in any hot water system. In practice, weather's irregularities and living patterns are better served with as few controls as possible. Two conditions must be anticipated and controlled: too much heat, and too much cold. In a simple system like the one illustrated, too much heat can be controlled by a simple thermostat which disconnects the power to the pump when the floor gets warm enough. If the panels boil from too much sun, steam vents through the panel's pressure relief valve. Too little heat is a more likely eventuality. A wood-fired water heater or a small fossil-fuel-fired batch water heater can provide enough standby heat for cold spells.

Laying out a hydronic floor, you get to put the heat right where you want it: by the kitchen sink, beside the bed, in front of the toilet . . . . Exterior walls and windows in particular, even

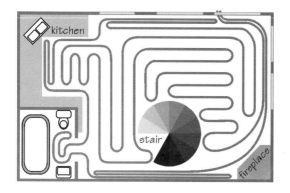

With a hydronic floor, heat can be placed where it is needed—along the walls, in front of the sink, and in the bathroom.

the best-available glass, still import cold, so extra runs go there. You can keep heat away from the pantry, woodpile, closets, under the couch. Larger households are "zoned" so that different areas are heated according to occupancy patterns, thereby reducing heating costs while improving comfort.

Now for a caution. Radiant heat is not "instant on." Heat injected into a slab four inches below the floor takes four hours to reach the floor and be felt, so turning up the thermostat is not as satisfying as standing over the forced-air system's floor register. Heat travels through a slab about an inch an hour. This is perfect for diurnal heat shifting: Late afternoon sun's heat is transferred from the tubes to the bottom of a four-inch slab, making its way to the floor and into the living space four hours later, just when needed. Accustomed as we are to instant solutions, however, this latency can be irritating. We loathe waiting four hours after turning up the thermostat to feel any warmth. This frustration can be minimized with one more control: a setback thermostat programmed to anticipate the family's needs.

Radiant flooring can be retrofitted anywhere an existing floor can be strengthened to support the concrete's additional weight, and where enough headroom will remain after the floor level is raised a few inches. If the floor cannot be reinforced to carry the new weight, special lightweight low-mass concrete can be used, but since heat storage capacity relates directly to mass, this is less satisfactory. Pouring a slab inside an existing building is an art form that should only be attempted by professionals. But the effort is worthwhile: There is no doubt that delivering heat from below, especially if the heat comes from solar panels, is more energy efficient than other heating tech-

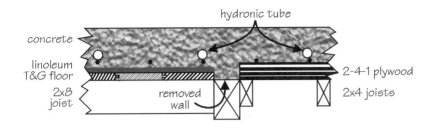

concrete

linoleum
T&G floor

2x8
joist

hydronic tube

removed
wall

2-4-1 plywood

2x4 joists

Renovating an old floor by "Chernobyliz-ation," covering previous mistakes in freshly poured concrete.

niques. Switching from forced air to radiant floor will repay the dramatic effort many times over.

When renovating an old building, new flooring will often supercede two or more unequal predecessors. Restoring a pre-1900 building in Caspar I encountered floors differing in elevation by an inch, and in thickness by another inch. Ripping up and regularizing such unequal floors with a conventional surface like hardwood flooring is costly, and also yields a mountain of toxic waste. My alternative solution was to "sister up" the floor joists by interposing a strong new joist between each old one, doubling the floor's strength. Next, leaving the linoleum, pine flooring, and asbestos tile in place, I laid down a grid of 6-6-10 reinforcement, the kind of steel mesh used to make driveways and slabs stronger, which then served as a grid for tying down the serpentine Hepex tubing. Then the concrete artist helped me pour in a lovely flat slab: four (or five) inches of entombment. In our climate, this simple addition to the area's thermal mass decreased its range of unheated temperature extremes by 50%, and reduced heating costs by 70%.

Stephen Heckeroth proposes that we achieve peace in the struggle for control of the thermostat by installing and retrofitting more radiant floors. Surprisingly small adjustments in air handling can have an amazing effect on personal well-being. If our feet are warm, our heads are happy in ten-degree-cooler air. In hot weather, we cheerfully endure higher temperatures if the air is moving. Skin is the best clothes. There is little doubt that our skin and hair are designed to function best when uncovered. My house achieves a tropical climate with woodstove or under the westward sun's kindly glare, inducing a holiday mood. This mood could be extended indefinitely and affordably with adequate thermal mass, ideally in the floor . . . another one of the many ways I would improve my house if I had only known . . . but then there's always the possibility of a retrofit.

By dedicating roof space to capturing the sun, the solar panels are incorporated neatly into the house's profile, while protecting and helping to insulate the roof.

Building in Harmony with Sources

By considering energy needs throughout the planning stages of a new, independent house, you will find many ways to integrate good energy practices and vastly improve your homestead's performance. The solar fraction is the percentage of a household's energy requirements that can reasonably be expected to be generated by harvesting solar energy. In some places, where cloud cover is rare and the sun beats down unimpeded, the fraction is easily 100%. An old government map of solar resources showed such spots in red, while less gifted sites ranged through the spectrum to blue where, at one time, it was thought that no solar capture was feasible. Vermont was all blue, but this estimate, like the report of Mark Twain's demise, was premature. Richard Gottlieb, perennial candidate for governor of Vermont, has been cheerfully harvesting solar electricity for almost two decades, and reports that in his part of the state (the south) the solar fraction could be as much as 70%. Deficits, of course, may be made up with more modules and a bigger battery bank, but it is more sensible to develop a second energy harvesting method if possible.

The solar energy falling on a south-facing roof can be either a curse (as an undesirable heat source for the space within, in which case we recommend the attic fan arrangement in chapter 2) or a blessed source of electricity and hot water. Low-voltage electricity does not travel long distances efficiently, but the area required for solar panels—photovoltaic modules plus solar hot water panels—often corresponds nicely to half of the house's roof area. This serendipity is helpful, however, only if the roof faces in the right direction. By dedicating roof space to capturing the sun, the solar panels are incorporated neatly into the house's profile, while protecting and helping to insulate the roof. As already noted, one progressive utility, the Sacramento Municipal Utility District, enlists customers in their "Give Us Your Roofs" program for precisely this purpose. Ratepayers have been rushing to sign up, willingly paying a 15% premium for the renewable energy they use.

To prepare for this design economy, a house's roof must align with the site's solar window, so the sun's energy may be captured most effectively. The sun's path across the sky varies in two ways, by time of day and time of year, and reaches its highest point at noon on the summer solstice. At noon on the equinoxes, spring and autumn, the sun can be found due south at the angle of the site's latitude below the zenith—the center of the sky directly overhead. On the summer solstice, the sun reaches its annual high point in the sky at noon: due south and at an angle below the zenith equal to the location's latitude minus 23 degrees. At noon on the winter solstice, the sun appears due south at an angle equal to the site's latitude plus 23 degrees. (Remember, the earth's 23-degree angle of precession causes the seasons.) Through the rest of the year, the sun commutes sinusoidally between these points. The proper angle of exposure for any solar device corresponds to the site's latitude plus or minus the seasonal fraction of the sun's annual 46-degree range. Some solar devices are racked or tracked so their angle can be adjusted, but seasonally adjusting a roof's angle is impractical, so a fair compromise must be achieved. Choosing the site's latitude as the roof's angle from the vertical varies from the optimum by approximately 23 degrees at the solstices, but is direct enough for most of the year. If maximum electrical loads are expected in the dead of winter, as is typical in the northern tier, the savings realized by mounting modules on the roof can be invested in additional modules to compensate for the inefficiency of slightly misaligned modules during part of the year.

Please note that compasses point to magnetic north, which coincides with true north in

Compass corrections reflect the earth's magnetic field. For approximate corrections in the lower 48, use this map. For final alignment, a large-scale local map should be consulted.

only a very few locations on the planet. Get the local correction before using a compass to orient the house, then verify your measurement by making careful observations using the time-honored method: With a tall stick (the gnomon) and pegs, mark the shadow's end over several hours; a line from the end of the shortest shadow through the gnomon points due south.

Carpenters express roof pitch as the inches of rise to twelve inches of horizontal run; a typical commodity house has a rise of three in twelve. To get the best out of Caspar's winter sun, I designated a section of roof for solar harvesting, and angled the roof perpendicular to the sun about a month before and after the winter solstice, or eighteen degrees higher than my 40° latitude, resulting in a precipitous pitch of fourteen in twelve. My best solar aperture is slightly west of due south because of trees I do not care to cut and typical wintertime cloud patterns. Solar energy experts have special tools for predicting the solar aperture and its changes over the solar year, and can help get the roof oriented precisely. Getting the array angle wrong can easily subtract 20% efficiency from your array.

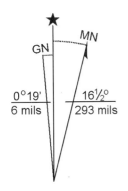

Good maps show magnetic north (MN) and the direction to the North Star as well as geographic north (GN). This compass rose comes from the USGS map for Caspar.

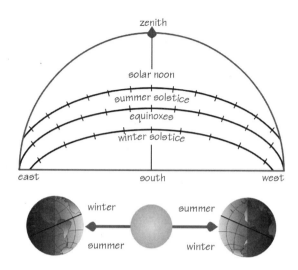

Face south and envision this image as a quarter sphere going into the page and large enough for your head to fit inside. East will be over your left ear, south right in front of your nose, the zenith right overhead. The sun will move along the paths shown on the equinoxes and solstices.

Coincidentally, increasing the roof pitch with latitude serves the need to employ steeper roofs in snowier lands. Since snow will cling to a glass surface at or below a 50-degree pitch (particularly if the surface is dirty and the snow is wet), some thought must be given, if the homestead relies on PV for snow-season power, to convenient ways to clear the snow. If you live in snowy parts or very far north, you might consider mounting your modules vertically in winter. When the air is cold and there is snow on the ground, vertically mounted modules perform better than under any other conditions.

In 1997, photovoltaic shingles and panels became widely available at very reasonable cost when considered as energy sources as well as roofing material. Architects have integrated solar modules into roofs for more than a de-

cade, cleverly solving the problems of sealing delicate electrical connections away from the elements, but owner-builders can employ these new roofing materials very cost effectively.

Photovoltaic panels become less productive when they get hot. This perverse behavior cannot be corrected by the usual means—you do not want to shade your array—so they must be mounted in a way that keeps them as cool as possible. Unventilated, the temperature behind modules will be as much as thirty degrees above the ambient on a sunny day, decreasing module performance by about 10%. Mounting modules on vertical rails creates a chimney effect, drawing cool air in at the bottom and allowing heated air to escape at the top.

When strong winds blow, modules are expensive, fragile, sharp-edged kites waiting to fly. Aficionados of the electrical code don't worry as much about this as I do; they worry instead about another unlikely but potentially devastating event, lightning, and insist that every module and metal part be individually grounded.

There is a brisk argument in the solar community about the efficacy of racking and tracking as compared to solid installation. By seasonally tinkering with the angle of the modules, small gains in efficiency can be made; by keeping modules pointed at the sun as it makes its daily transit from east to west, productivity may be improved by as much as 50%. Active trackers using electronics to follow the sun are accurate to within half a degree, and get on the job first thing in the morning, but they are also subject to the failures and costs of any high-tech equipment. Passive trackers that use freon (a CFC, but safely encapsulated, we hope, and not prone to

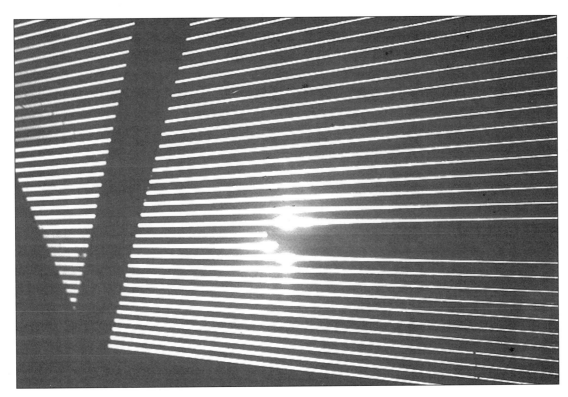

Looking up through the photovoltaic skylights in the Heckroth barn roof. Photovoltaic bands (dark) in these architectural glass panels convert part of the sunlight to electricity. Sunlight streams through the untreated glass between bands.

escape) are less accurate, and take half an hour to wake up in the morning. I hasten to warn that any use of CFC must account for eventual decommissioning, when it is important that the equipment be relieved of its ozone-eating molecules without further damaging the stratosphere. Racks and trackers are typically rated to withstand 120-mile-per-hour winds; in Caspar, or wherever the weather gets intense, that is barely adequate. As a veteran of winter's lusty gusts, and having seen the effects of serious storms and weathering, I value the advice of those who favor fixed mounting of the solar array. Having lived now with a tracker in the front yard for several years, and having repaired it only twice, I also see merit in arguments favoring trackers.

In 1993 I wrote, "To my eye, a tracker full of photovoltaic modules in the yard is just slightly less offensive than a powerline swooping in from the street: It bespeaks a lack of forethought and a hint of impermanence." I now admit I was wrong. The powerline is a blight, but the tracker and its modules are a sort of rectilinear sunflower productively following the sun's great eye across the sky. If we wish to make a sculptural statement about the source of our energy, a tracker does so very nicely.

Economic factors may help resolve the controversy in many cases. Trackers usually cost about 30% of the modules that populate them. At 40 degrees latitude, trackers can add more than 40% yield to a PV module's summertime power harvest assuming the whole day-long solar path is available. In winter, when the sun shines from a lower angle, a tracker's improvement to the daily yield drops to less than 20%. If year-round power production is important, then adding one-third more fixed modules eliminates moving parts, costs the same, and harvests the same energy. If excess summertime energy can be used to power fans, pumps, or sell energy back to the grid, the tracker makes sense.

Many homesteaders are attracted to photovoltaics because of their elegant passivity. If it is impossible to orient the home's roof correctly, or if a better solar window is available at a distance from the home site, it may be better to incorporate the array into the appropriately sloped roof of an outbuilding than to plant a large tracker in the yard. This shed may become the power house, sheltering batteries and power-conditioning equipment. Take care to account for energy losses due to distance between source and loads.

Typical single-crystal cell modules convert about 15% of the sun's energy to electricity. Since the photovoltaic effect has a theoretical maximum efficiency of less than 30%, and the most efficient cells are more expensive to manufacture, another obvious way to improve energy yield is by using lenses and mirrors to concentrate sunlight from a large area onto smaller high-yield photovoltaic cells. New modules employing this strategy have been in the field for a few years now, and work satisfactorily where available space is at a premium.

But modules with optical concentrators must be kept pointed directly (to within a quarter of a degree) at the sun, and existing active trackers are barely capable of pointing this precisely over long periods of time. Concentrator modules are likely to be a high-maintenance solution perfect for folks who like to tinker. In order to harvest the most summertime energy, perhaps for agricultural pumping, trackers improve yields in a benign climate. For most households, over the twenty or thirty years of life we expect from our modules, I still believe it is more cost-effective to buy a third more modules and mount them solidly on the properly oriented roof.

One photovoltaic module may provide enough energy to keep the battery in an Emergency Callbox charged, but most homes need several modules. (How many modules do you need for your life style? You will find a worksheet in the appendix.) When mounting modules, remember that they can be expected to operate for thirty or more years, and so the mounting structures and hardware should be correspondingly durable. Since "up" for an electron is away from a negative charge, modules can be aligned in any way that is convenient so long as their faces are unshaded and perpendicular to the best sun. We have already noted that modules perform better if kept cool, and so this also suggests mounting strategies that enhance air flow.

Experienced installers arrange an array of PV modules into conveniently wired subarrays. This arrangement permits the use of smaller-gauge wire for connecting the arrays, which is cheaper and easier to handle, and also simplifies troubleshooting. If you suspect that your array is not producing as it should, you can test each subarray with a multimeter until

207

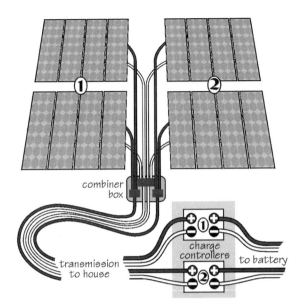

By arranging the PV array into subarrays, smaller-gauge wire and lower-capacity charge controllers can be used. Redundancy is good.

you determine which subarray is performing poorly, and then start unwiring modules and testing them individually. Carrying the subarray notion on into the balance-of-system equipment, using two or more small charge controllers, is often cheaper and more reliable than using a single large controller.

Roofs capture another precious commodity: water. In most of the world, roof catchment is the primary source of domestic water, but in conventional American homebuilding, roof run-off is considered more a nuisance than a resource. Roof-captured water requires treatment, particularly in a suburban or urban setting, but as our shared water resources approach their limits, we ought to put these pennies from heaven to good use. Thanks to gravity, it takes very little effort to divert roof run-off to a holding tank; it takes even less when this intent is incorporated into your plan early on. If the run-off is to be used for drinking or to water edible plants, care should be taken to select a roof surface that does not spoil the water with leachates or unfilterable impurities. The manufacturers of the roofing material should be able to provide reliable information.

Renewable Energy Gotcha!s

Things that spin, sing. Photovoltaic panels bask in the sun while quietly producing electricity, but most generators convert rotational energy into electricity—wind-spinners, hydro-turbines, and internal-combustion-powered generators—and this conversion makes noise that may be disconcerting and will certainly intrude on rural peace and quiet. Owners of homes with nearby micro-hydro turbines and wind machines told me that they found the sound reassuring in the night, but I wonder if this sentiment is shared by other family members and neighbors. These devices are best at a distance—wind machines, in particular, as far away as possible. Since electricity loses energy when it travels, transmitting the energy back over a long distance requires heavier-gauge wire, or conversion to higher voltage. As we saw in chapter 4, distance costs power, and so does conversion. Whenever possible, we want to generate the same form of electricity that we intend to use. Low-voltage electricity loses more energy over distance, and so wind-spinners and hydro-electric turbines sited far from the home usually spin high-voltage alternators or generators. Furthermore, moving parts, particularly those in contact with hostile forces like water, wind, and heat, inevitably require periodic attention and replacement. Compared to other generators, PVs look expensive. But when long-term maintenance costs, together with the cost of resolving the conflict between electricity's unwillingness to travel and our need to keep spinners at a distance, are considered, PVs are cost-competitive. PVs are likely to decrease in cost and increase in efficiency more than competing technologies, and so there is a compelling argument in favor of building on a good site for PVs even if you do not expect to use that source right away.

When possible, houses are built out of the wind. Wind strips heat out of a house, and wind-chill can make a place very uncomfortable, so good home sites are seldom the best places to harvest wind power. Since we avoid building houses where the wind resource is best, we naturally distance ourselves from the insistent *thwocka-thwocka* and whir of these noisy devices. Any hollow structure (like a house) in proximity to a wind-spinner can act like the hollow body of a guitar, amplifying the sounds, so wind-spinners are best kept by themselves atop a tall tower as far as practical from the home. Hydro-turbines need less distance, as their high-pitched whine is easily contained within the massive cement structures that house them best.

Planning for Maintenance and Change

Plan to install extra electrical outlets. There never seem to be enough, and electricians report that adding outlets is their commonest and least favorite electrical job. As with placing windows, anticipation of the tasks to be performed in a particular area allows for better planning. In some areas you will probably install twice as many as the minimum required by code (which typically calls for one every twelve feet), paying special attention to areas where you expect to work or concentrate electrical equipment. Underwriters Laboratories reports that most electrical fires result from overloaded circuits and faulty or damaged extension cords. This danger can be circumvented by anticipating where extra outlets might be needed, and wiring accordingly. Where electrical use will be particularly intensive—the home office, computer station, entertainment center, and kitchen counter—using plug strips or building wiring into counters and furniture may be more sensible than a surfeit of wall outlets. Trade electricians usually charge by the outlet

MOBILE HOMES

From an energy standpoint, it is hard to imagine a more miserable shelter solution than a mobile home. Using aluminum for lightness (a material embodying enormous energy because it is refined in electric furnaces from bauxite, and that is further cursed with good heat-conduction properties), these thin-walled, barely insulated, single-glazed monstrosities are serious blemishes on the landscape of our nation. They are "independent" in the sense that their occupants can, at a whim, pump up the tires, kick out the blocks, and be gone, but they are horribly wasteful. In fact, this impermanence commits the occupants to a life of unwitting disconnection from the land their trailers infest.

To heat or cool a mobile home requires roughly triple the energy of a properly insulated house of equal size. The residents, who invariably pay the energy bills, are condemned to perpetual servitude to the energy mongers, because of their abode's extreme inefficiency. It is practically impossible to operate anything but the smallest mobile home on an alternative energy system anywhere near the realm of acceptable comfort. Moreover, trailers are firetraps; heating a mobile home with wood is suicidal as well as illegal. Mobile homes are built tight of toxic synthetic materials, so their indoor air is dreadfully polluted. Evaluated by the freedoms enumerated in an earlier chapter—rational construction costs, use of sustainable materials, reasonable costs to operate and maintain, and responsible options for recycling the structure at the end of its useful life—the mobile home is a costly mistake from beginning to end. Significantly, elsewhere in the world, not even the most disadvantaged people consider them even remotely habitable.

Apologists for mobile homes offer the excuse that it may be the only kind of habitation that can be afforded by low-income folks. That is absurd on its face: So much goes into making the damn things mobile that equal energy and equal care put into a stable habitation would provide a much more functional and beautiful home at much less cost to the resident and the environment. To what extent have the relaxed health and safety standards applied to trailers made their existence possible, while similar relaxations have not been made available to solid, responsibly designed low income housing?

The advice of many who have moved onto undeveloped land and lived in a trailer while building their home is, don't waste your time and money. Camp in a tent and build, instead, a guest cabin to weather the first winter while you learn the site firsthand.

To heat or cool a mobile home requires roughly triple the energy of a properly insulated house of equal size.

box, not by the hour, and bid their jobs based on the box count; on my travels I heard prices as high as twenty dollars a box. Naturally, if our focus is the job cost, we may be tempted to cut costs by eliminating outlets. As always, remember that you will pay for such false economies the whole time you live in the house. It is much cheaper to install an extra outlet when the walls are open than to add one afterward.

Design for easy maintenance. If simple tasks such as changing lightbulbs and water filters are inconvenient, we are likely to leave them undone. Satisfied homeowners showed me many innovative arrangements for making periodic maintenance easier, from built-in light fixtures with hinged translucent covers held closed by magnets, to sturdy built-to-size trays under the sink so that the welter of household chemicals kept there could be easily set aside while changing filters or working on plumbing. Traps and pipes inevitably drip when being serviced. By building cabinetry that easily allows us to slide in a plastic tub or tray before the dribbling begins, and out again without spilling, and by designing the plumbing so that connections are located above the places where such a tub can be set, we make the chore quite a bit more pleasant.

Placing lights in the dark places where someone must occasionally work is a novel idea that often occurs too late to homebuilders, even if they expect to work on their own systems. The same goes for outlets in places where power will be required only occasionally; by wiring with such foresight, our tool-gathering task is lessened by at least one extension cord.

The Light of Experience

The preceding three chapters have explored the physics and mechanics of homebuilding. In the next chapters, we move from theory to practical experience, starting with the experiences of several pioneers. Their successes and failures provide advice I wish I had heard before I began to build.

CHAPTER 9

PIONEERING AND SETTLING THE NEW ENERGY FRONTIER

From the point of view of efficient use of energy, the most successful homes I visited were built in the last twenty-five years from the ground up in accordance with a builder's vision. Many of these homes seemed to have grown from their conceptual hearts outward, adding layers as a tree adds rings.

Primeval dimetrodons were solar-powered dinosaurs that heated and cooled themselves by circulating blood through their sail-like backs. Dimetrodon is also a quarter-century-old architectural experiment, a collaborative cluster of homes designed by a group of "outsider" architects on Prickly Mountain near Warren, Vermont. This residential Dimetrodon was conceived by its builders as an innovative marriage of unified structure and diverse expression. Dimetrodon's skeleton provides structure and services, and defines eight spaces that the architect-residents fleshed out according to their personal visions. Every residence is different. One of the founding builders, William Maclay, lived with his family for two decades in their small but spacious-feeling apartment on Dimetrodon's western side.

As one of the original co-housing projects, Dimetrodon remains a success, but its dominant technical aspect, the solar south wall, was disconnected after a couple of winters, a failure I found in many early re-inventions of the independent home. Standing with Bill on a snowy

Clustered together within Dimetrodon's skin, the small residences conserve heat, and constitute a self-contained, supportive neighborhood.

William Maclay and Dimetrodon, on Vermont's Prickly Mountain.

winter day and feeling disoriented, I asked if I was mistaken, or did Dimetrodon's windows really point the wrong way, to the north, away from the sun? Bill shuffled his feet in the snow and agreed: "Well, we had to make a few mistakes. . . ." Then he explained how the south side's active solar heating system was meant to work. Most of the building's southern side was dedicated to collecting solar energy; water trickling over a black plastic surface would absorb the sun's energy and then be circulated to heat the building. Sadly, sound theory proved not to work well in practice, and the collector was abandoned. Someday, perhaps, this sloping south wall will be covered with sparkling, sapphire-blue photovoltaic modules and state-of-the-art solar water heaters, and the founders will be hailed as solar visionaries, the failure of their original technology forgotten.

All other aspects of Dimetrodon work well: With the solar wall turned off, the large, home-made wood-fired heating plant operates efficiently. Residents told me they look forward to their turn at cooperatively stoking the furnace through the winter. The heating plant looks like it belongs inside a dinosaur, but its exposed pumps and pipes make it easy to understand, troubleshoot, and service. Clustered together within Dimetrodon's skin, the small residences conserve heat, and constitute a self-contained, supportive neighborhood organized around shared indoor and outdoor space. Maclay's home has five distinct levels with playful interior windows and archways between them, and so it seems much roomier than its modest square footage suggests. From within, the dense Vermont woods seem much nearer than Dimetrodon's seven other domiciles. Compared to other nearby dwellings, Dimetrodon hums with life. None of the founders are still in residence, but Dimetrodon continues to shelter young families efficiently and enjoyably.

Pioneering new technology or employing it in the real world for the first time is always chancy. There is always the risk that the technology, or one of its essential components, will fail (as did the plastic membrane at the heart of Dimetrodon's solar space-heating scheme). Better technology and better applications in-evitably come as a result of the efforts of pioneers. Newcomers to independent homesteading do not come to an unexplored frontier, but to a well-developed and safe place where most of the big mistakes have been made. The challenge for you is to avoid making our mistakes over again.

Stepwise Refinement

Innovative builders laugh when someone suggests that a new technique will be perfectly applied on the first try. "Third time's the charm," they reply. Alternative energy innovation is not exempt from Murphy's Law. Building inspectors and other vested upholders of convention tend to regard self-sufficient electrical systems suspiciously: experimental, haphazard, temporary, armed, and dangerous. In the frontier days of low-voltage residential systems, before "clean" and well-integrated systems were feasible, inspectors encountered rough-and-ready hookups, and concluded that alternative electrical practices are treacherous and shoddy. Still, no electrical fire or loss of life has yet been attributed to a renewable energy system.

Knob-and-tube wiring was the experimental, haphazard, temporary, and dangerous technique used by builders until the 1950s. Repairing or changing a house's knob-and-tube system is difficult. The wire's insulation ages badly in hot attics and moist crawlspaces, and fails under normal stresses such as rainstorms and cats in the attic. In contrast, safe and robust modern (1970s-era) wiring is easy to install or edit. Even so, considering that 98% of homes built before 1955 used variations of the earlier knob-and-tube wiring strategy, surprisingly few met with catastrophe.

Compared to kerosene and candles, electricity, even when primitively deployed, is extremely safe and fails benignly. Circuit breakers replace fuses; conduit, Romex™, and boxes supplant knob-and-tube; home wiring continues to evolve under pressures from the housing, insurance, and now the renewable energy industries. Today's fail-safe hardware and belt-and-suspenders techniques are the product of incremental refinements based on new materials and needs, failure analysis, and feedback from practitioners. This hardware and practice suits renewably energized homes as well as conventional ones on-the-grid. Working with modern materials and techniques, you can add an outlet or replace a light switch in a four-decades-old system and know it will continue to work safely for decades. But the changes can also be fickle: The best available circuit breaker in 1973 is a valuable antique today, if it is in working order. Any home's electrical system ages while technology evolves, causing core devices such as circuit breakers and the "breaker box" or the distribution center they populate to become expensively obsolete.

Conventional electricians and inspectors are baffled by low-voltage wiring and systems incorporating the whole electrical cycle from generation through distribution. They were trained to apply standard procedures, not understand basic principles. Confronted by a complex assemblage of unfamiliar devices and control circuitry, naked conductors, heavy-gauge wires, yet not wanting to appear stupid, an inspector might mutter noncommittally while inventing an excuse for withholding approval. After 1993, finding such inspectors in the field became rare. Low-voltage and whole-system electricity is technologically appealing to most electricians and inspectors; their theoretical grasp of whole systems and unfamiliar voltages may prove shaky, but they are generally interested and willing to learn. The building inspection establishment is reluctantly integrating renewable energy into its knowledge base, especially in regions where renewable power is more common. Well-informed inspectors can be excellent advisors, and are right to hold us to a high standard of safety. Conversely, we should expect building officials to be tolerant of new technologies, and eager to learn about them.

The wiring in older independent homes may still look rough; owners of vintage systems often ask "Why fix it if it ain't broke?" Such elemental systems, built before metering and safety devices were available, are still producing abundant power after many years of demanding use because low-voltage electricity is so easygoing, and the technology is fundamentally solid. Nevertheless, I must emphasize the importance of good overcurrent protection and control devices, while agreeing that even the most unsafe alternative installations are safer than the high-tension utility distribution systems that run over our heads and through our neighborhoods. The discovery that DC is gentle compared to AC can give a false sense of safety, but our trust in low voltage's benign forgiveness does not justify sloppy habits. Even with low-voltage DC, hasty or temporary repairs can cause equipment failure, electrical fires, battery explosion, and potentially harm.

Evolving Standards

A second generation of homesteaders is now building their homes and energy systems beyond the grid's reach. What pioneers made possible, these new settlers make permanent, safe, and trouble-free. Looking at the energy systems now in use, I see a second standard emerging. The pioneers' experimental first-generation systems, like the knob-and-tube era of grid electrification, were necessary because innovators were forced to mix and match primitive components in unusual ways to provide for their basic needs: lights, pumps, refrigeration. The eccentric, brilliant, and practical installations of first-generation operators look rough to those who do not understand electricity. Many first-generation system operators continue to test new methods and refine system performance, often keeping heritage equipment in operation; because this process is never-ending, why should the systems be covered up? Insulation fatigue, corrosion, loose connections, and other problems are more easily found and corrected, because pioneers' systems are continually being monitored, evaluated, repaired, and maintained. Second-generation systems, begun less than a decade later, integrate new, well-designed equipment that is designed to fit together according to the conservative standards that govern modern house wiring. Settlers, second-generation users, have called for sophisticated systems that can be installed, covered, and, except for periodic maintenance, treated as robust and trouble-free utilities . . . better than the grid.

Settlers, second-generation users, have called for sophisticated systems that can be installed, covered, and . . . treated as robust and trouble-free utilities.

FUEL CELLS: WHY I AM UNIMPRESSED

Fuel cells consume hydrocarbon-rich materials and produce electricity. Within the next decade, we are told, fuel cells will begin appearing in homes, where they will be fueled by a range of fossil fuels, primarily natural gas, a substance delivered through buried pipelines, which are more reliable than electric lines draped through the air. Natural gas, like all petrochemicals, is a heritage fuel. In 1998, experts concurred that about 40 years of crude oil remained, along with something on the order of twice that much natural gas at *current rates of consumption.* Experts also concede that consumption is increasing. Shifting electrical generation to fuel cells has no real benefit (except, in the very short term, to the makers of fuel cells and suppliers of natural gas). I view this development as a distraction from the true issue at hand: How can we thrive using no more than our current solar income?

While second- and third-generation low-voltage and residential power-handling equipment improves rapidly, pioneering and settling strategies appeal to different people. For innovators who value technical sophistication and enjoy tinkering, new renewable energy equipment is irresistible: improved efficiency and reliability, better system tracking, bells, whistles. . . . Do-it-yourselfers and energy hobbyists who love to assemble our own systems find that fail-safe equipment is easily available and cost-justifiable. We find a supportive community of willing and knowledgeable advisors. Proceeding deliberately with common sense and caution, anyone with a solid grasp of the elements can build a renewable energy system that will function for years with routine maintenance and minimal intervention. For settlers, for whom electricity is a means to an end, modern renewable energy technology provides sensible and responsible ways to live comfortably without ever having to plug in to a grid that draws power from suspect or threatened sources. In earlier days, carpenters installed knob-and-tube wiring, and tinkerers installed their own early renewable energy systems. Now homesteaders in every region where the need for off-the-grid energy exists can find competent and experienced experts who can install and service robust, long-lasting systems. With either approach, residents who use renewable energy can rest assured that their electricity is safe and sustainable.

Beyond Experiment: Low-maintenance Systems

Many energy pioneers continue to devote their lives to experimenting with energy technologies, but renewable energy providers report increasing demand for home systems that require less from their operators. Those who live in experimental homes, and who demonstrate their pioneer systems to curious newcomers, confirm the trend: The new wave of independent homesteaders express a clear preference for durable, undemanding, self-regulating systems that incorporate hard-won improvements, need no further refinements, and require little maintenance.

Native Americans, many of whom have never lived on-the-grid because their homes are too remote, and others who might return to the land if they could have electricity, could benefit greatly from remote energy. Electricity has been both carrot and stick in coaxing them to assimilate into Anglo civilization. Ethnic Polynesian Hawaiians find themselves caught between a desire to resettle land spoiled by foreign incursions and the need to keep the amenities of modern life. In 1920 the State of Hawaii set aside substantial plots of land on each island for the "rehabilitation" of the *Kanaka Maoli*, the original Hawaiians, but until 1998, health and safety laws decreed that no building could be allowed on these Hawaiian Home Lands until the resettlement area was outfitted with subdivision infrastructure in the form of wide roads, water, sewage disposal, and fire suppression. What will it take to convince these *haole* health and safety watchdogs that a home can harvest all the water and power it needs, manage its own wastes, and rest lightly on the generous Hawaiian land?

Navajo home.

Ancient passive solar co-housing project at Mesa Verde.

The second generation: Lafayette Young's story

On the island of Maui, a confluence of potential new home builders, a water shortage, and the energy crisis has created a vortex of second-generation independent homebuilding. Laf Young is a builder and perfectionist who has applied his theoretical expertise to alternative energy systems. He lives with his wife Beverly in an owner-built, off-the-grid home on the ocean side of the twisty Hana Highway near Haiku. Both Laf and his son Cinco (Lafayette Young the Fifth) recently finished houses just down the road. Laf's new shelter is intended to be a vacation home for eco-tourists. While building for himself and immigrants from the mainland, he is always aware that his robust, self-managing electrical systems and self-sufficient catchment and residential water systems prove that a Hawaiian home is truest to its land when it lives without most of the burden of infrastructure.

After teaching vocational school in American Samoa, I came to Maui to administer the Community College, but after seven years I put myself back into vocational tech—homeowners' courses—so I've been exposed to all the building trades over my adult life. After a while, you learn the right way to approach projects. I'm bringing in our new vacation rental at a ridiculously low figure, considering how well-built it is: under forty dollars a square foot, finished. I figure it will be a great vacation place for the guys who work with alternative energy all day, then go home to their all-electric homes. They need to live off-the-grid, catch their own water, to really understand the dream.

We raised our two kids here, and when they got home from school, and finished homework and the usual rural chores, they had to calculate, based on wind speed and battery charge, whether we were in a net positive power situation before they could ask to watch TV. A guy came over from SERI (the government's Solar Energy Research Institute) for his honeymoon, and looked me up at the college. I couldn't come out to the house with him, so I told him the kids would be home from school and could give him a tour. Cinco gave him the rap, and we didn't hear anything more about it until two weeks later, I got a letter. Apparently Cinco blew the guy's mind, because the letter said he knew more about energy and water systems than any adult at SERI.

Most of our power comes from an old Jacobs 32-volt wind generator. Do you know why they chose 32 volts? Delco was the market-maker then, and they were looking to electrify the dairy farms in Wisconsin. You get good transmission above 30 volts, but the farmer standing ankle-deep in cowshit doesn't want too much voltage when he gets a shock. Turns out he can let go of 32 volts.

Deadly reliable machines, the Jacobs, turning 225 times per minute. The low rotation speed keeps it from destroying itself in the bigger winds. I've seen a Windseeker go into runaway mode here, and spin so fast it melted the high-tech teflon tape on the leading blade edges. The Jacobs was easy for me to work on, but I rebuild motorcycles, too.

We have a perfect wind resource, more power than we can use. Photovoltaic complements wind perfectly here, because the wind always blows when the sun doesn't shine.

In the 1970s when I had to go overseas, I'd call home and ask how's it going, and Cinco would say, "Shit, Dad, it's pitch dark in here!" Because the system was jerry-rigged, and we didn't have adequate controls. We didn't get balance-of-system equipment until the 1980s. The best part of the Hawaii experience has been finding responsible designers, and keeping after them until we get the equipment we need. We've driven the industry to provide controls that just weren't available. I go away for weeks now, and the system just hums along.

It makes us feel smug, when Maui's grid goes down, and we're still living normally. For three years in a row, Maui Electric has gone down during the Superbowl; of course, the towers up the mountain lose power, and the TV signal stops, but friends call up to see if they should come over with their coolers.

In my original installations, now thirteen years old, the battery management isn't as good as it would be now, and the wiring is more California-style, wires looping cans together. Nowadays, I'm more concerned about managing the ferrous/nonferrous metals in advance, before they corrode and fail. Increasing the wire size between array and controls improves efficiency: There's no point in heating up wires. In the past, stuff was so expensive that we installed the minimum, but now we should sell people the system they need.

I like to create bulletproof residential electrical systems that are transparent to the grid, so occupants and guests don't have to know. My philosophy is that I want everybody's experience with alternative energy to be a good one. The system should be self-maintaining; if you draw the batteries below a certain point, the generator should cut in automatically. There are no black eyes in this trade, but the first house that burns down because of crummy DC wiring will be a tragedy. I don't mean it to be one of my houses.

Architecture in Hawaii is hideous. No one is considering the energy future. Even the energy present is bleak, because we import all our electricity as petroleum. I like to get customers before they've been spoiled by the

Laf Young and a serious battery bank.

architects. We tell them about how Hawaiian houses can cool themselves by letting the trade winds in. Fans don't like inverters, and speed controls and inverters are a nightmare, so it's better if we let the air move itself. Usually, the guy says, give me a 12-volt system; I'm ready to go, but when I ask the wife if she wants a washing machine, they both look horrified. Of course they do! By then, I'm thinking about a 24- or 48-volt system. I get people to come with a list of their present appliances, and we work backward from that. I get some of the best questions from the women, because they really grind on the concept until it gets clear for them. Beverly's helpful in the summer, when she isn't teaching first grade, because she can show folks around, the Maytag and all.

When I've worked with a *malahini* (a newcomer to the Islands) and put in a new system, I figure on two call-backs after they're launched. Either there will be a few operational questions, or they'll crash their batteries through inattention. The hardest part is keeping the system balanced, and that's because it's hard to measure the battery's state of charge. Voltage is no good, and neither is the hydrometer, because unless you're doing something to stir the electrolyte you've got stratification. My experience is that if a system is well thought out, it seldom fails, and so people don't lose refrigerators full of food. Then I'll see them on the street or in the supermarket, and they'll come and give me a big hug and say, "Thank you, you've changed our life!"

It's important to understand the whole scope of the system if you want to live inside it. If you're already used to an electrical lifestyle, you need to retune, and ask questions you've never asked before: Where does this come from, what does it cost, and do I need it? When you understand that you are catching your own electricity and water, you start to practice water and energy conservation naturally. Instead of being upset when you get a rainy day, you say, Yes!—because you're catching water and filling your tank.

The most serious mistake to make in the tropics, especially in the low altitudes, is to build on a slab right at grade, so there is no air underneath and all sorts of insect and critter problems ensue. Centipedes are a specialty (one got in my bed just the other night), and higher up, there are scorpions. In our climate, we want circulation over, under, and through. The climate is so hard on wood that the roof should carry well over the decks, which should be wide enough to use.

The tropics are a difficult environment, but you take the same care in building you would elsewhere. We foam the holes drilled for electrical wires to control infiltration, but it's insects, not cold, we're controlling. For the same reason, everything below the roof is treated. If there's an open or cathedral ceiling, it makes sense to insulate to keep it cooler in the daytime. Cross-ventilation and orientation to the trade winds are critical. I used stainless steel screws on my rental's new roof because aluminum screws back out, and I'd rather spend a thousand on screws now than several thousand on a new roof after a storm. I like C. W. Dickey's split-pitch roof for the Islands, a seven-in-twelve pitch over the peak, then three-in-twelve over the decking. I've found that, with a little blocking, I can do it without breaking the metal roofing where the pitch changes. We're at 21 degrees of latitude, and so a four-in-twelve pitch would be better for solar panels, either water or electric, because you need to optimize for winter here. At the new home site, we get stellar afternoons, so I oriented the house about 10 degrees west of south.

C. W. Dickey's graceful proportions work well in the tropics. Rain falling on this roof is piped to the cistern, and provides ample household water.

On Maui, if you need remote power, you need remote water, too. On the windward side, there's plenty of rain, so a good catchment system will do. I gutter the roofs carefully, and bring the downspouts to a manifold which comes up inside a good-sized ferroconcrete tank. Residential water should be reliable, so I don't like the lightweight, intermittent-duty pumps.

Of course you use solar hot water. In Hawaii, the natives (whose land all this was, originally) have gotten land grants, called the Hawaiian Home Lands. There are two big plots here on Maui, and every native is supposed to get a parcel. But the bureaucrats won't let them build because there's no infrastructure. They've got to have water meters (and there's a moratorium on those), fire suppression, paved streets, electricity. I'm trying to build the case that we work around those problems for rich *haoles* all the time, why not for the original residents?

I've also been lobbying the banks to make loans for alternative energy. I've finally gotten First Hawaiian Credit to agree to finance a responsibly installed alternative energy system. It's very hard, maybe impossible, for most lenders to write loans in a rural environment when they encounter alternative energy, because all banks resell their notes in a banker's market that requires conventional housing, you know, streets with curbs, powerlines. If they can't resell, they have to hold the notes themselves. The Community Reinvestment Act (CRA) requires them to do just that, to keep a percentage of local notes. My approach worked, because I explained to them why alternative energy projects were good candidates for their CRA portfolio.

I spent thousands of hours and dollars lobbying the whole state to allow 6-liter—.6 gallon—flush toilets instead of the wasteful three-and-a-half-gallon models, and they're finally legal. At one point in my campaign, the plumbing honcho told me he could get them legalized if they came in some color other than white. I asked what colors he had in mind, and he said, "Green . . . ?"

The PURPA (Public Utility Regulatory Policy Act) program here is a joke. They'll buy your excess power for five cents per kilowatt-hour then sell it back at thirteen cents retail. Then they charge you a fifty-dollar monthly metering fee because they say it takes a second meter reader with special training. So it doesn't happen much. I understand the utility's point: On an island, where it is the sum total of its own grid, if it can't dispatch a resource, it's a frill.

Committee work can eat you alive. Friends of mine spend hours on IRP panels (power company Integrated Resource Planning advisory groups) and it's all voluntary. I'm on sabbatical from social involvement, and Beverly has

Independent homesteaders no longer need to invent or jerry-rig the equipment they require. Off-the-shelf components are readily available, yet it is critical to design and maintain a renewable-energy system carefully. Laf would call this a "loops and cans" instal-lation—workable, but not elegant.

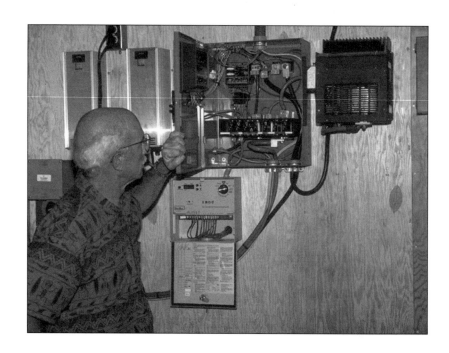

taken over. She's working to reorganize the education department. I'm worried that technology has divorced us from knowing how things work. Like semiconductors: What do they do? And is there anything we can do about it when they don't? My son-in-law is trained to troubleshoot equipment seven levels deep, but when the problem is found they still swap out whole modules, not the failed component.

We trust that technology will take care of us, but actually that means we need to have (and trust) an elite, who do understand, to take care of us.

1999 Update from Laf:

My commitment is still to my clients. I want them to know I'm there after the sale. Like Cully Judd says, it's about service, or it's about price. If you plan to stay with your system, you need to build it right, and you should choose people who will be with you year in and year out. I'm going to be fifty-six this year, and I'm going to stay with what I do. I want people to have a successful experience with renewable energy. I'm always disappointed when I see decisions being made on sticker price alone. I have a reputation for being expensive, but that's just judging by original costs. Since my systems work better and last longer with less tinkering, it seems clear to me they're a bargain. Since the island environment is so hard on electronics, installers have to be extra careful to install systems so they won't corrode and rot. Of course this means spending more money at the beginning, to make the system robust.

In my business, equipment sourcing has really changed, particularly in Hawaii: Solar Depot, Alternative Energy Engineering, and Real Goods dominate the market. With toll-free numbers and credit cards, every solar "expert" can buy direct. These wannabe experts have no clue about building and electrical codes and island conditions, little or no background or training, but they're so enamored with the technology they plunge ahead. They don't know how to include value-added features like corrosion prevention, for example, and so they put their customers' investment at risk.

The renewable energy industry has been taken advantage of by a parade of back-to-the-landers, boat people, groups that include a large percentage of surface miners and bottom feeders. Right now, there's a group of bottom feeders and scroungers who can't resist a good deal, so they're bringing in pallets of used lead-calcium batteries, military batteries, golf course batteries, and they make these the heart of systems which they bid, in competition with me. Of course their systems fail, and that gives us all a bad name. You can't build a deep-discharge system around used batteries,

Since my systems work better and last longer with less tinkering, it seems clear to me they're a bargain.

or lead-calcium batteries; they go down the first time you put any stress on them. But what do these guys know? I keep losing jobs to people who are selling purchase price over service and durability.

On the other end of the scale, I am getting requests from people on the grid for independent systems with big arrays and battery banks, sine wave inverters, propane gen-sets: people who want reliable, renewable electricity. There have also been lots of system upgrades: sine wave inverters, more batteries, a few wind turbines. I've been doing more water system development. For quite awhile I stayed out of the water tank business, because everybody wanted a 15,000-gallon tank for $1,500. These tanks are open to the environment, covered by shade cloth . . . when the tank is full, it's a birdbath, a tea strainer, a place for rats to party before they drown. No cement pad, so the tanks shift and fall apart; the sun's ultraviolet light rots the shade cloth. Not something I wanted any part of. Now I'm putting in nested galvanized silos with liners and solid roofs. I paint the inside of the tank, because a condensate forms between tank and liner and the silo rots from the inside out. We've put in tanks from 5,000 to 72,000 gallons. Of course some of the cost is site-specific, but if you can get a cement truck to the pad, we can put tanks on the ground for 70 cents a gallon with a ten-year warranty.

Most systems we do now use inverters, and so we like to use Grundfos's JQ series—Great pumps! with a DC motor controller, so they start soft and finish soft, which is where most centrifugal pumps tear themselves apart with water hammer. These pumps have another feature mind-boggling in its simplicity: The inlet and outlet fittings have built-in articulation designed to improve the performance of their O-rings. The problem with all centrifugal pumps is that if they aren't bolted down tight, and you use plastic fittings, they eventually start leaking at the threaded joints because they vibrate. The Grundfos articulated fittings allow an installer to get the alignment perfect.

Matching load to source is always the best idea. In marginal systems, if you can avoid an inverter, you're amp-hours ahead. We do some deep-well work using Solar Jacks. Now that Jim Allen is giving us a range of pumps and controllers, we can match the circumstances exactly. His systems are bulletproof . . . and, of course, it's all DC. Two hundred twenty-five feet is about the limit for direct use, and you have to include a linear current booster or a control. We have several systems running at one hundred eighty feet, and their performance, once we got it tuned right, has been marvelous.

I install Bergey wind turbines. The 850 in a trade wind environment is a great performer, and the new blades are much quieter. Of course all the power comes from the outer third, and the noise comes from the tips. High wind is not the issue for the new blades; the problem now is at cutting windspeeds, when the blades make a noise that carries upwind. A true airfoil, like my old Jacobs, never makes any noise. I don't know why windmill companies think that by using a cheap constant-cross-section airfoil blade, they're going to save money. If they'd invest the money in a true airfoil, they'd get more energy and make less noise, and the objection to wind would go away. Right up the street from me, there's a guy making fins for sailboards using a CNC forming machine. My guess is that if Carl Bergey put blades like this on his 850, it would cut in earlier and make electricity in lower winds, it would be at least a kilowatt machine, and it would be silent. He wouldn't be able to keep the simple hub he uses for his cheap stall-regulated, pull-extruded, constant-cross-section airfoil; he'd have to use an articulated hub, but if he'd take a look at the old Jacobs, forget the old spruce blades, but check the way the rotational speed is controlled. . . .

Our rental house's systems have been very successful. The off-grid system and water have been flawless. I spent thousands of dollars advertising in solar rags, and didn't get a single bite. All I can surmise is that our industry pays people so poorly that no one can afford a Maui vacation. So we put it in long-term rental. People pay a premium rent to drink rainwater and run without electric bills on off-grid power.

Deregulation on the islands is going to be tough. If the sugar producers decide to stay in bed with Maui Electric (MECO), there won't be any competition. For years mill-operators accepted whatever MECO gave them, but that was because they were producing energy only when they had bagasse. When they figured out they could bring in coal or anything else that would burn, and get better year-round contracts from MECO, they got more serious about making electricity.

I've worked a lot on Maui's backside, where they say there's so little rain. When I drive out there early in the morning, there's a surprising amount of water on the roads. There must be more nighttime rain than people think. The Chinese up in Keokea figured this out a hundred years ago: All their roofs are steep, pitched 7 in 12. Why? Because every night there's a heavy dew which condenses on steep surfaces and runs down into the gutters. The humidity swing goes from 40% to 85% between 7 PM and 11 PM. That's where the water is.

The lack of energy consciousness on the Islands remains stunning. It starts with the crappy tropical architecture. I am really concerned about the Hawaiians at Kahikinui. If the folks out there could settle on a typical system, a standard voltage, a reliable supplier, they'll be able to put together nice systems for a reasonable cost. I'd like to see them put together a design team, get all the relevant experts together, and specify a generic bulletproof system. That way the folks living out there can talk to each other about their systems, share experience and insights, swap parts. I'd like to be sure the gounding and bonding is good enough so no one gets hurt. I don't care where the modules and batteries come from, but they ought to use a standard one-line drawing. There should be agreement and consistency about water pressurization, catchment surface, and water tanks. . . . We can get way out in front of the curve if we can get these people's attention, but not if the folks in charge don't ask for expert help. ⬌

Integrated Response

The secret of an independent home's success is the integration of separate systems into a coherent, responsive machine. Where a conventional state-of-the-art device (like a heat pump, which incorporates electrical, plumbing, and mechanical subsystems) offers an opportunity for the electrician, plumber, and mechanical engineer to point their fingers at each other, the independent homesteader seeks a more holistic approach. The dependent home's conventional systems are meant to be stuffed inside walls and ignored, but attentive independent homeowners expect to be involved in the day-by-day and hour-by-hour management of home systems. This hands-on approach to shelter thrives when humans enthusiastically incorporate themselves in the management loop. Mod-

LAF YOUNG ON GAS REFRIGERATORS

Gas fridges—propane or natural gas—work by absorption: They don't make cold, they remove heat by exposing the chamber to a mixture of liquids and gases that draws heat and then radiates it through fins on the back of the box. The evaporator in the ice cube tray must be exactly level, because the droplets of refrigerant run down by gravity. Some units had built-in levels: That was a good idea. The newer models are sometimes so poorly made that levelling the case isn't good enough, because the tray is off level, so be sure to level the evaporator inside the box.

If a gas refrigerator stops working, see if it will run on electricity. If it does, you know the burner's dirty or the chimney's clogged. If it doesn't, and you verified that you really have given the unit the right kind of electricity, check the temperature of the refrigerant reservoir at the bottom.

If it's hot, turn the unit off, quick. When it gets too hot, the buffering chemical in the hydrogen-water-ammonia mix cooks into a solid that plugs the little tubes in the evaporator permanently. The old idea, that you could turn the whole box upside down, doesn't work with the new units. If you find green goo around any fittings, or you've smelled ammonia, then you've had a leak and the unit needs recharging.

You have to service a gas fridge once a year. Do it yourself with a little homemade manometer with the right kind of fittings on the ends. You're looking for eleven inches of pressure. Clean out the furballs, the industrial ruby orifice, blow out the burner, and take a shotgun brush to the heat exchanger. If you're not up to all this, call the service guy. It's

ern power-gathering, heating/cooling, lighting, and control systems work well enough without much supervision, but they run best when they are being steadily harmonized and adapted by a critical and interested user. It is hard to know if Laf Young is prouder of his house or of his family's ability to operate it happily in his absence. In my view, only part of the credit for the success of renewable energy goes to improved systems; the pioneers have been heroic in their dedication, and their years of conscientious caretaking show that a family can learn what it has taken generations of on-the-gridders to unlearn through neglect.

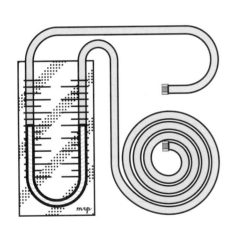

Laf Young's homemade manometer consists of clear plastic tubing fitted with the right brass connectors for your propane refrigerator's plumbing.

Several conditions contribute to "bulletproof" residential systems: The system must be well designed, durable, and intuitively understandable, and the home's resident collaborators must comprehend what must be done to optimize energy production and conservation. This design should include metering that clearly shows current system status. Parts requiring maintenance should be presented in a sensible arrangement. Seasonal adjustments and periodic servicing requirements should be documented, self-explanatory, and manageable, so the operators will be eager to take an active hand in operations. And the owner-operators need to know that competent, dedicated help is available, so they are encouraged to be curious, ask questions, and develop the sound judgment they need to negotiate new situations. Good design, lucid metering, and decent transfer of technology from installer to operators will satisfy many of these requirements, but it may take another decade before responsible local support is available everywhere it is needed. Many pioneers take their expertise into the marketplace, adapting their own techniques to the needs of more recent settlers, resulting in happy and thriving energy-self-sufficient communities.

Knowing what I do about the trials and travails of independent homesteading, the homes that appeal to me most were built with serious, stubborn consideration for the long term, not the quick gain. In these places I could tell without asking that important, easily overlooked details had been well handled: details such as comfort in all seasons (minimizing infiltration, cost-effective heating and cooling, reasonably mud-free access). I feel at home in these places, because I know that health, safety, and disaster planning have been carefully considered. I can see that the home is sustainable, and that family members are cross-trained and the system documented, so the principal operator can go away overnight without the house crashing.

A pioneering shelter may be little more than a lean-to for protecting dry goods from the storm. Jim Loomis, who lives in Maui's jungle and considers himself a water ape, asks only this of his shelters: that they keep most things mostly dry most of the time. For explorers breaking trail into new territory, it may be this kind of deprivation that keeps the pioneering edge honed, and I admire them for their fortitude. I consider myself a settler, and demand for myself and my family a more convenient standard.

As I visited pioneers on the technological frontier, I noticed again and again that the work was so consuming that it inevitably spilled over into the life, to the point where work threatens to overwhelm life without an almost epic effort to find a balance with more traditional endeavors. As these energy frontiers are tamed by the pioneers, however, the renewable energy life becomes possible for settlers like myself.

CHAPTER 10

HOME ENTROPY—
IMPROVEMENTS
AND REPAIRS

INDEPENDENT HOMESTEADERS COME TO THE LAND TO STAY. CHANGE
is inevitable: more batteries and photovoltaic modules, better metering,
maybe even whole new rooms . . . the end of every appliance's life-cycle
is an opportunity to improve overall system efficiency. Those who settled
far from the grid may even connect to it if the powerlines creep close
enough. But we seldom retreat back to the city. For us, home is more than
place, it is process: land, people, and shelter growing together.

It would be wrong for you to think that the living is easy; this independent life is demanding. My editor, a dedicated off-the-gridder, urges
me to remember that our many friends who live under their own power
experience a steady, long-term wear-and-tear which is especially trying
for those who commute, or whose adolescent children go to conventional schools.

In a sense, people who live in extraordinary houses are trapped. How
can one leave paradise? If you have an investment in being rootless, be
very careful that you do not live someplace too perfect. If you must work
in the city, or if your children need a cohort of friends, then this independent life I regard so highly may not be for you . . . yet.

Over the years, those who are content to be rooted to a nurturing
home ground develop a good sense of what it takes to maintain a homestead, along with a concern for improving its workings, adapting to the

inevitable changes time brings. With experience, most of us formulate strong conclusions about how to put a home together so it stays together.

Living beyond the fringes of civilization requires—or develops—patience. Anyone who lives in a rural setting must acquire tolerance for those inevitable delays between the failure of an appliance and the arrival of the service technician or replacement part. At the end of the road, and relying as we do on unusual equipment to generate our energy, who can we call when the lights go dim and the music stops? As more and more people locate themselves in the backwoods, their presence promotes change. Energy equipment reliability is improving, as is the service infrastructure of trained technicians with the right equipment to maintain and repair whole energy systems. A far-flung community of supportive independent homesteaders is evolving, so that when professional repairers are reluctant to come to the end of the road for a triviality or on a Sunday, friendly help is available. On the frontier, the community survives because it is mutually supportive.

Spreading the news: Richard Gottlieb and Carol Levin's story

We are often called upon, in this age of rapidly changing technology, to cope with equipment and concepts for which we have no formal preparation. Electricity is an exceptional example of this: We all use it in multiple ways, and take it completely for granted; its absence is considered an emergency. Yet for many who rely absolutely on electricity for the practice of their livelihood, electrons are mysterious and frightening. Independent homeowners, many of whom have been exclusively trained in the gentler disciplines, are glad to demystify electricity, and embrace the straightforward realities of providing their own utilities. They assert that to live fully in this complicated technical world, it is crucial that we never stop learning. Until a century ago, when the pace of innovation was slower, many journeymen finished schooling or apprenticeship and then practiced trades that scarcely changed during the remainder of their working lives. Nowadays, most of us are called upon to learn perpetually outside our areas of expertise, whenever we buy a vehicle, communication device, or other new high-tech gadget. In this time of proliferating microprocessors, we must at least learn what the most important controls do. Some of us succeed through inspired trial-and-error, or by reading the user's manual, while others need personal trainers or classrooms and teachers.

Richard Gottlieb and Carol Levin live in a rural valley near Brattleboro, Vermont. Their solar-powered house and workshop gets its backup power from the grid. For years they have offered classes and coaching in remedial technology to their friends and neighbors.

Richard Gottlieb and Carol Levin.

Carol: I got into all this solar energy business because I met Rich when I was running a coffeehouse in Brattleboro and we got married.

Richard: What galvanized us into doing solar was our first experience with a re-windowing project, where we grossed half as much in a week as we had in the previous half-year's farming. We started doing solar hot water, back during the tax credit era, but there was lots of competition, and then the credits evaporated.

Carol: Solar is like magic, you put up panels, and electricity comes out. We traveled some, visited with friends in Tempe, Arizona, who said, "You can't do solar in Vermont: That's in the blue zone."

Richard: You know, referring to the map that shows solar potential, it's red in Arizona and yellow part way up the Atlantic Seaboard, and blue says there isn't enough winter sun to make it work. We thought, "Solar is a neat idea anyway, and there's nobody doing it in New England," so we started doing hybrid systems.

Carol: We just got into our niche, put in our own PV system, started gathering data, and

discovered you could get plenty of power here. We were petrified when the tax credits went and hot water went down the tube, but then we found that where people wanted solar hot water for the tax break, they wanted PV for its value. And they were wonderful people! Where else would you find people who invite you over for a champagne party when you finish putting in power to run their home?

Richard: We make people happy. Some of the people we work with are making a philosophical statement. Maybe 20% of our customers want to face their grandchildren when asked what they did to keep from ruining the planet. They come to us and ask, "What can we do?" So we tell them. More and more of them are really having fun by making a statement.

We are most pleased with the work we've done with Ben & Jerry's bus and truck, which are seen by thousands of people every day. They've got solar panels on their roofs, which crank up when they park. They keep their ice cream in Sun Frost freezers powered by energy from the sun.

For years we provided solar amplification for Pete Seeger's Clearwater Hudson River Revival Festival. We'd set 300 watts of modules on top of our van, take along the 200-watt tracker array, and plug in a 500-watt sine wave inverter to the batteries that power the old-timey stage.

We run one-day, hands-on classes for eight people every four or five weeks from spring through autumn right here in our workshop. Theory in the morning, then we assemble a four-panel system in the afternoon.

Carol: About half our students are homeowners, a quarter professionals—architects, electricians, contractors—and the rest are students and retired folks who are interested in seeing something new. Most of them have never done any electricity before, so we start out slow.

Richard: Some have had more experience than others, who may hang back, but on the panels there are eight connections, so everyone gets a chance for hands-on experience. By the end of the day they see a whole system, panels, transmission, batteries, load center, and equipment doing something useful. And they get confidence that they can do it.

Mostly, we do off-the-grid projects around southern Vermont. There are 500,000 people in Vermont, and six solar dealer-installers, and as many PV systems per capita as in Arizona. Here in Vermont there are lots of regional power companies, and the electrical source may be 60% hydro, 30% nuclear, and 10% other, much of which is bought from neighboring power companies through the New England Power Pool.

Carol: Most people come to us after finding out how much it costs to bring in a powerline. They just want lights, a refrigerator, a pump, entertainment, a computer, and they find out it will cost five to eight dollars per foot. For that much money, it's easy to break even on an independent system if the extension is half a mile or more.

Richard: People come to us with the line extension cost as their price: "What can I get for fifteen thousand dollars?" With help, most people get the installation right the first time, although we often see the systems grow as soon as people see how easy they are. There's a constant tuning process, as new appliances get added, or they add panels, or the trees grow and the tracker needs to be moved. We really try to anticipate where they will want to be three, five, eight years after installing their system, and put them into the right equipment to begin with.

When we put in a 500-watt system, in-

stalled, it costs about thirty dollars a watt now. If they assemble and install it themselves, it comes in at about half that, fifteen dollars a watt.

Carol: Or we'll preassemble it here for something in between, and they do their own installation.

Richard: When we do the whole system, we make sure it conforms with the spirit of the National Electrical Code, and we walk the new owners through the maintenance, and give them a manual: a nice package. We've done a few nice systems for people who can get grid power easily, but prefer to be their own power companies. [Berta Nelson tells about one such system in chapter 5.]

Carol: We also get people who have been living away from the powerlines, and are tired of kerosene lights, or they've got kids coming up, and they want proper light for homework. There's one guy, a welder from Putney, who empties the change from his pockets every night when he gets home, and every year or so he's got enough to add another module. And then there are some long-time customers who have been running their systems for seven or eight years, and it's time to change the batteries.

Richard: Nobody wants to water their batteries properly. When the Ben & Jerry's Solar Roller came back after two years, all the batteries were practically dry, but we'd sized them conservatively and they were still working. I put two gallons of water into them. But with proper training and attention, batteries are not much of a problem.

Some people are starting to worry about electromagnetic radiation (EMR). I don't know why, but we started working with DC lights from the beginning, and have gradually switched most of the lighting in our house over. I just don't like AC lights.

One thing that comes with experience: We find ourselves doing prettier systems, with cleaner work. The equipment is better, easier to work with, and the code keeps changing. We closely follow the work that John Wiles, the safety guru, is doing with proper fusing and grounding, and incorporate that in our systems.

Carol: In Vermont, the state cares about septic systems, but there's no building inspection outside the big towns, and no agency cares if you build things right. Our own system is always under development. We'd like it to be state-of-the-art, but customers' needs come first, and so something's always under construction here.

Richard: But the days of alligator clips and jumper cables are over. Independent power systems are in.

1998 Update from Richard:

Five years later, there are almost 600,000 people in Vermont.

The New England Sustainable Energy Association asked us to be part of a new Towards Tomorrow Fair in July of 1998. The original TTF happened at Amherst in 1975, and was very important in the thinking of many of us here in New England. We taught a three-day workshop for thirty students, soup to nuts. A lot to cover, so I had to talk fast. They also asked us to PV-power the two-day conference and fair, so we provided renewably harvested power to all the vendors and all the music.

Golly, there are lots of projects: We still get a regular flow of people wanting little 300-watt back-to-the-land systems. These guys are so young! they look like my grandchildren. But there are plenty of big projects, too. We just finished a 3-kilowatt line-tied system in Concord, Massachusetts. We're working on a YMCA lodge in the Catskills, and keep doing projects out on Cape Cod. Jobs like these are getting bigger, which is very encouraging.

I finished a straw bale house up at Dummerston, Vermont, which combines PV with wind—they're trying to create a non-toxic environment, so all the batteries and equipment are in outbuildings away from the house. They've been running well for two or three years with *no* back-up. I'm getting called to do more and more systems with integrated wind generators. If you put your spinner up high enough, the potential is good here; it flies in fall, winter, and spring, and the PV flies spring, summer, and fall—a good mix.

I keep talking to people about cogenerating: using the waste heat from their propane generators for domestic hot water and in-floor heating. Since it looks to them like another pricey plumbing job, I have no takers . . . yet.

People come to us with the line extension cost as their price: "What can I get for fifteen thousand dollars?"

Nevertheless, the trend seems to be toward more integrated systems, not just a few PVs for electricity: whole-house energy systems using all their energy efficiently.

A few years ago, a lady up in Wilmington asked me put in a little 600-watt system on the two-hundred-year-old off-the-grid house she's been summering in since she was a girl. Of course there wasn't a single suitable roof, so we put in a tracker one hundred feet away. Now the woman has gone electroholic on me! She called and asked if she could run a thirty-cubic-foot SubZero refrigerator on her system. I told her probably not, but when I got up there to take a look, she'd already bought it—looks like a Sun Frost with no insulation—so we put some better insulation around it and doubled her PVs. Now I hear she jacked the old house up, put a proper perimeter foundation around it, blew in insulation, and is now talking about living there year-round . . . amazing how all this energy conscious-ness got started as soon as she experienced the magic of a little PV.

Around our house, alternative energy systems come and go. The hot water system blew over in a windstorm last year, and so we renovated it. It's been working well for twenty years now. In fact, all the household systems continue to work great.

I'm planning to run for governor again this year; I skipped last time. I was thinking to run on the internet. That's where the next campaign is going to be waged, I think. I'm going to try to avoid debating Howard Dean in hamlets all over the state of Vermont.

Richard Gottlieb explaining the balance-of-system components to students.

We got called to a conference about the president's million solar roofs. When they asked for feedback, I stood up and said "Here's one thing that would make a big difference: Tell that Bill Clinton fellow to get some hot water panels and maybe some of those photogalactic things up there on *his* roof. Y'know, if you're gonna talk solar, you've got to live it." Afterwards the official fellow said he was for it, but that it isn't as easy as it was in the Carter era, now that it's a historical structure and there are guys on the roof with rocket launchers. . . . ⟶

Thinking Like an Island

When the power fails in your independent home, for whom does the bell ring? That's right: you! You are the power company.

It is a well-documented fact that grid power fails more often than independent power; agencies such as the Federal Aeronautics Administration and the power companies themselves use reliable independent systems for backup, for equipment that cannot be allowed to fail. Many who are designing their remote homes so that essential systems such as wells and communications are immune to the vagaries of grid power are doing so because they expect a crash. Outages are increasingly common around the globe, because growth and urbanization are out of control, reservoirs are silting up, aging distribution systems are getting knocked down more often in storms, and competitive cost-cutting among utilities has thinned out emergency response as reserves of equipment and workers are cut ever thinner in pursuit of immediate profit. Underscored by the experiences of city-dwellers in Florida, California, and Hawaii, and under the threat of Y2K and the larger notion that our

far-flung utilities are fragile and subject to unforeseeable natural and man-made disasters, many homeowners and institutions are planning for autonomy. The path of least resistance still leads through the gasoline generator, but many people are seeking hardier, more self-reliant earth- and human-friendly solutions.

Independent home systems seldom fail, but their power production often dwindles. This can be the result of reduced sun, rain, or wind, or worse: If periodic maintenance is skipped, and users are inattentive to their meters, the system's capacity may be affected. The first time or two this happens, we are surprised, but then we usually learn to pay attention, and head off trouble before it happens. Since our expensive flooded lead-acid batteries are damaged by being seriously drained, this is something we are wise to avoid.

Over the life of a house, mechanical features—anything involving wires, pipes, or motors—will require periodic maintenance. The practical rule, a corollary of Murphy's Law, is that the probability of failure is directly proportional to the difficulties involved in correcting the problem. Cascading failures encountered or created while repairing other problems can complicate our lives in spectacular ways if major reconstruction becomes necessary. I have spent enough time standing on my head working with my arms twisted uncomfortably to learn to plumb and wire for accessibility. Distribution panels and other equipment requiring troubleshooting and maintenance should be placed as comfortably as possible; the code is very specific about leaving room for the electrician. I have learned (the hard way) to cover mechanical, plumbing, and electrical connections with removable panels, so that systems may be revealed and repaired

ISLANDING

Utilities, especially those in New England, the upper Midwest, and the Far West, where alternative energy production systems are concentrated, are beginning to direct some thought toward what their systems will look like when their ratepayers are also their principal suppliers. One analogy they use to analyze this phenomenon is "islanding." This is their scenario: Residents of a small town like Caspar make the solar commitment and cover their south-facing roofs with energy-harvesting equipment until the village finally attains the point where its community energy production from PVs and other sources equals its base load (minimum) domestic consumption. Caspar's energy customers would then be producing something like 50% of the power they are using. California is one of a dozen net-metering states, which means that a customer may generate electricity and send any excess back into the grid, spinning the meter backwards; at the end of the metering cycle, the customer pays for the net amount of electricity consumed. The sun shines in Caspar by day, so it is likely that Caspar would be a net *exporter* of energy by day, "storing" energy in the grid for nighttime use. Of course, the grid stores little or no energy; urban, industrial, and agricultural users would gobble up the excess power during the day while Caspar's producers watch their electric meters spin backwards. The utility would be able to reduce its reliance on expensive thermo-electric peak-load generators. At night, when the sun goes down and big users retire, the meters on Caspar's homes would spin in the conven-

tional direction, powered by the utility's base-load generators.

But what happens when a winter storm, a squirrel, or a drunk driver takes out the tenuous powerlines that connect Caspar to the greater grid? Caspar could become an energy island. By day, Caspar's small individual systems would still be able to power essential equipment even for neighbors without energy-harvesting roofs. Even today, by carefully conserving battery-stored energy, most of the essential services for my home and my immediate neighbors can be maintained for days. The whole little village could easily provide the generation and storage required to maintain essential electrical service in times when grid power is unavailable, if we can all agree to be conservative. Our utility would have to install automatic local equipment which would allow Caspar

Natural "islands" of self-sufficiency (shown here as unshaded boxes) may exist beyond the grid's tendrils.

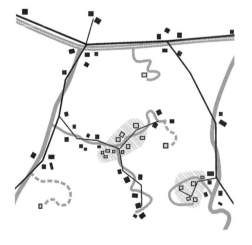

to "cast off" its ties to the failed grid and steer its own course until grid service is restored.

Utility safety officers envision another scenario: What happens when the powerlines serving Caspar require maintenance? Workers could easily cut the main connections to the grid, but to the independent systems of Caspar this would look like another grid failure, and they would automatically go into "island" mode, keeping the local grid energized. To make islanding practical, all constituent power producers must be able to be told, "go local," whereupon they would discontinue pumping excess power into the community's distribution lines so workers can do their work safely. As more and more independent power producers come online in net-metering states and good renewable energy regions, this becomes a serious management challenge to utility planners, and no final solution has been determined.

The implications for future power sourcing are apparent. As centralized, fossil-produced power becomes scarcer and dearer, local independent systems powered by cogeneration, photovoltaics, wind, and hydro will proliferate. The grid serves distributed and centralized sources equally well: It is an instantaneous power broker. The production end is already well developed, and convenient domestic-scale power-management equipment can be purchased for less than most homes pay for electricity in two years. In the next decade, the problems and possibilities of islanding will be widely explored and developed.

and the undamaged panel replaced as neatly as when new. Screws are much less expensive than nails when they must be removed.

There is no substitute for an intimate, up-to-date knowledge of a home's systems. Experienced homesteaders catch and correct problems before they turn into outages or equipment is damaged. To avoid the "it only fails when I'm gone" syndrome, routine inspection responsibilities should be shared by everyone in the family tall enough to read the meter, evaluate the water flow, or note any other signals that may reveal changes in the health of energy systems. In many independent homes, an amp-hour meter, the instantaneous indicator of battery-bank balance, is mounted in the kitchen where all may see it.

By refining our sense of what should be happening at any given moment in the operation of a system, and comparing this with what is actually happening, troubles can be seen coming before they bring the system to its knees. Experienced users have a finely tuned sixth sense of how much sun fell on their panels, and how much energy should have been accumulated in storage, either as hot water or battery amp-hours. For example, if the actual battery state is less than expected, then a short, seasonally adjusted litany of causes is checked. In the fall, as the sun's apogee falls lower in the sky, new growth in trees may shade the photovoltaic array where half a year previously the array stood in full sun. A failing module, loose connections, a faulty charge controller, a power-hungry task performed earlier, or a small energy-pilfering load left connected—all these can affect the battery's state of charge. A single variation is worthy of note, not concern, but if the system continues to perform below expectations, the causes must be found and corrected before equipment is damaged. This implies vigilant attention on an almost-daily basis. Since this reckoning is complicated by season, climatic variations over many years, and changing supply and demand, new operators should not despair; the intuitive sixth sense of the system may take years to develop, but it surely will come.

The amp-hour meter has revolutionized the way we track our power systems by showing us the balance between incoming and outgoing power, letting us know exactly what energy we have made and spent. Modern meters reset the zero point when they deem the batteries to be full, usually at the end of a charging day. As we use power, the meter counts down; when power is generated, the meter climbs back up from whatever low figure it reached during the current cycle. Experienced operators set limits on various activities, since it is unwise to operate batteries "on credit" by pushing them into new low territory. With my

Cruising Equipment's E-meter, the simplest system monitor available.

system, during the winter, an outage brings an immediate halt to vacuuming, ironing, and use of the washer and dryer. If we anticipate a prolonged outage, we bring use of major elective loads—the big desktop computer and its printer, for example—to an orderly conclusion. When reserves fall below −400, we go into emergency conservation mode, quickly consuming any sorbet remaining in the freezer before disconnecting the refrigerator, and using only essential lights and the pump. Once self-sufficient homesteaders learn how much electricity a task requires, they can easily decide if there is enough energy in the bank, and if the task is worth the loss of other functions later on. Intelligent management of demand minimizes surprises.

When installing an electrical system, experienced homesteaders are unanimous: It is a mistake to economize on metering, switching, and lighting in areas that must often be repaired. This applies to plumbed systems, too: Careful use of valves and unions allows for repair rather than major reconstruction. During initial installation and under budget pressure, we may think it wise to leave these inessential parts out, but we are invariably wrong. If it can fail, it will fail. Time and money spent in making systems easier to understand, operate, and repair pays itself back quickly, often the first time the system has problems. Good systems are designed to work well, to be maintained and repaired easily, to recover quickly from failure, and to be as cost-effective as possible over their whole installed life.

In addition, well-organized systems are controlled, metered, labeled, and documented so that it will be easy to localize a failure. Once the failed device is found, we should know, or be able to find in the system manual, what procedures to follow. Before the system really fails, supervised "lifeboat drills" will help us know what to do. It is easy enough to separate essential circuits—minimal lighting, water pumping, telephones—and make sure everyone in a household who must operate the system knows how to shut off all of the inessential loads speedily and safely.

Critical components—computers, water systems, telecommunications—should have their own backup power or be on separate circuits so that we have a few minutes to collect ourselves, make an orderly retreat, and call for help after the lights fail. A delicate balance must be found between too many redundant systems (each of which requires separate, time-consuming maintenance) and too few self-sufficient systems. My priority is to let water pumping go at first so I can maintain computer power while I save my work—pressure in the big tank keeps us wet for a few hours if we are conservative.

Water systems have similar failure modes, and designing for maintenance is equally important. In a gravity system or a system with a big pressure tank like mine, when the source has failed, pressure decreases slightly for awhile, then reduces to a trickle. In pumped systems, pressure decreases from normal to a dribble as water is used. In either case, hours may pass between a failure event and its recognition, and so some means for monitoring performance of the remote source may be wise. Salamanders periodically stop Frank Dolan's system, and earwigs in the pressure switch accounted for 80% of our water outages until our resident plumber—that would *not* be me—placed a little neon lamp in the pressure switch housing to keep the photophobic little creeps away. By using each failure as an instruction on

how to make a system more robust, we may expect failures to become rarer and more interesting.

When a home utility outage occurs, low-voltage DC lights cast a useful, reliable glow in places where troubleshooting may be required—the metering station, the distribution center, the battery enclosure. Batteries are seldom so weak that they cannot make a halogen bulb glow. Independent homes should also be well endowed with functional flashlights, easily found, near places where people spend time or where darkness would be disconcerting. As above, so below: Solar-rechargeable flashlights are especially gratifying.

Since emergency-only systems are found to be inoperative only when we have an emergency, I like such systems to be in everyday use.

Understanding Failure: Darkness or Dimming?

For inexperienced operators, problems with a new system are surprising and confusing. One reassuring difference between electricity and water systems is that electrical systems, when they fail, do not fill rooms two inches deep in escaped electrons. Installed safely, power fails safely: It stops, and the lights go out.

When the lights go out, the first step is to take a few deep breaths while reviewing causes and consequences. Grid-connected folks send their imaginations ranging far: Did another drunk hit a power pole? Did another exotic tree fall on the transmission line? Has there been a terrorist attack? What have the squirrels done *now*? The independent homesteader need only think as far as the local system and the way the power failed. Did the lights slowly dim, or did they go out as if switched? Are all systems down, or only the lights? Were there any accompanying sounds or smells? Where is the flashlight?

Electrical system failures come in two types: device failure (which is rare) and empty batteries. The first step in successful troubleshooting involves reviewing the clues leading up to the failure and the circumstances at the time. To track down the problem, we need an understanding of how energy flows through the overall system, and the ability, through well-conceived switching, metering, and overcurrent protection, to isolate circuits and test them. The troubleshooting challenge is to think of the system's flowing particles, and follow that flow until the stoppage reveals itself.

Imagine the system as streams of energy cascading from source through storage to loads. System protection devices normally interrupt the flow to all downstream constituents, which go dark. If some appliance was just plugged in or turned on, it is the likeliest culprit: Either it is defective or it drew so much current that an upstream protective device, a fuse or circuit breaker, shut down the distribution branch or system. A sure way to cause this kind of failure in my house is to turn on the oven while the washer is agitating. Before restarting the system, the offending device should be unplugged or switched off. Whenever there is an outage, the breakers and fuses are a good place to start troubleshooting. In independent homes with electrical systems designed to look like the grid, and in which house current is delivered through conventional plugs and in-the-wall wiring, failure may be so infrequent that householders forget there is a home system present, and commit a momentary lapse in prudent load management.

Sometimes system dimming takes place over such a long period that we do not notice.

⎯

Sometimes system dimming takes place over such a long period that we do not notice (or adduce our perception to presbyopia) until an inverter or other voltage-sensitive device notifies us by invoking its low-voltage protection and unceremoniously shutting itself off. This strong message generally needs to be received but once, whereupon we resolve to monitor the system's meters more carefully in the future, to be sure to be the first to know about low system power.

Whichever the failure mode, the place to start troubleshooting is the affected system's center and likeliest trouble source: the battery bank for electricity, the pump for water. We should be able to check battery voltage from the metering station. If it is low (in a 12-volt system, that would be below 12 volts), there are two probable causes: A big load is drawing the voltage down and should be disconnected as soon as practicable, or the battery bank has been overdrawn. If a big load is causing a low-voltage brownout, removing it may solve the problem temporarily. Even when the problem is a depleted battery bank, the batteries having drained themselves below safe operating levels, it will generally be possible to get through the night by minimizing loads until the batteries can be recharged. If the big load is important or stubbornly unproductive weather continues, a prudent system manager starts the backup generator.

If the batteries show plenty of charge, then the problem is downstream in the power- conditioning and distribution systems. Inverters are the most complicated technology in the system, and so are next in line, after the batteries, as candidates for failure. If alternating house current is out, but low-voltage DC loads still run, check the inverter. Most modern inverters have built-in over- and under-current and -voltage protection, which may have tripped; remove all AC loads and try resetting the inverter, then test the AC output with a small, expendable AC load like a trouble light. If all seems well, gradually bring the rest of your AC circuits online.

Optimizing Performance

As soon as we find ourselves within new walls, either new-built or new to us, we begin to find innumerable ways to make the house work better. We seek to achieve a transformation that is alchemical, to turn the house into a home. By lavishing intention and love, by investing ourselves in the walls, fixtures, furnishing, and all the fine-drawn details of a shelter, we can work that alchemy.

Planning is ever confounded by the fact that we planners can seldom start a process with clairvoyant confidence right through to conclusion. Attempting to visualize results and predict complexities, we rarely get all factors right. As a room grows from paper plan through siting, excavation, foundation, flooring, frame, roof, electrical, enclosure, finishing to—at last!—livable space, we see it wax and wane like a living thing coming to life. Daily, unexpected views, delightful nooks, lighting effects, textural and spatial juxtapositions surprise us. Our vision of the home shimmers between intention, potential, and reality the whole time it is being built. Some alterations can be accomplished better and less expensively during construction, yet mid-course corrections are the most frequent cause of cost-bloat. It is wise to plan for what we need, not what we want. Time and money are always important factors; we balance costs and benefits as wisely as experience allows. Some enhancements can wait;

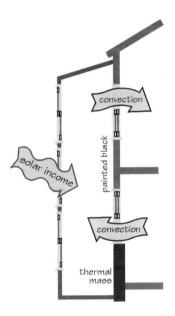

Sunlight turns to heat inside the narrow "greenhouse" of a trombe wall, and heats the house by convection through open windows.

certain puzzles resolve only after daily life makes solutions moot or obvious. We and the house grow together.

Many of my own first efforts were waylaid by inexperience and by my willingness to plow ahead uninformed. Despite my inexperience, my early results worked as well or better than conventional solutions. I was innocently unafraid of failure, willing to try anything, enjoy limited success, refine my technique, and try again. My pioneering wife laments to this day the unfinished nature of the space we occupy. A completely happy medium seems nearer every year, but not yet attained.

Learning to live in new space, patterns develop. We (and our cats) find the best places to sit, work, and move through. The furniture settles into apt locations. Currents and circulation paths become apparent, and enhancements begin to recommend themselves.

The membranes of a house, the walls, roofs, and floors, need not be thought inviolate—set in concrete, as the saying goes, unless, of course, they *are* made of concrete. Punching a window through an existing wall may not be such a daunting task, especially if the window is small and fits between the wall's framing members, or if we have had the foresight to frame for an added window already. The resulting framed view or solar task-lighting quickly repays the effort.

A window is an appliance we use for getting light, views, heat, and sometimes ventilation through solid walls.

A window is an appliance we use for getting light, views, heat, and sometimes ventilation through solid walls. Like any other appliance, a window can and should be replaced when a better-performing version becomes available. After 1993, single-glazed windows became almost impossible to buy, but many "legacy" single-thickness windows still exist in older houses. First-generation double-paned windows, plagued with seal failures and low-quality aluminum frames, are also good candidates for retirement, because modern glass is at least twice as resistant to thermal loss. The cost of re-windowing repays itself very shortly, often in less than four years, and super-glass windows continue to give efficiency and satisfaction indefinitely thereafter.

A retrofitted trombe wall in Vermont.

Improving the circulation of conditioned (heated or cooled) air is a simple and effective energy-management project. Designing good convection patterns into the home from the planning stage is best, but retrofitting registers, archways, interior openings, and other passive measures can improve ventilation when the original plan does not quite work. Small fans, if possible directly connected to solar modules or running on surplus energy, can help redistribute trapped pools of expensive warmth. Ceiling fans can add a little motion to the air with only a small energy expenditure, thereby efficiently moderating temperature extremes.

HEDGING WITH WINDOWS AND WIRE

One optimization strategy is to plan for more, but install less. Windows, for example: Too much glass or too little affects the cost of comfort in a room. Designing for fewer windows, but framing for more, avoids future drastic house surgery to add a window: The headers and framing are already in place. It becomes useful to remember where the framing is buried. Take pictures.

Consider *wiring*! If you might ever want DC current in the walls, put the wire in the walls before the skin goes on.

Modest openings between rooms and floors and between hot and cool places can accomplish as much, though less showily.

Carefully thought-out energy features can be added to existing homes easily and cost-effectively. Maria and Arnie Valdez (whose story will be found in chapter 13) showed neighbors how to add passive solar collectors to their existing homes in the cold but sunny high desert of southern Colorado's San Luis Valley. Hardier original builders might complain that these retrofits are ungainly, but the inhabitants are warmer, their energy bills are lower, and another declaration of energy independence has been made. Partly in response to the Valdez's pioneering examples, new buildings are being planned to incorporate these features more aesthetically from the beginning.

Understanding the arrangement and purpose of pipes, ducts, and wires in our walls, and adjusting them to our needs becomes easier. If lighting or electrical outlets are located inconveniently, a solution more suitable than extension cords or deprivation can be found. Installing permanent outlets and lights is sometimes more complicated than adding a window, because walls may need to be removed (or worked

behind) to expose existing wiring and connect to it. We always prefer to anticipate such needs and put in adequate fixtures or extra wiring before the wall is paneled. In areas where change is likely, a change-tolerant paneling scheme makes alterations easier: plywood panels rather than gypsum wallboard, screwed to the studs rather than nailed. Built-in cabinets and counters can be wired to plug into existing sockets. Surface-mounted wiring hardware is an acceptable last resort. As a veteran of these kinds of campaigns, I was astonished to see how tolerant adobe buildings and earthships are to this kind of change: To add an outlet, chip out plaster and adobe to about twice the depth of the wire and box, affix wire and box in place, then replaster and repaint. Being your own electrician makes this whole electrification process better and cheaper.

Never Finished: Enlarging on the Theme

Driving past a row of New England's traditional "big house, back house, outhouse, barn" compounds recently, I was reminded that real homes are seldom finished. Despite the best efforts of architects and home-makers, homes are organic entities, always growing, always undergoing substantial change. I have heard this constitutional incompleteness mentioned as benefit as well as regret by many who live independently; for some, who lived amidst construction debris for too long, the best argument in favor of a conventional home is the attraction of living in finished rooms. Neither proves to be exactly right for everyone at all times; what we must decide is where between these polar opposites we wish to steer our course.

Even solidly put together with wood and stone, copper and glass, houses are nevertheless alive . . . and growing! Homes designed under inflexible time and budgetary constraints rarely allow for space beyond immediate requirements. Homes built for reasons other than shelter tend to bloat and impose. Still we fail to anticipate social and familial metamorphoses that, in a surprisingly short time, can alter the way we relate to our home space. While I visited no home where space was being removed, and in most of the homes visited some expansion of enclosed space was just completed, being actively planned, or underway, I also note that many folks close off drafty rooms and the unoccupied rooms of departed children in winter. In fairness I must also admit that I visited many places, especially conventional homes, where space was used poorly. Slash-and-burn building—build it, use it up, and move on— is no longer possible. With the western frontier thickly settled, it may be time for us to learn to add new space only when existing space is being used to its best advantage.

I do not want you to fear that an independent home can never be complete. More than

half of the homes I visited were substantially finished; additions or improvements were small and being undertaken with the continuity of ongoing life very much in mind. I merely want to propose the idea of a home as a dynamic and uniquely protean envelope for living in, thus allowing us to adapt to unforeseen changes in ways that enhance life in the future without putting too much strain on living in the present.

Whenever an addition or improvement rends the exterior membrane or requires heavy indoor construction, we must once again survive the hardships of living in a construction site. Savvy homesteaders learn to time renovations carefully—winter is fine for small indoor projects, but miserable for re-roofing. Experience proves, and my partner Rochelle reminds me, that well-planned temporary containment for any building project repays the time spent on its conception and execution. Easier cleanup, less dust on horizontal surfaces throughout the house, less grit tracked into bed, fewer indoor gales and deluges, and an acceptable level of domestic concord during construction: These benefits are valuable to builder as well as his living partners. Taking the time to spread dropcloths and separate building zones with plastic sheet or tarp, and cleaning up after messy phases such as sawing and sanding, while the dust is still a local phenomenon, is the hallmark of a conscientious renovator.

One way to manage our need for ever expanding space is to build another building. In realms of harsh winter weather, direct connection through a breezeway or attached hallway works well, providing additional storage as well as shelter. David Palumbo's shop communicates with his big house through a storage hallway, and the glass greenhouse connecting the power room and privy to Frank Dolan's main house also serves as a canopy over the front door.

Starting over with a new building recommends itself for a number of reasons, not the least of which is that we get to do it right based on previous experience; no longer exactly naive, we may look forward to making a whole new set of mistakes.

Homogenize or Customize?

Quick shelter for quick profit is the imperative of the modern housing industry, but in homebuilding, speed is usually the antithesis of quality. No amount of mass-produced gingerbread can conceal the fact that generic buildings are equally inconvenient for all. As we have already noted, residents are not fooled by this supposedly democratic trend to mediocrity, but willingly abandon or replace quick-built shelter with dismaying frequency. Because of the tempo of hurried development and short-term occupancy, a homogeneous building style now pervades America and has been exported to many other countries. These structures share a common parent: the wartime warehouse. Independent homebuilders, building what we hope will become ancestral homes, are not afraid to appropriate the best tools and techniques of mass-production building (nail guns, chain saws, rigid foam insulation, quick-connect electrical sockets, high-tech glass), although we wield them, at first, warily, but with a craftsman's pragmatic spirit. The sensitively wrought, slow-emerging results fit so comfortably that a family may root and hold for generations.

Two repeating themes impressed me as I visited independent homes and their satisfied owners. One theme, customization, was found in every home. Built-in conveniences such as my self-closing storm doors, David Palumbo's baseboard night-lighting, and Stephen Heckeroth's "baseboard heater" harken back to homebuilding in earlier centuries, craftsmanship that lives on among fine furniture makers and custom boat builders. In many homes, no effort was spared in fitting details of the house to please its inhabitants. The other theme is restraint, a minimalist's insistence that form must stand the test of function, and less can be enough. Taken together, these two divergent themes produce personal, whimsical, comfortable places to live, homes that, surprisingly, cost less than their equivalent mass-produced counterparts.

Independent homesteaders and the keepers of ancestral residences patiently adapt their homes to themselves and their families while decades go by, testing, rebuilding, retesting, and refining their places. My hosts, when asked where they might be in ten or twenty years, answered promptly and unhesitatingly, "Here. Right here." Of 50 off-the-grid homesteaders interviewed in 1993 for this book's first edition, 49 still lived on the same land in 1999, although two have connected to the grid. Studies of a comparable "normal" sample of the populace predict 25 address changes in the same period.

To accommodate this brisk occupancy turnover, standardization is the watchword of the housing industry, and customization is a bugaboo: "I wouldn't advise that," contractors counsel homeowners who want to install energy-efficient renovations. "You'll never get your investment back at resale." For successful independent homesteaders, a home is more than an investment, and resale is an unlikely consideration. Launched children, at college or working in the city, communicate their continuing attachment to and reliance on their safe ancestral abodes. We are to steward these places, they insist, until they return.

In old houses as well as new, on-the-grid and off-the-grid, I found evidence of loving customization everywhere I looked. Energy consciousness is consistently accompanied by willingness to make a long-term commitment to a home and invest ourselves in it. In some older houses, the original structure could be seen only as a venerably papered wall, a comfortably unsquare doorway, or a rounded brick hearth peeking out beneath the high-tech woodstove . . . new layers growing, emending, and improving over the old. This approach was most natural in houses built before the assembly-line values and techniques of commodity construction took over. I was heartened to learn that fine old homes could be adapted to work as energy-thrifty dwellings. From within, through the eyes of the family, these were built to be treasures, remain so, and are worthy of preservation. Sadly, I also found a large number of corrupt and unsalvageable buildings, mostly built during the "post-war building boom": a tragedy of energy embodied in rotting expediency.

Building cabinets, furniture, and other details to fit their purpose makes a house more livable while often saving space and minimizing the use of wood and other building materials. Designers of commodity housing discourage custom-built features because tenant's preferences differ, and accommodations should therefore be provided by modular amenities reduced to low common denominators—so that no one feels very comfortable, as one happy realtor expressed it. And yet it is

surprising how much time and effort can be saved when the kitchen matches the cooks and the food they normally prepare, for example. Standard counter heights fit standardized people well enough, but if one consistent judgment can be made about owners of independent homes, it is that they are not standard at all. Why should a short or tall cook spend years of discomfort and resentment hunching or stretching to reach workspace at the wrong height—to conform to the prejudices of a standardized housing market? The same question can be asked about work surfaces throughout the house, as well as about the way clothes, personal articles, tools, and supplies are stored, work and play spaces are configured, services and living areas are accessed. Until all points where we interact with the solidity of our home places as we go about in daily life serve and please us, we keep planning to make changes.

Independent homebuilders innovate and invent with mixed results. In most cases the outcome is adequate, but, being newly coined, may not be as refined as solutions that have evolved. In one friend's home, three generations of spiral staircase reveal the way his craft has grown with time; each is lighter, stronger, simpler, more compact, and easier to climb than the last. Much can be learned from earlier craftsmanship and other crafters, and so the most successful customizers tirelessly study, compare notes with each other, and borrow from other cultures and disciplines. Innovations are second only to children as favored topics of conversation, so the question "What do you like about your house, and what do you wish were different?" can let loose a torrent of useful information. Customization is most beneficial where space must be optimized, and inspiration is sought in domains where personalization, spatial economy, and grace have been elevated to the level of high art: boatwrights, Japanese carpentry, fine furniture and woodworking, and well-crafted homes built before 1930. I hate to admit it, but mobile homes and recreational vehicles are triumphs of miniaturization; the materials are deplorable and execution may be trashy, but they offer useful lessons. Thorough research can save several cycles of building and rebuilding.

Building into the Future

In the quarter century since I closed in my house, dramatic changes in material availability and energy consciousness have transformed the way houses are built. While many of these changes are still made in the interest of quicker construction, even at the expense of long-term durability, many of them have come about for reasons of failure in early applications, resource efficiency, and the development of new materials. Fewer

---

THE BRAZIL EFFECT

This term was originally coined, I believe, by Seymour Papert, an educational theorist, to describe the coming of modern telephony to Brazil. Our North American telephone system has grown gradually and almost organically over a century, and some of its farthest-flung branches still rely on early technology: copper, relays, mechanical plugs and switches. Gradually, the fastest growing parts of the system have been upgraded to fiberoptics and semiconductor switching, while the most far-flung are now reached by cellular service rather than hard wires. The work is often very technically sophisticated, and repair specialists in many regions must be trained in, and carry tools for, three technologies. Our system's performance is sometimes hindered by awkward interfaces between old and new. In Brazil before 1975, the phone system was practically non-existent, and therefore required building from scratch using the best available technologies; the repair people there only need to be trained in, and carry tools for, the more modern technology, which was then fiberoptics. Brazil's telephone system cost less to build and maintain, and works better, than its older northern counterpart. Does this imply that it is wise to wait until a technology matures before taking the first step? Of course not: far better to risk being cut by the leading edge of technology, to pay the price of being the first, and have the advantage of months or years of experience.

houses will be built in the next two decades than in the last, but increasing pressure to improve performance in important areas—especially heating, lighting, and other energy-intense applications—will keep the pace of innovation brisk. Clear-eyed, possibly clairvoyant analysis on my part might have convinced me in 1973 to importune local suppliers for better windows, but the best even in 1993 would not have been as good or as inexpensive as windows retrofitted into my home today. This so-called Brazil Effect may be expected to bite us as well as gift us many times over the life of anything we build.

A home exists in four dimensions, three in space and the fourth over time. An ancestral home is a monument to all who live in it; it will outlive us, and serve our children well. Its surfaces, nooks, and spaces connect with surroundings that in turn shelter the home and lead us outward to village, fields, and forests that shelter human life. The home changes just as the trees in the garden grow, and the family's members mature, all manifesting patterns of space and time, seasons and textures. The hearth's at the heart of all. Caring for this *genius loci*, polishing and tending it and being nurtured by it in turn, is the homesteader's happy lot.

CHAPTER 11

GROWING WITH THE LAND

OUR HOMES GROW IN BEAUTY AS LIVING THINGS INSIDE AND OUTSIDE them grow, and as we work to create and care for our microcosm. The grounds surrounding the home, the strand between home place and the rest of the world, enfold us in an immediate source of the energy and materials we need. This microclimate profoundly affects the house's energy budget. A garden demonstrates an understanding of our bioregion, and reveals the appropriateness of our adjustment to our environment. If we garden and build with sensitivity to region and microclimate, our works further engage with the natural world, making possible a continuing fruitful relationship with the land.

Slash and burn is still a common human approach to nature. Yet who are we to ignore and negate our land's particular gifts? During construction, especially when contractors and heavy equipment are rampant, we are called to serve as the land's steward and voice. Only clear, far-seeing vision and vigilant intervention can protect the land from lasting damage. In many temperate places, life is tremendously regenerative, and may be drastically cut back without being killed, so long as the web of interdependencies on which it relies remains intact. Sadly, brutes with diesel-powered tools and arsenals of poison still dominate land development, and scorched earth is their favorite building site. For those who intend to stay on the land for decades, it is worthwhile, even if demanding, to imagine the new home amidst its naturally regenerated flora and fauna before whacking the first bush or turning the first spadeful of earth. The

Trees are the land's leading citizens, and our relationship with them tells an instructive story about our overall responses to the land.

best adjustments are found to be slow, gradual, gentle, and life preserving.

Trees are the land's leading citizens, and our relationship with them tells an instructive story about our overall responses to the land. It has been said that where a tree thrives, a human can also. Trees often occupy the same settings we prefer for houses, and must be asked to yield, especially where space and sunlight are limited. If trees must be felled to clear the home site, we can transmute their energy into shelter directly by milling them and building with their lumber. The timbers in my house were milled by my own hand from wind-felled white fir, which lumbermen in these parts call "piss fir" and consider trash trees unworthy of salvaging. A friend and I spent days sawing their storm-uprooted trunks into timbers with a primitive Alaskan mill, a chain saw with rollers. Salvaged from commercially undesirable wood, these timbers are rough and oversized for their purpose, and hanging them was a challenge for this solitary builder, but twenty-five years later they still confer authority, solidity, and a powerful sense of place to my home.

Compared to the way settlements grow, the forest moves incredibly slowly. Where I make my home, redwood trees (*Sequoia sempervirens*) rule the oldest and most stable communities. A fortunate few have tended the same ground for more than two thousand years. These serene and stable neighborhoods contain nothing more disorderly than a bluejay yacking near the stream. Redwoods grow fastest near flowing water, but the stream's incessant undermining of their shallow roots ultimately defeats trees growing too close to the unstable bank; when these fall, they dam and slow the stream while releasing a long, narrow meadow for smaller life forms. On the hillslope the scar of a fallen tree, perhaps caused by human roadbuilding or weather's more gradual work, shows how thin the soil is here: a hydroponic forest where the landscape holds itself together with a fabric of roots to resist time and weather. Unlike most trees which have a natural lifespan, redwoods never stop patiently enlarging their massive structures as they nurture the supporting network of sorrel and fungi on the flats between stream and slope. After centuries of accommodation to terrain and erosion, the remaining old-growth redwood forest is a harmonious community of ancient spirits.

The edge of the forest, the tree shore, is where life is most abundant. We may wish to select this place for our home sites, but it is wise to preserve as much of the original tree line as we can, so its existing community can enrich and naturalize the homestead. The precise relationship we choose to the tree line will be defined by climate, kind of trees, and perhaps the energy source we select.

Open space is desirable, and if our whole plot is wooded, we will want to clear. Although the ancestral or regenerated forest may present us with raw material, the land that begs for release may not be in marketable trees but in young or trashy cover that will provide only firewood. If we are unfamiliar with the species of any large natives on our land, it is important to find someone who knows them better.

Much of the land available has been mistreated by our civilization's avarice for timber, and is now covered with opportunistic pioneer species: Fred Rassman's thornapple, Dave Palumbo's poplar, or the willow, gorse, and blackberry that infest California hillsides. This abused land may require centuries to heal itself without our help. Several landholders reported surprise when vistas and features opened as they cleared their sites. At the very least, I expect someone buying wooded land to climb the tallest available tree, or a tall ladder, to get a sense of the topography. Arranging a photographic overflight in a small plane provides a bird's-eye view and baseline record that will be useful as long as the land is held. Two acres, especially if hilly, is a surprisingly large parcel, and can easily take a half day to explore thoroughly. Without thorough exploration, preferably conducted over several seasons and in different kinds of weather, how is it possible to find the best house site?

While thinking of vistas, consider "viewshed" and solar access as well. Unless we have come to the remotest of locales, any changes we make, any clearing, building, or planting, may obtrude on our neighbors, their cherished views, and the way the sun falls upon them. It is one of the rudest kinds of aggression when an unfriendly or heedless neighbor erects structures or plants trees that spoil our views, steal

SELECTING A HOME SITE

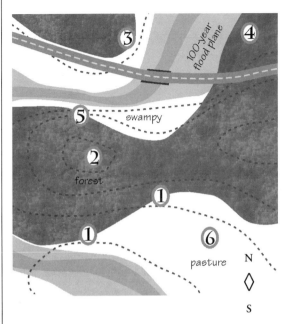

Look to the land when selecting a home site. Here, sites at the forest's southern edge (1) are best. A home on a forested hilltop (2) may be windy, hard to reach, and may damage the treescape for others. A site near the road (3) may be more accessible in snow season, but will be noisier. Sites on north slopes are typically dark (4) and damp (5). It is always a shame to give up good agricultural bottomland for homes (6).

our sun, or impair the value of our holding. Just as we should consult with knowledgeable natives about the habits and suitability of trees, we should make forestry planning a neighborhood activity, working for a stable beauty that suits all. If we admit the trees themselves and other life forms to our homecoming celebration, we will find ourselves treading gingerly on sacred ground until the land itself shows us our place.

Twenty thousand trees: Frank Dolan's story

Frank Dolan lives on one hundred acres of meadow and woodland at the end of a long dirt road high above California's Lost Coast. As I listened to his story, I recognized a pattern of naturalization that I had noticed in others who settle remote land: Seeking a place to live peacefully, they find that they are inspired and informed by the land. Where conventional development holds nature at bay, independent homesteaders work with the land, and refine their efforts as the land's lessons are apprehended.

There's a Chinese saying, "the wise man chooses the place nobody else wants," so in the late 1960s I thought I'd seek that. This is the first time I've been part of a community that isn't deteriorating, in fact it is still getting better thirty years later. My land requirements were simple: I needed reliable, gravity-fed water, good sun exposure, good soil, and good neighbors. And it had to satisfy my spiritual needs but be perceived by others as not attractive. I was looking for a place with a grim economic future, because I knew that I could always create a way to make a living. I prospected all over the West for two years, finally narrowing it down to southern Oregon and northern California, which had the temperate climate that fits my nature.

I started living here before there was a road, and so I packed everything in. The roads, as far as they went, were pretty primitive. It was sad to see the prairie oak go down a few years back: We used to tie off to it and winch cars up its hill. When I first drove through here in the early 1960s, the last of the old-growth Doug fir logging was going on, and they had murdered the landscape. It looked worse than the pictures of Vietnam on TV. I didn't see how it could ever recover from the scalping they had given it, but in the early 1970s, when I returned, it had recovered enough that you could see its potential. Thirty years after, it has healed (except there's little old growth, of course) because of our incredibly regenerative climate. You can still see the scars, many of which have been caused by post-logging, misguided, government fire-suppression efforts. It's easy to see that the land is better off in private ownership, with individual landowners cherishing their own holding, rather than big chunks in the hands of corporations or national forest. In my twenty-two years here, I've planted nearly twenty thousand trees (of course not all survived); it wasn't that hard, and didn't cost that much. There are *Sequoias* [redwoods] that I planted that are as big around as me and forty feet tall.

The first business I tried here was an organic farm. In the third year, after successes with specialty peppers and eggplants, we grew twelve thousand eggplants on three-quarters of an acre. Closely followed by twelve thousand beetles, which taught us the limitations of monocropping. That attracted a good group of Pacific moles, for whom I've found the best remedy is a stick of Juicy Fruit gum down the active hole.

I got an early start with emus: For a pair that cost five hundred dollars, I was offered thirty-three thousand. Now there is a wonderful small flock of Black Welsh Mountain sheep on my steep upper pasture.

I still grow food, a good portion of what we consume, but for a cash crop eggplants gave way to hookbills: parrots, cockatoos, and other members of the *Psittacine* family, intelligent, opportunistic, and manipulative birds, well suited to being human companions. I pioneered a live diet of sprouted grains, automatic watering, and thermostatically controlled misting—an early masterwork of 12-volt spark-and-arc design (I do my electrical work with a pocketful of fuses; I call it the spark-and-arc method)—to create a humid rainforest atmosphere in the summer. I had long-lived, happy birds, but no more babies than other breeders. Like humans, hookbills breed in adversity, and often require stress to get them to reproduce. It's commonplace in the business to sell an unproductive pair, and have them breed soon after arrival at their new home; we even had success crating a pair up and driving them to the city and back.

I've moved from birds to bird hardware and supplies, because the birds make nonstop demands you can't ignore, while a mail-order business fits life here better. I've developed a state-of-the-art incubator for market. And here's a trunk full of feathers, one of my byproducts, that I must edit and send off to the Native American Church and to a project in Panama

Frank Dolan's thrifty Black Welsh Mountain sheep thrive in the rain. Frank's home and parrot nursery are powered by a hybrid system that draws energy from sun, wind, and water power.

that supplies feathers so the natives won't have to kill wild birds.

When the house construction has reached a resting place (it's never finished), and there's enough stability to the business, I plan to get off the place and play.

The land here exerts a powerful selection process of its own, and we're lucky it does. Not long after I got here, I remember the morning our regional paper, the San Francisco *Chronicle*, ran the headline that made us famous: "Humboldt County's Outlaw Growers," with a front-page map with an arrow pointing right at us. The article said you could get rich quick by moving to Garberville and growing dope. There followed exactly what I had tried to get away from: a four- or five-year boom. But by the end, the get-rich-quick guys gave up and got out, and left those the land selected, a tight community dedicated to the idea of taking care of ourselves, healing the land, and getting government out of our lives. The myth created by the media and the police, that this is a place of outlaws and dangers, spoiled the real estate market.

We are an endangered species. We started as beatniks—I did, in any case—then became hippies. People around here don't categorize, and there's a diverse cross-section of styles, radical elements, wild-west cowboys, new agers, computer programmers, organic farmers, and some that combine these opposites. . . . Style is not important. We're eclectic, and spirit is at the center of everything. Everybody, of whatever style, agrees: Get out of my house, government! The land selected us all, a long time ago. When you have sanity and health, why do you need government?

One of the things that made it possible for us to stay was the absence of the intrusion of government. We end up doing what should be everyone's right anyway: sheltering ourselves according to our individual needs. Fortunately, government leaves us alone for the most part. I have no building permit, and won't get one until it is a meaningful process rather than a method for social control and revenue generation. There are two conditions remaining in my dialogue with the county: First, a certification program where citizens (remember them?) can ask Planning and Building for advice on health and safety issues without being pursued for violations. The citizen pays for the services, but is not liable, and the inspectors are bound not to prosecute anything they see that doesn't conform. Second, there should be a grandfather program that lets us enter the permit process without paying fines. As long as the alternative owner-builder isn't re-spected, innovation will take place outside the system. "People should have the same freedom to choose their shelter as they have to choose their clothing (as long as it doesn't endanger the health and safety of someone else)." That's my credo. The burden of proof should be on the government. The way it is now, if it doesn't specifically conform to what's gone before, it's illegal.

I assume that bureaucrats are essentially lazy. When they go on the attack, immediately ask them for documentation. That sets them back at least thirty days. Meanwhile, you get into their backyard, and find their dirty underwear. It's always there. My strategy is, only get angry when

"Show me the plumbing." Frank's full-disclosure policy makes the plumbing on his bathtub (a recycled wine vat) easy to service.

appropriate, and tell the media that you're doing it for free, to make the county a better place.

KMUD, our community public radio station, is a great resource and source of strength. For example, somebody will get on the air and tell folks about the current outrage the county's hatching: They want to rename rural roads. They'd destroy the history of the county. KMUD lets us generate the kind of energy that's required to back off the bureaucrats. It's the electronic link for our remote homes.

There's another Chinese saying, "a man who finishes his house, dies." If I was forced to finish this house under time pressure from Big Brother, it would have been a very different, inferior house. I have to be able to solve problems in the time frame of my own life. I can't finish a part of the house until I can visualize it.

I collect several springs into a 12,000-gallon ferroconcrete tank a third of a mile away, with 115 feet of head. I run two 2-inch plastic lines to the house, one from the top of the tank, and the other from the bottom, so I know if the tank is full or not. I chose 2-inch pipe because it was all off-the-shelf, and quite a bit less expensive, but I have enough water most of the year to justify 3-inch and possibly a second generator. From Thanksgiving in a wet year until June, my two-nozzle hydro generator puts out 22 amps twenty-four hours a day; I know when a salamander has plugged one nozzle because the output falls to 7 amps. This rig has paid for itself many, many, many times over, even at the utility's subsidized rates. But you can always use more power.

I have three composting toilets here of my

Frank's volunteer grass lawn thrives on soil created in the composting privy.

own design, but the design doesn't really matter. To succeed with a compost toilet, you've got to get good at making compost in your garden first. Master the art of getting kitchen and garden scraps to compost (which is simply making microorganisms happy), and you'll be able to design a compost privy for your circumstances. My luxuriant green lawn out there used to be unfertile Laughlin-type soil, but I gradually covered it with the output of the privy, and encouraged grass to colonize it. The cuttings in turn are used in the privies, and fertility increases as the years pass.

My greywater system took a couple of tries. The first one failed utterly; if I'd been doing it with Big Brother looking over my shoulder, he would have declared it a failure, and would have made me go back to the conventional model. I fine-tuned the idea, and now it works perfectly. ✐

Fitting the Landscape

Where fire is a threat, our relationship to the forest or range must respect the way these ecosystems sustain themselves. Healthy dryland forests and grasslands thin and maintain themselves by tolerating small, periodic, lightning-set fires. Homes are not as tolerant as the surrounding flora to this periodic fire release, and often do not perform well even in the mildest grass fire. By choosing fire-resistant roofing and siding, and by replacing the nearby forest understory with plantings that do not burn well, we may be able to hedge on the firefighters' bare earth requirement somewhat, but, as has been repeatedly learned in California's urban and suburban forests, the relationship between homes and dryland forests is precarious. Besides careful planting, those who build in such a place must plan several emergency escape routes within the house and from the property, and during fire season keep their possessions elsewhere or in perpetual readiness for flight.

Both sun and storm should figure into calculations about neighboring trees. Anyone who has heard a tree fall on a stormy night, or has seen the wreckage, will want to be sure that climax trees—trees that have reached their maturity and have a propensity for falling—are kept at least a tree-length away from the house and any important or valuable possessions. A car under a fallen tree is particularly forlorn, since the fall could have been foreseen and the car moved. Balanced with this risk is the fact that a line of evergreens may offer, as it does on my land, the best defense against nagging winds. If trees are still growing, we must plan our solar exposure with their eventual size in mind, or plan someday to cut them. Cutting is an inevitable aspect of forest management, so if a windbreak is important to a home's microclimate, tree stewards must consciously plant replacement saplings well before the older trees attain their full growth and are ready for harvest.

Deciduous trees and bushes shade us from too much sun in summer, but winter sun shines through leafless wintertime branches; this seasonal shade will suit our seasonal heating and cooling requirements more perfectly than anything we can build. In addition to being deciduous, orchard trees offer up other seasonal delights: blossoms in spring and fruit in summer.

When planting trees, we should look first among the indigenous regional flora. Gardening arrogance is at its worst when plantings are taken wholly out of their regional context, and require water, fertilizer, and protection to survive. Exotics are a good choice only if they come from a similar environment and can be controlled effectively. Japanese black pines, a low, slow-growing species from the salt-blown headlands right across the ocean from Caspar, settle well among our native shore pines, and thrive without driving the natives out, adding healthy diversity. Their cousins, western yellow pines that thrive just a few dozen miles and a couple of ridges back from the ocean, near enough almost to be considered natives, find the salt and fog stressful; they bolt, get leggy, then top-heavy and brittle, and fall prematurely.

For a demonstration of bad species management, take a trip to Hawaii. Here in these most isolated of land masses, tropical growing conditions and pockets of good soil allow imported flora and fauna to overpower what once was a wonderland of unique species. Generally for gain, but often just for sentiment or a misguided sense of fitness, exotic species were imported from every continent and broadcast by humans and birds. Now native species, which never before needed to develop competitive strategies, are all but gone. It is estimated that less than 20% of the original Hawaiian species survive, and native species cover less than 10% of the land area.

Only recently have we become aware of the magnitude of this tragedy, which involves not only plants but also insects, animals, and especially birds, as aggressive exotics wipe out native species and ruin well-balanced ecosystems right around the globe. Introduction of any exotic species may set off a far-reaching and irreversible environmental change affecting every species in an ecosystem. Hawaii has been overgrown by the banyan, a large tree native to the Indian subcontinent, which produces fruits much loved by birds, many of which are also imports. The fruits contain tiny indigestible seeds which are neatly packaged in fertile guano

When planting trees, we should look first among the indigenous regional flora. Gardening arrogance is at its worst when plantings are taken wholly out of their regional context, and require water, fertilizer, and protection to survive.

before being broadcast as far as the birds fly. It is supposed that the banyan, which dominates Hawaii's humid forests by choking out dozens of more useful natives, including precious native sandalwood, came to the islands just a century ago as a house plant.

Ark Maui: Wittt Billlman's story

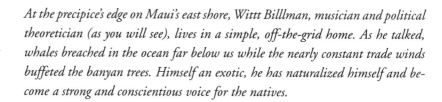

At the precipice's edge on Maui's east shore, Wittt Billlman, musician and political theoretician (as you will see), lives in a simple, off-the-grid home. As he talked, whales breached in the ocean far below us while the nearly constant trade winds buffeted the banyan trees. Himself an exotic, he has naturalized himself and become a strong and conscientious voice for the natives.

As I get more serious about my endeavors, the more I am considered a comedian. Or a karmedian. You can spell karma with the H or without, depending on how Tibetan you are feeling. Spelling is a very important subject, so be sure you get my name right, triple T, triple L. I favor everyone adding a personalized twenty-seventh letter to the alphabet, so we can play with the jokers, too. Fifty-two letters doesn't seem to be enough in these times.

I came to Maui on a triple: I was looking for my high school math teacher, Jim Loomis, and I knew he was here; I was in a ten-piece Brazilian dance band then, playing a gig in Portland, Oregon, and a woman came up to me and said I was going to Maui, she could see it in my eyes; I still didn't take it seriously until I met another woman who said she was going to Maui, and asked me if I would come with her. It ended up all three of them lived on the same road, within a short distance of where I've been living nine days a week ever since (I bought the land from George Harrison, and I know it took him eight days a week to get it, and I'm sure I work longer hours). We're all from two different galaxies, you know: the Pleiades . . . and the Work-days.

The first ten years I lived on candle power. Water comes from rain and gravity flow. I've experienced duress, expense, and time-loss keeping the utilities out of our neighborhood. My energy needs evolved, and I got one solar module, although I've recently been considering getting grid of it. It powers my electric piano, but I'm thinking about going acoustic, and using the power to dehumidify the baby grand instead. I like that image: a row of bulbs inside the piano keeping it warm and dry, and at night, when we need light, we open the top. I've written a new song for the Billary Allegory in the White House (isn't that just a primer coat?). Would you like to hear the song? I call it, "Inhale to the Chief."

Wittt Billlman watches whales from his east-Maui headland.

Politics is very important to me. Right now I'm running from office, with the Party party, and have been making all the right moves: I have a political insultant, and a full-time champagne manager. Originally I had no platform, because I think everybody should have both feet on the ground (assuming they have two feet, of course. I sincerely don't mean to insult anybody). My supporters have pressured me to get a platform, so I've adopted one with a big hole in the middle, so I can stand on the ground, surrounded by platform.

When Hawaii becomes a non-nation, those are the offices I'm especially interested in running from. We hope that Hawaii sets such a prezident (that's with a Z, because we're near the end, and should be using more Zs than ever). We're working very hard on Ark Maui. Did you know that 87% of the endangered species are in the Islands, 60% on Maui, and that 30% of America's endangered species live only on Maui? We are pleased that more people are starting to kNoah-bout the Ark.

Money is a symptom of centralization, and I believe we have to decentralize to survive. Since we went on the uranium standard, I've come to favor hundred-dollar bills over ones, seeing as how both really cost the government two cents to print. Lots of bananas and papayas grow on our land, and that's how we survive . . . until lunch. We have a little salad garden, some sweet potatoes, so with a little fish, we don't have many food needs. Our taxes started at two dollars twenty years ago, then twenty, two hundred, three thousand! That's 15,000% inflation. We pasturized our land, dedicated it to agriculture, and the taxes rolled back to one thousand.

Except for that, the county doesn't bother us much. I was working on a house for a neighbor the only time I ever saw the building inspector. By coincidence, I was just gluing the last joint in the water system, and a big rain cloud was coming up over the mountain when he showed up. He got out of his truck and looked around, so I went over and said, "May I help you?" He looked me over and said, "Some folks call you hippy, but I call you pioneer." He got back in his truck and drove off. ✈

Slash and Burn Gardening

Starting in 1880 the inland redwood forest, coastal meadow, and shore pine community that is now the village of Caspar was scalped, flattened, settled, scalped again, poisoned, and replanted with fast-growing cash trees, to be abandoned in 1955 to the ground squirrels and a new wave of settlers. Starting about the same time, New England was being scalped in order to pasture sheep. Across the country wherever trees grew, the land was scarred by careless use of the tools, machines, and, more recently, chemicals employed in the war between humans and nature. Fragile topsoil built up over centuries is still being carelessly scraped away or abused, then allowed to wash into the ocean. California's native salmon, which rely on clear, cool streams, are nearly gone, and New England's rivers, declared dead decades ago, are just beginning to recover.

Plants and building styles transplanted from other climates survive, but seldom thrive. In California, one of the worst immigrants is the eucalyptus from Australia. According to one story, eucalypts were imported by an igno-

rant but predictably greedy speculator who had been told Down Under that gum trees were fast growing and made great hardwood for cabinets. He was not told, or was in too much hurry to understand, that there are many different eucalypts; he brought home the fast growing ones that make miserable timber and are only marginally useful for firewood because their ribbon grain makes the wood unworkable when dry. Only in the last few years, land stewards have discovered that eucalypts are allelopathic: the oil in the bark, leaves, and seeds that the trees shed in abundance contain a toxic substance that kills everything but the eucalypts, and persists in the soil for decades. Because of their habit of quick growth, however, the infernal eucalypts now infest every temperate continent. Along California's coast, first-time tree-planters, myself included, thought one pine as good as another, so they planted fast-growing mountain species. Cypress, another misfit exotic, is a Mediterranean native that here becomes brittle, sheds limbs, and finally falls after a fifth of the years it enjoys in its homeland. All these exotics, unhappy with coastal fog and rain, bolt like berserk weeds in their forlorn search for sufficient sun; in less than a decade they are leggy and unstable, and fall to the buffeting winter winds. In a storm in 1997, powerlines were downed in over a thousand places along the north coast by falling trees, of which more than eight hundred were climax exotics. Yet these trees grow quickly and handsomely in their early years, lending a brief but reassuring pretense of green healthiness, which is all the developer requires. As a consolation, all this downed timber feeds our fireplaces.

Only lately have we come to understand that other trees, the original natives, are more fit for this place, and require less from us and from the fragile soil. When I came to Caspar three decades ago, I too planted mountain pines and cypress, and have lived here long enough to see them steal the soil's vigor and fall before they attained maturity. The California State Department of Beaches and Parks, in a rare burst of environmental heroism brought on by the drought and the resulting fire hazard among the stressed and dying exotics, has instituted a program of removing them and replanting with the original inhabitants, the shore pine. Eucalyptus abatement will cost the people of California more than a billion dollars, but the expenditure is justified and urgently needed, according to silviculturists, because eucalypts poison the land they grow on, rendering it unfit for all other life forms. It is hard to imagine us summoning up the political will and hard cash required to correct the mistakes of our forefathers.

In three decades I have learned a great deal, and fortunately my land is forgiving and resilient. A Sitka spruce, one exemplar of this resplendently upright, pointy-needled species at the far southern edge of a natural range extending north all the way to the arctic tree line, is pushing out the neighboring western yellow pines as if to say, Be gone, this is my place. Competing on merit, unhindered by pushy human pretensions of control, natives prevail here. In the spruce's shadow, the indigenous coastal forest micro-flora is regaining a toehold and pushing out the periwinkle that lived in the unhealthy shade of the unhappy pines. The deer have returned, and my roses are their favorite food despite my most patient explanations that native flora is better for them. Recently I have learned to protect my edible exotica, lettuce, leeks, sorrel, and other crops the deer deem tasty, by building cloches of rebar (the steel used to strengthen foundations) and chicken wire.

The deer have returned, and my roses are their favorite food despite my most patient explanations that native flora is better for them.

Beneath the Trees

Houses and trees are equals, but the understory is more our size. Shrubs and other greenery soften harsh weather, humidify and cool us on hot days, and slow winter's gusts. Deciduous plants, especially vines trained to a trellis, extend living space and offer blessed shade in summer but retire graciously thereby permitting access to added warmth in winter when the sun comes in more weakly at a lower angle. Seasonal plantings draw our attention outside, and remind us that we are a part of a larger picture.

Native plants thriving on a home site may suffer when a house takes their sun or changes a feature of the microclimate on which they rely; by

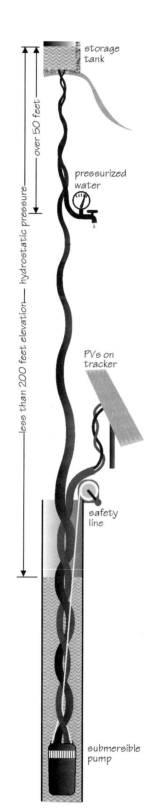

studying their habits before construction begins, and re-establishing them as soon as possible, we help to preserve and even improve the natural relationships that originally drew us to the site. The north side of a house offers precious shade and a moister microclimate for delicate species; here in Caspar, native violets, rhododendrons, and fuchsias thrive if they are also sheltered by trees from nor'westerlies. Such northern gardens, well protected from the wind and selectively viewed through small super-glass windows, can brighten rooms on the utilitarian dark side of a house.

Storm run-off, including that from roof drainage, paving, or hardscaping, is one of the most devastating long-term effects of development. Nature holds water and slows its progress across the land; too often, development hastens its passage. If rainfall, which is easily ignored because it takes place when we are safely indoors, is anticipated and managed, we can create or improve habitat. By capturing, redirecting, and even storing run-off, we can care for plants that normally require more or less moisture than the local climate provides.

Roof catchment of rainwater in cisterns or ponds, coupled with direct solar water pumping, provides appropriate and satisfying ways to lengthen growing seasons and encourage wetland plants on a normally dry site. At one farm, a subarray of photovoltaic panels contributes to the home electrical supply in winter when there is less sun; when the rains stop, this array is switched over to pumping water from the river at the property's low point to a large tank on the ridge above for gravity irrigation.

Getting above Mud

A bird's-eye view plan of the house, its approaches, outbuildings, and enfolding garden, adds important patterns when mapped to a home's site rose. Our daily comings and goings around the house, our traffic patterns, pass through the four dimensions of space and time in a context with seasonal weather patterns and the special qualities of the home site. If walkways, plantings, and clearings are planned effectively, the home and its pattern of habitation may be sustained by surrounding nature.

Ideally, the water storage tank sits atop a hill fifty feet above the homestead. Dedicated PVs directly power a submersible pump that sends water to the tank during daylight hours. Any overflow is diverted to irrigate the garden.

Using this map, preservationists can declare vehicle-free zones and build walkways before undertaking development, thereby minimizing impact on fragile terrain.

Repeated travel along unplanned paths breaks important natural connections. When nature's web sunders, mud happens. In 1993 I was surprised by how many independent homesteaders laughed mud off. "Mud season's like this; it will dry up soon."

"To dust!" I thought to myself. Having lived in a rainy spot for a long time, I have learned that a good walkway is always better than mud machismo in a quagmire. Nature answers mud with a delicate seasonal wetland garden of pioneer groundcovers, mosses, and grasses, supporting a whole community of busy native gardeners. A few years later, all but the rawest sites are showing a growing awareness of the importance of healthy groundcover. Successful mudders learn early to avoid disturbing their tiny underfoot neighbors, whose continued vigor augurs for the welfare of the site's topsoil. If groundcover fails, we may expect to waste heroic energy scraping mud, stuck in the mud, muddy footed or worse, muddy watered, waiting for mud to go away. There may be a few spots on the planet where mud is truly unavoidable; elsewhere, I suggest that a modest but comprehensive web of pathways and driveways will save more energy than can be told. In Caspar, in order to preserve a healthy flora and fauna in summer, we seldom drive across the meadow, and never after the first rain. The approach to the house is by walkway; a wheelbarrow is provided for difficult loads. The inconvenience is more than compensated by mudless feet and a summer meadow full of abundant insect and bird life.

Communicating these preservationist values to big-footed workmen is easier when they see you suit your walk to your talk.

When planning to add outbuildings, we must ask important questions about the impact on our present home place. Will the new structure improve wind shelter or will it create a wind funnel during bad weather? How will it change garden exposures, favorite views, and traffic patterns? Being careful not to fall, one good way for us ground-dwellers to envision these changes is to take to the roof on a sunny day.

The Kitchen Garden

It is a poor home indeed where no fraction of the food we eat comes from our own garden. A good place to start is by visiting an experienced local botanist or forager, who can tell what native plants are edible, and how best to encourage them in your garden. Neighborhood gardeners, too, can offer a wealth of specific knowledge about what grows well, and how favorites can be coddled. If you time your visit properly, you may come away with enough starts to begin your first year's experiment.

Small gardens succeed when more grandiose plans collapse from their own intensity. Smaller garden plots, a mere four or six square feet, can be maintained, protected from predators, and watered, and will be astonishingly productive given a small amount of effort. By adding a few square feet at a time, we can adjust the size of our gardens to personal gardening energy. If we intend to live for decades on a plot of land, successful establishment of a single perennial or permanent planting should be considered a good year's progress.

In an exceedingly benign growing environment like coastal California's in spring, the struggle is slightly different from that in a wintry environment. Whereas in Vermont the gardener will go out after the final heavy snowfall to see what favorite plants were able to overwinter, I go out with jungle-cutting implements to see which of my favorites has not been choked out by spring's onslaught of riotous weeds (mostly, alas, exotics imported by livestock). Wherever we plant with intention, we find our plans may not coincide with those of the Greater Gardener, who always plants first *and* last. By starting modestly, with one bed at a time, we can be heartened by the balance of success and failure.

Nature abhors single crops and straight lines, and attacks them with all the pests and pestilence in existence. Independent homesteaders will resist bringing chemical treatments near their homes, and particularly near their food crops. As with energy and building materials, rejection of conventional petrochemical wisdom leads to rediscovery of traditional techniques, in this case integrated pest management. Similarly, seed saving and planting of heritage crops in preference to unreliable and single-season hybrids allows us to naturalize a garden that thrives across years. The result is tastier tomatoes and peaches, hardier regional crops, and reduced load on the planet's energy resources. We see that traditional agricultural practices—companion planting, attentive care and watering, integrated biological pest control, and acceptance of the whims of nature—are more satisfying and effective than aggressive, expensive technological interventions.

In the garden as in the house, the goal is independence. Homesteaders prefer systems that require minimal maintenance once established. By studying the home site we can learn about the garden, where our sensitivity to microclimate resolves to a focus on increasingly finer features—rocks, trees, and shrubs—and a concern for the effect of person-sized and house-sized impacts on nature. Over seasons of firm yet gentle urging, we can tune the garden to be our most productive energy source, converting sunlight, moisture, and nutrients into fuel for bodies and minds. By learning to follow the natural rhythms and tendencies of the land, we will further decrease our reliance on imported energy. A completely sustainable garden requires only what comes to it naturally—light and water—and produces no waste. Paradoxically, the largest single contributor to American landfills is yard waste. Only nature does it perfectly, and we will do well to observe, learn, and imitate.

Applying whole-life cost analysis to soil amendments, planting materials, and gardening hardware is just as reasonable here as in the house, and you will find that when these measures are applied, some highly innovative technologies become economically justifiable. If the garden is the right size, we might well enjoy the time spent watering it by hand in the old way, with a watering can filled from the rain barrel. Reinforced hoses, durable brass fittings, smaller plots, appropriate groundcovers, and sensible planting patterns are sustainable over the long term. In our appropriation of energy efficiencies in the home and here again in the garden we must not be too quick to embrace all offerings presented by the marketplace. Drip irrigation pinpoints water delivery, wasting less water and demanding less of our time. This method uses petrochemicals reasonably: Instead of burning them up, they are embodied in a plastic irrigation system that will be useful for many years if well planned and cared for. My experience with drip irrigation exactly parallels my findings indoors: There is no free lunch. Quick-and-dirty solutions such as slip-on pressure-fittings and fussy little sprinklers allow for quick gratification, but stand the test of years poorly. As with any project, take time to plan and assemble the parts with care, thinking not of this year's garden but of the garden in future years. Since coiled poly pipe has a mind of its own, allow it to relax for a few sunny days in the garden before making final adjustments and connections. Design the system so it can be dismantled and redeployed in a different configuration; try to minimize the use of parts that only work once, like slip-on connectors.

Cold frames, greenhouses, and other techniques for lengthening the growing season can be readily cost-justified in any productive garden. Common sense judgments about materials and their whole-life cost should still be made. Customized solutions take longer to plan and erect, but they usually work better, require less fussing, last longer, and are more easily reconfigured.

The kitchen garden should be awarded the most favorable solar exposure. Within the garden, sensitivity to the solar requirements and growing habits of each crop will be repaid: Low-to-the-ground eggplants and squash require as much light as they can be given, while lettuce prefers a smaller portion. Peas, beans, corn, tomatoes, and other tall crops happily climb in the back row without shading lower-growing plants. Home energy harvesters rapidly discover that such considerations are intuitive.

Discussing home food production from a garden, Richard Gottlieb suggests a ratio similar to the solar fraction (which, as you may remember from chapter 8, represents the amount of electrical energy that can be reasonably expected to come from the sun). In the solar blue zone of Vermont, he says, it is very difficult to push the local food fraction above 10% of homegrown food. His New Hampshire neighbor Leandre Poisson disagrees, and other growers in the region, employing season-lengthening techniques and legacy seeds adapted to cope with local conditions, report results of up to 50%. Eliot Coleman reports that his gardens in Maine are so productive in autumn, winter, and spring that he takes summers off.

Growing Energy: Renewable Fuel

We have learned not to heat using electricity, but how will we cook when the dead dinosaurs are used up? Direct methods such as solar ovens are delightful, but provide at best a partial solution. Yet sustainable biofuels

Over seasons of firm yet gentle urging, we can tune the garden to be our most productive energy source, converting sunlight, moisture, and nutrients into fuel for bodies and minds.

A solar oven.

are nearly ready to replace some of our fossil fuel needs, particularly our need for fuel for cooking. Hydrogen and methane can be renewably harvested using the sun's energy, and offer us an efficient way to store heat. These technologies are on the brink of practicality, and some pioneers are using them already.

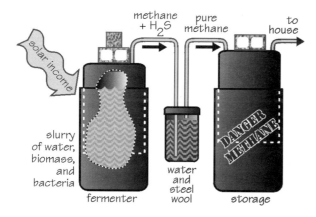

Household-grade methane can be produced using low-tech means. A soup of water, methane-producing bacteria, and high-cellulose garden clippings ferment, and the output gas is bubbled through a steel-wool bath to remove the sulfur dioxide. The whole operation runs at very low pressure – one to three concrete blocks work fine.

The idea behind biofuel is that a suitable crop is grown for its energy content and its ability to improve (or at least not deplete) the soil. Fast-growing, nitrogen-fixing crops, including *Gliricidia* or fava beans, are ideal candidates. Mature plants are chopped up, mixed with water, placed in a closed tank, and anaerobically digested by a community of microbes already in residence in the digester. Methane bubbles off and is captured, dried and desulfured, stored, and used as a fuel.

Hydrogen generation is simplicity itself: Nudge water molecules with sufficient electricity, and they split into their constituents hydrogen and oxygen. In the most direct process, hydrolysis, solar electricity is used to split water into its atomic components: oxygen, which we can always use, and hydrogen, which can be captured, stored, and used for heating tasks.

Hydrogen is a particularly desirable fuel. Our shared horror at pictures of the exploding Hindenburg notwithstanding, hydrogen is easy to store, and a far less aggressive explosive than most petroleum-based products. It burns cleanly, producing no chemical gotcha!s, and can be conveniently stored, as long as the storage container doesn't need to move. Methane is not quite so well behaved during storage or combustion, but is manageable on a small scale. Neither hydrogen nor methane lends itself to efficient liquification, and so the problem of transportation remains a puzzle, because vehicles need a light, concentrated fuel source. Until the techno-wizards figure out how to commercialize these two bio-fuels, they will remain a curiosity for a few home-brewers to run their kitchen stoves, and for the Big $cience guys with their silly fuel cells.

The Garden Within

Planned for or not, the whole south side of a building is a solar collector.

(We apologize to our southern hemisphere readers; for you lot, it's the north side.)

Using glass and thermal mass, we can keep living space within a comfortable range of temperatures. In many of the houses I visited, the area between the glass and the mass is used for indoor gardening. People who live or have lived in greenhouses rhapsodize about the joys—Robert Sardinsky and his midwinter carrots, Amory Lovins's iguana and banana tree—and testify to the delights of the indoor jungle.

Using the energy-capturing capabilities of glass and mass, indoor gardens and season-extension techniques using cloches and cold frames work everywhere there are seasons. In addition to adding healthy home grown produce year-round, these methods keep reminding us that the life force is directly related to the sun, which always returns. Many independent homes bring plant cultivation right into their living rooms.

Wes and Linda's greenhouse: even in the coldest climates, a sun space provides abundant heat and growing things.

"I've never met anyone who isn't moved by a garden."

⟶

While the concept is natural enough, the execution is challenging. Indoor agriculture introduces into the compact terrarium of a house a wide variety of pests, petals, and pollenizers requiring energetic, attentive, and even inspired management. Some jungle vegetation is generous with its spores and pollen, so an allergic gardener will choose tropical species carefully and introduce them cautiously. Often we find that devoting the sunspace either to space heating or agriculture, but not both, is more practical. When adding any element as seasonally dynamic as sunspace to a home, we must remember that such additions will be present and will affect the overall microclimate of the home not just in the season for which they were designed to be beneficial, but all year long. A space that generates welcome extra heat in winter may also produce undesirable extra heat in summer.

In cold weather sites, the greenhouse can add humidity to air that has been freeze-dried outside, with beneficent effects for the house's inhabitants, but less predictable results in the library, where books and papers may not appreciate the added humidity, and spores from the garden may initiate a bloom of mildew. In temperate and humid climates the greenhouse's added moisture and odors can be even less welcome, and good ventilation is important. At Rocky Mountain Institute, a PV module coupled directly to a fan (like the direct-solar powered attic fan described in chapter 2) provides air circulation when the sun shines, a load matched perfectly to its source. Sun-powered greenhouse devices such as exhaust fans and window openers adapt nicely to the indoor garden space.

Including gardening space inside the home's envelope reconnects us with growing things. Growth—flower, fruit, and the annual cycle of reproduction—makes us glad. As Robert Sardinsky says, "I've never met anyone who isn't moved by a garden." Watching Arnie and Maria Valdez in their greenhouse living room, pinching dead leaves and tending the tomatoes as they talked to me, I could see how naturally such space becomes an expression of independent living.

CHAPTER 12

GENERALISTS AND SPECIALISTS

MANY OF US WHO CHOOSE THE PATH OF ENERGY INDEPENDENCE CAN offer plausible economic reasons, but at heart we follow a deeper imperative. We feel called to care for ourselves and our surroundings in the best possible way, keeping foremost in our minds the interests and birthright of our children and our children's grandchildren. We find we cannot resist the challenge of constructing and managing our own life-sustaining systems. We imagine that some of the greatest rewards come from consciously negotiated arrangements with the *genius loci* of the place we inhabit. We even believe we can improve on conventional ways of caring for the biosphere, the local ecosystem, and our family. And so we take up the fool's cudgel, and become jacks of all trades.

It normally takes a mob of specialists to develop a home site, starting with realtors, planners, and architects, then cat-skinners, back-hoe jockeys, formers, framers, plumbers, and electricians, until finally the finish carpenters, appliance installers, and landscapers wrap up the details. If all these tradespeople were to work together as a team on my house, with me as their captain, we would all benefit—but when has this been your experience with the planning and design apparatus or the building trades? Amory Lovins, who has identified twenty-five different development specialists, describes the building process as a relay race, not a team effort. But hold! This house is meant to shelter my family, keeping us safe in times of trouble, encouraging us to laugh and make love. Our home, the domestic constellation of family, shelter, garden, and land, is so close to the center of our lives that we ought to give nothing less than the best of our attention and the greatest part of our time and substance to its realization.

Renewable energy pioneers are often specialists, skilled at collecting knowledge and getting disparate technologies to cooperate. Many of us (especially the men) are unworldly eggheads who are willing, even eager, to spend a thousand dollars on a solution that will save a nickel in a lifetime, because we like technological elegance. We energy devotees strive tenaciously to widen our vision, and are insistently curious about the edges of our specialties, from which we may peer over into other fields of endeavor. And we share the fool's fearless willingness to attempt the impossible and unknown.

Whereas specialization tends to lump assumptions together and treat them as established facts in order to get to the product quickly, a more encompassing intelligence takes time to pick apart lumped assumptions, looking into the cracks between them for alternative solutions, missed details, and hidden opportunities.

Both specialist and generalist tendencies are manifest in the minds and practices of independent homesteaders. No matter how well disciplined in our original specialties, in our earth-oriented, home-centered lives we incline toward animism, ceremoniously attributing consciousness and intention to photons and rocks, rat snakes and people, storms and sunshine. Thinking about particles leads us to thinking about waves. Particles help us understand our place in the midst of things. Waves tell us about grander currents and movements.

Think, for instance, about particles of air: The warmer, more energetic ones bumble to the top of their pile over their less energetic neighbors, who sink. We can take advantage of this by encouraging convection patterns inside our homes, by closing doors or opening drapes, by helping the currents of particle motion work in ways that benefit us.

Driving a nail through a piece of wood, we are assisted in the effort if we visualize the waves of wood fibers straining to hold together while the nail strives to wedge them apart, the grain parting in advance of the driven nail-point. Drive a nail hastily too near the cut end of a board, and as surely as particles have a finite ability to stick together, the wood splits . . . and a nail, a piece of wood, and the time and energy it took to get them where they were meant to go is wasted.

Phyllis Lindley, a veteran independent homesteader, explains, "We like hand tools for most work. They take longer, but mistakes are smaller and easier to fix."

Independent homemakers delight in tools. The cultural bias that determines which tools girls and boys pick up is directly short-circuited in a homestead where the women have a crucial role in maintaining the energy systems, for example. Lafayette Young, master solar teacher and motorcycle mechanic, could scarcely contain his pride while telling me how his daughter knew without being told, and at quite a tender age, whether he needed a Phillips or a straight blade when asked for a screwdriver. We make great gains for ourselves and for our children when we unwrap the mystery and work with the nuts, bolts, buttons, and lights of technology. We learn that brute force is seldom required, and that mind, intention, patience, and the right tool will prevail. We learn that women can do anything men can do, and some other essential things besides. We learn that two heads and four hands can lighten tasks, and that ears and minds as well as mouths and hands must be engaged for serendipity to work its magic. Watching nature, we learn the joy of cooperation and the certitude that competition is seldom productive. We learn to take very good care of our fingers and our eyes, because we only get one set. All these intricate lessons combine to teach us to pay attention, to connect, to care just as passionately about the tiniest particles as we care about planet-sweeping waves.

For land and family: *Wes and Linda Edwards's story*

Linda and Wes Edwards have lived for a quarter century in an off-the-grid solar house they built on a heavily wooded ridge a couple of folds back from the Pacific Ocean near the hamlet of Ettersburg in southern Humboldt County, California. A daughter grew up there, and still lives in the community. These three and an extended family of apprentices help each other with their cottage industries—Linda packages mustard, and Wes's specialty is carefully crafted home energy systems. Once restless travelers, Wes and Linda now find themselves staying closer to home these days, and their land flourishes under their attention.

Wes: We were born and married in Rochester, New York. In the late 1960s we had a revelation, that we should be growing our own food, and having something to say about our environment. We thought we would like to build our own house, and if we were to move, we thought we might as well choose a better climate. We arrived in Pleasant Hill, near San Francisco, in 1970 with the idea that we'd like to move to the country, but when we told the realtors we wanted rural property, they laughed: It was very costly. So we started looking for property in our camper, mapped out California, and started our big adventure every weekend.

A friend of a friend had eighty acres in Humboldt County. We were trying to comprehend what eighty acres must feel like! We came up and looked at his house, and were taken by the area. It was a hot summer day, and he told us where to find the swimming hole. When we found it, there was a nude baseball game going on, and we met some nice people. We decided we'd found a wonderful place. We went into town and found a realtor who really seemed to listen. We told him we needed lots of water, a southern exposure, trees for firewood, and a site to build on. He immediately showed us this land. We didn't want to say we loved it too fast, but we camped here that night, and left our deposit with the realtor on our way back through town the next day.

We camped here every weekend, and built a ten-by-ten cabin with

We were awed by the land: It was so much more, and there was so much more of it, than we had ever imagined.

lumber scrounged weekday nights from the dumpsters in the city, and hauled in a trailer behind our camper. We worked with the land all weekend, trying to figure out how to use it. The winter of 1973 was a killer, one hundred twenty inches of rain; we're glad we didn't come up here to live until spring. We were awed by the land: It was so much more, and there was so much more of it, than we had ever imagined. We sold all our stuff, our appliances and conventional possessions, and ended up with all our remaining furnishings and belongings in a pile under a tarp. Our daughters were thirteen and fourteen when we moved. They'd really been a big part of our decision to move, because things weren't good for them where we were. I remember really late one night, we'd just finished a new roof on our city house, and Linda and I were sitting up there, at two-thirty in the morning, and this car drives up, a door opens and a couple of beer cans roll out, and then this lout staggers up the walkway. "Can I help you?" I ask him from the roof, which doesn't seem to strike him as too strange, because he says, "Can Patty come out?"

Moving to the country was a philosophical decision for us: getting away from neighbors in our faces, where we couldn't be private with our space. The land came first, and it came without power. At the time I remember thinking, I can deal with that! We found out that power is a survival issue.

We fenced three and a half acres so we could have a garden, increased the size of our natural pond, started studying house plans and doing electrical things.

Linda: I hated kerosene fumes, and we couldn't read with kerosene lighting.

Wes: Our first power source was the car. We'd drive to town and back, which charged an extra set of batteries, then plug the house into the car when we got home. In the early 1970s there wasn't much equipment available for remote living. We got a wind machine, set it up on the ridgetop, and moved batteries back and forth between the house and the wind machine.

Generator-based systems are expensive. The units are costly to begin with, and even with intensive (and expensive) maintenance, they have a finite life. Their real problem is that they consume fossil fuels at an alarming rate. I figured in 1986 that, all costs included, I was paying seventy-five cents a kilowatt-hour to run a generator. We only have to run ours in the fall, when it's cloudy and the hydro system hasn't started yet. We run it as little as possible, and when it's running we load it up with the battery charger, washer, dryer, and anything else we need to use. Running just to charge the batteries is not efficient.

The Edwards home makes the most of the sun.

When we first came here, our plan was to have a neighborhood school on the ridge, and we even poured the foundation. But our girls could already do what high school students needed to do. . . .

Linda: And there was so much for them to learn here, homesteading, how to grow things, build things. It worked very well.

Wes: At first, the girls acted like we were going to a foreign country: "Dad, we're not going to meet any boys."

We lost Patty, our older daughter, in 1976, to complications of cystic fibrosis. She was sixteen.

Linda: We took some time out then.

Wes: I was non-productive for several years. It felt like my heart was ripped out. That was a setback. . . .

My true love is being out in the field, installing systems. PVs: I love to handle them. What a concept! Sun falls on them, and out comes electricity. I get to work with wonderful people, houses, and equipment. There's not a day goes by but I say, and I get paid for this? I'm getting more people who have sold suburban homes and are ready to build serious solar electric systems. In five years, they'll certainly be competitive with the utilities.

With utility power, you pay for convenience. If you run your own system, you must get involved with it, learn about meters and maintenance. And you must know and stay within the limits of your system. It's like *Zen and the Art of Motorcycle Maintenance,* you've got to get involved with the machine. Otherwise, you'll be calling somebody like me a lot. I like to see the batteries clean, and the terminals greased. Of course you can pay to have the system serviced, but that's not the idea.

Everybody uses more power than they've got. I make sure every one of my systems has good metering. Here, we live (and die) by the amp-hour meter. And education has to be part of every installation: How many hours can be used before you run the generator, how do you equalize the batteries. Batteries are the weakest link, and the least understood, and they get abused because they're not understood. Install good metering, and believe what it tells you!

In my experience, it's the women who know how to listen and learn. Maybe men have a block about listening to technical instruction. If I get a sense a man's not listening, I explain to the woman. I get calls all the time, and I can't believe what I hear. The guy says, "The meter's broken; I didn't use that much electricity, just a few lights." I talk to his wife, and she tells me the TV was on all day, and this and that, and he spent 900 amp-hours. Hmmm, not bad, 900 amp-hours out of a 720-amp-hour battery bank. I work with her so she understands the batteries, charging, equalizing and all, and now I'm confident it will all work right.

The Information Grid

Popular electronic culture urges consumers toward a hands-off approach to the world. Lights turn on and off when people enter or leave a room, computers apprehend and transcribe our speech, the waters flows into the lavatory when we put our hands beneath the spout, and the toilet flushes when we walk away. TVs, stereos, VCRs, and even ceiling fans are always powered on so we can control them with "remotes." Are humans preparing to evolve out of their bodies, or emigrate from Earth? Independent homesteaders are torn between fascination with new technology and preference for hands-on, do-it-myself, manipulative and physical interaction with the world. A quarter century ago, some of the mind-tools and electronic advantages we cannot imagine living without today were barely conceivable. Yet we rejoice in our bodies and their attachment to the earth, and readily take our values and rhythms from nature.

As human particles in a large and crowded Gaian organism, it is surprisingly easy to disconnect from the brute force of grid-supplied electricity, even as its suppliers attempt to convince us that energy addiction is in our cultural bloodstream and distinguishes us from the animals. In scarcely a century, we have also become dependent on much broader and possibly more justifiable electronic grids: telephone, radio, television, and the internet. Of the nearly one hundred homes I visited in 1993, two were without telephones, by choice of their owners. Every other home, no matter how remote, was connected to the world by phone. I conversed with one independent homeowner who can only be reached at his island off the west coast of Vermont by marine radio-telephone; only one of us could talk at a time, but we communicated and exchanged intelligence, despite the inconvenience. Returning to visit in 1998, I found almost every homesteader connected to the internet. By then, cell phones were affordable, common, and service reached into places far beyond the reach of the telephone grid. Cordless telephones are now capable of transmitting more than five miles, a fivefold increase in five years. In 1993, many remote homes were connected to the telephonic universe with "farmer phone lines" that delivered party-line service over wires draped through the trees for miles. Under competitive pressures, rural phone systems have been upgraded, and the same wires now bring a dazzling range of caller-ID, forwarding, and central-office services to homes beyond the end of the pavement. Telephone service can be brought the extra miles more readily and inexpensively than electricity, because information can be transmitted so much more efficiently than power.

Work-at-Homers

Income is one grid that most of us cannot disconnect, and one common question of those contemplating the move to the country from a well-paid specialist's job in the city is "What can I do for a living." As soon as the telephone is connected, many homesteaders are able to conduct business as well or better from the edge of the grid than in an office cubicle miles from home. The telephone and its partners fax and modem enable independent homesteaders to "move electrons instead of protoplasm," as Amory Lovins says. Vacationers often wryly remark that they can perform their jobs from

the beach or poolside as well as from behind a desk, and work-at-homers merely carry this idea to its logical conclusion. Homemade electrons run sophisticated tools more reliably than those supplied by the grid. If our work requires a lab or workshop, we build them; we love to build. If work calls for collaborators, we expand our workspaces and sense of family to accommodate co-workers, or employ a new breed of teamwork tools developed to allow groups of workers to manage shared projects. Modern commerce demands more knowledge workers than ever before, and the internet enables us to tele-commute and collaborate without regard to distance. In a twinkling we can refer to resources halfway around the globe and exchange works-in-progress instantaneously with others in our work group, and so the computer and telephone join to become an essential tool for many who thrive under their own power.

Conventional houses are designed to tolerate neglect, to provide automatic comfort without the intelligent participation or even the presence of the inhabitants: barracks for an army of dispossessed workers who are gone half the time and exhausted the rest. Homes powered self-sufficiently reward our attentive presence: We open and close doors and windows to take advantage of changing breezes, adjust the photovoltaic array to catch the best sun, and work our gardens, and are rewarded with a more comfortable and efficient place to live. Our homely tasks fit seamlessly among our money-making chores, and the texture of living is richer. Instead of mall-crawling for lunch, I forage in the garden then walk behind my battery-powered lawnmower for a few minutes before starting the dinner soup. Independent homesteads perform optimally when attended, and I find myself leaving home reluctantly. Homes left alone while we commute to work almost seem to pine for us, like abandoned pets. My yen for independence from civilization and for rootedness to a place endows my home with a creative, familiar, and supportive atmosphere that I now find inextricably woven into my ability to create. Yet living independently runs so much at right angles to conventional patterns of employment that work-at-homers are forcefully encouraged to undertake innovative forms of work.

For an increasing number of workers, staying home simplifies lives and reduces economic needs. Time is one of the biggest gains: During the time it takes to dress for work and travel to the office, commuters must often employ others to perform tasks they might prefer to do themselves—child care, food gathering and preparation, as well as energy management. Commuters are subjected to the constant siren call of consumerism, and frequently compensate themselves for the indignities of commuting and working in a corporate environment, and for their time estranged from family, home, and hearth, by indulging in costly and wasteful "retail therapy." Work-at-homers avoid these and other inefficiencies because their homes provide entertainment and pleasure as well as gainful productivity. By investing ourselves, our time, and our savings in a home place and improved energy equipment, work tools, and work places that serve us better, we liberate more time and energy for ourselves, our families, and our community.

Shifting the workplace from the conventional office cubicle to a desk at home requires a dramatic change in direction: Bosses must be educated, tools improved, policies formulated, and we must each find our own best balance between homework and time in the presence of our office colleagues. Even as bosses and forward-looking companies encourage workers to "call it in," thereby saving on overhead costs at the office, most of us who work at home feel a need to spend some time face-to-face with our colleagues. Video-conferencing and other telepresence techniques can only partially substitute for "face time." Fortunately, we can approach a home-centered livelihood gradually as we make the home place more self-sufficient and fulfilling. Successful work-at-homers told me that their transitions began when they detected a faint but insistent urge to stay home, which led them to a long, fascinating inquiry into their lives, habits, talents, needs, and ambitions. Most started with the realization that too little of their time was their own, and that their accomplishments at work did little to enhance their world. Many off-the-gridders begin their quest when sickness, an accident, or a personal crisis forces them to recognize that no salary can compensate for the anxiety they feel about their unwitting contribution to the global despoliation resulting from a century of profligate energy mismanagement and corporate greed. For some, the response began with a few compact fluorescent bulbs, a recycling bin, a compost bucket, a course in sustainable living shared with a spouse, and rededication to love and family. For one friend, a single take-home project allowing her to "call in well" liberated enough time to launch a little kitchen garden and a long-postponed do-it-yourself solar experiment. At some point, the cumulative small changes combine, escape velocity is attained, and the independent adventure is launched.

A few discover that the sometimes solitary and intensely familiar way of the work-athomer is not as productive as the commuting life. Even in this extreme case, the dedicated commuter usually discovers that the focus once reserved for career and office broadens to include home and family, responsible practices become enjoyable habits, and consumptive patterns come under control.

Phone It In

Many of us are unable to make a living without leaving the homestead, and only a few have been able to settle within human-powered or mass-transit reach of work. And so many who live at the end of the road still commute. Most long-time off-the-grid commuters regret the rush-hour time they waste, and the wake of burnt dead dinosaur fumes they leave behind. Since phone, fax, and computer have made it possible to work effectively from any place with telephone service, why not leave the slippers on and phone it in?

For people who do their work with a computer, the biggest trend in the last five years of the 1990s was the rapid evolution of the internet. In 1993, the first edition of this book barely survived submission by floppy disk. Online resources were scarce, and email-enabled collaborators were few. Corporations and others who might require electronically transmitted services were afraid to grant off-site access to their systems for data acquisition, and few were capable of receiving electronically transmitted reports. The 1994 Northridge earthquake destroyed highway access between parts of Los Angeles, and nudged acceptance of telecommuting in that rush-hour-choked city—phone lines could be added far more quickly than freeway overpasses could be rebuilt—and this acceptance rippled across the country. By 1995, forward-looking executives were deciding that having a website and a corporate identity preceded by http:// was the ultimate in modern chic. After that, corporate resistance to telecommuting evaporated in most industries. Companies routinely justify the cost (and risk) of outfitting valued employees with home computers and telephone connections to the corporate mainframe by reckoning their savings in office overhead. Resource efficiencies and other newly recognized benefits motivate such schemes, and the transition away from commuter culture could be hastened by a nationwide program of tax breaks and other incentives for work-at-homers as well as employers. Some areas with air pollution problems, like the Los Angeles basin, already offer pollution credits to employers who "farm out" their work to work-at-homers. A trained staff is every corporation's most

Since phone, fax, and computer have made it possible to work effectively from any place with telephone service, why not leave the slippers on and phone it in?

valuable resource, and telecommuting now allows specialists with new babies, disabilities, or other compelling reasons to stay home to practice their professions from the end of the road. Thanks to the internet, remote data communications, in 1993 an arcane specialty, became in less than a decade a utility as transparent as the electric grid, and a thousand-fold more efficient.

Some of the people telling stories in this book have solved the remote income puzzle. In 1993, most of my off-the-grid families included one or more commuters; in 1999 most still speak longingly of the day they can work at home. Disconnecting is harder than it looks. Old-paradigm bosses still like to see bodies at desks. Certain tasks require personal interactions with co-workers or specialized equipment. Some of us are such social creatures that we wither without face-to-face interaction with customers and co-workers. Others, borderline agorophobics and hermits like myself, yearn to be home.

Most work-at-homers made the transition from office work to home work gradually. Changing their regimen from a five-day-per-week commute to three days, then two, allows for better adjustments on both ends. Family responsibilities and patterns may require rebalancing; in my household, contention between the cat and me over who gets the rocking chair on sunny winter afternoons continues to this day. Bosses may experience an easier withdrawal as they come to appreciate the increased creativity and productivity that work-at-homers generally experience. Many independent homesteaders are entrepreneurs and creative heads of companies who "escaped the office." David Katz told me after one extended weekend, "I should do this more often. I get five times as much done when I don't have to talk on the phone." Surrounded by the simplicity and quiet of their independent homesteads, they claimed renewed clarity and insight as well as the expected increase in productivity. They have founded new forms of decentralized industry, self-reliant in terms of energy and in other ways attentive to their possible environmental impacts. Their new ventures in some cases created alternative, dignified, and productive livelihoods for other at-home workers and cottage industrialists.

In parallel with the internet, the capability and affordability of computers has enabled a work-at-home revolution. In less than two decades, breakthroughs in silicon technology have touched almost every kind of work, and it is quite natural that many who harvest their power with silicon photovoltaics would also express an affinity for doing their work using silicon computer chips. In the last decade of the 1990s, power

Disconnecting is harder than it looks. Old-paradigm bosses still like to see bodies at desks. Certain tasks require personal interactions with co-workers or specialized equipment.

generation and computer technology continued to converge. In 1997 PV modules appeared on the market fitted with inverters that immediately change sunlight into 120 volts AC onboard the module and are directly connected to the household distribution circuitry; they can report their productivity and status through polling serial interconnections which enable the power-harvesting equipment to monitor and adapt to local demand, storage status, and surplus energy opportunities. Power is increasingly interlinked and integrated with information equipment into the "intelligent home's" neural network.

It is hard for me to contain my enthusiasm for these breakthroughs, interconnections, and synergies. As an aging author, I find myself perpetually flailing for a fact or word. Just in the revision of this chapter, I have sought and speedily found two facts through the internet, and an elusive synonym using the thesaurus built into my word processor. In the time since this book was first published, I have seen my editor, publisher, and indeed most of the people I work with, testing, hesitating, then plunging into the electronic maelstrom.

I sometimes wonder to what extent I over-computerize my work. The skills I apply when making energy efficiency decisions help me with this speculation. Given that my time is the precious resource, and accomplishment of the task the only appreciable benefit, will I achieve a task better, quicker, more efficiently with the computer or with another means, perhaps pencil and foolscap or a walk around the yard behind the solar-powered lawnmower? Will an email—or, worse, an email storm—be more efficient than a conference call or, horrors! a face-to-face meeting at the office or over lunch at the mall?

In the home office, efficient ergonomics are as important as in the OSHA-scrutinized workplace. There is no question that workers in the conventional workplace are subjected to stresses and abuses directly attributable to the overpowering presence of computers: carpal tunnel syndrome, eye problems, exposure to ionizing electromagnetic radiation. Monetizing these externalities is as elusive, and as important, as monetizing the externalities of coal-fired thermo-electrical generation. As we at-home workers spend more time seated at our computer workstations, we run the same risks as our desk-bound office colleagues. As expected, I saw innovative solutions in independent home offices: standing desks, decentralized work centers, portable workspaces, low-energy-consumption equipment, including the use of powerful laptops as replacements for desktop computers. Off-the-gridders know how to build ourselves into something comfortable.

We developed-worlders are an enthusiastic and impatient people. We embrace new technology, the car, the telephone, the calculator, the fax machine, the internet, with childlike wonder, and too often without bringing our critical facilities wholly to bear on problems, adjustments, and dislocations caused by the new technology. Like any of these powerful tools, the computer brings with it unexpected dangers and limitations. Once again, the benign quality of the independent home buffers us from the worst kinds of exposure: When I am at a loss for a word or an idea, or am just feeling a little too intense, it is easier for me to break out, walk outside and talk to my neighbors, wash dishes, mow the jungle, or otherwise unblock myself than it would be at work.

Wild West: Wayne and Debbie Robertson's story

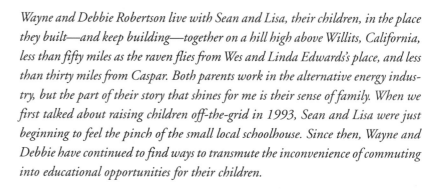

Wayne and Debbie Robertson live with Sean and Lisa, their children, in the place they built—and keep building—together on a hill high above Willits, California, less than fifty miles as the raven flies from Wes and Linda Edwards's place, and less than thirty miles from Caspar. Both parents work in the alternative energy industry, but the part of their story that shines for me is their sense of family. When we first talked about raising children off-the-grid in 1993, Sean and Lisa were just beginning to feel the pinch of the small local schoolhouse. Since then, Wayne and Debbie have continued to find ways to transmute the inconvenience of commuting into educational opportunities for their children.

Wayne: We got tired of working all week so we could afford to get away on weekends from where we lived—why not live where we wanted to be? I'm an anarchist: It's a threat to have my energy come to me at someone else's pleasure. I see Igor with his hand on this big switch.

We produce the energy we need, and there's a beauty to that. If I had to pay a utility bill, I'd feel punished.

Debbie: If I'm missing any modern conveniences, I've been living off-the-grid so long, I don't know what they are. . . .

Wayne: We have everything we need. Suppliers send me alternative energy equipment to test here, and I try to destroy everything I get. That helps the technology improve. A remote home is the toughest test there is

The Robertson home.

for electrical equipment, because there's no pattern to the loads. When we were building, we ran four worm-drive saws off PV using a Trace inverter. My friends, who were building with me, loved working without the noise of a generator. We've got all the modern electronic kitchen equipment, including a microwave. You should see the surge when the dryer starts up! Maybe we could use a freezer, except we're not big carnivores.

Debbie: But there's all that tomato sauce in the refrigerator's freezer; it'd be nice to have a better place for that.

Wayne: Here on Third Gate Road, there are no powerlines. On the other roads, the people that got plugged in have their freezers and refrigerators and other power suckers. If everybody generated their own power, they'd be acutely aware of power usage. We're more in tune with natural rhythms this way.

Debbie: Our kids were born out here, and they wonder why anyone wouldn't want to use free electricity.

Wayne: I remember walking with Sean in town when he was about five, and he pointed up to a power pole and said, "What's that, Dad?" I told him people get their electricity from it, and he asked, "You mean, most people don't use PV?"

Debbie: Sean and Lisa are great recyclers, always trying to figure out ways to reuse things instead of throwing them away. Their schoolmates are all off-the-gridders, too, and even the school was PV-powered until recently, so they've never known another way.

Wayne: They're used to outlaw housing with voodoo electricity. They are mostly aware of energy conservation, and usually turn off the power when they leave a room.

I expect this house will be finished when I die. I've got to have something to do with my hands, and my eyes are always seeing things that could work better or be better made. Before we start a project, we get all the books we can, and research it, and I visualize it to the end. Plans help us get started, and I talk to the County guys, but we change things as we go along, and always see better ways to go, and ways to go farther.

When we left the city, I took my retirement and put it into this house. We heat it on less than a cord of wood a year because it's so well insulated. I'm a perfectionist about foaming around the windows and sealing cracks and holes. I even foam around the outlet boxes, so this house is tight. I plan to grow old in this house.

It takes half an hour to drive to town, and so I'm aiming to move my work up here, and only go down the hill once or twice a week. I do all my work by phone anyway: With a laptop and a phone I can work anywhere.

Debbie: As long as the kids are here, one of us will probably commute daily for them or for work. There aren't many other kids around, so ours are real resourceful about finding things to do.

Wayne: When we aren't torturing them. [The kids giggle; they don't look very intimidated.] They've got the pond, an ever-evolving tree house, and the woods.

Debbie: After school they get plenty of time with other kids before we pick them up. As they get older, I can see that will become a problem; it already is for other families back here.

Wayne: We call this area the Wild West. People who live back here are real different from each other. The year-rounders have learned how to live here, but the springtime residents have no regard for the land. They just come to rip it off.

The Robertsons.

The more people who come, the more full-timers there'll be, and the better it will be. Right now, the people back here don't even pay their road assessment, so I had to get on the Road Committee to get any work done. Up until recently, the road ate a car a year.

Debbie: We've expanded our power system as our power needs have increased, i.e., dishwasher, three computers now, washer, dryer, microwave, etcetera. We've changed from a 12-volt system to a 24-volt system. We've added a new power center, a sine wave inverter, a larger battery bank, a new generator, and a new wind charger since we talked in 1993.

Once a week one of us cleans off the modules. During the winter we run the generator about once a week to charge the batteries, which requires regular gas, oil, and maintenance.

We also have a new solar pump on the water system, which has been a great addition, works well, and is very convenient. After about three weeks of cloudy, rainy weather we need to run the AC pump to refill the water tank.

We've constructed a shop with a guest house on the second level, and are converting the old tool shed to a stand-alone office. Both new dwellings will be powered by their own solar systems, but they are not installed yet.

We'd never want to be on the grid!

Our ranch road is still in bad condition.

Wayne: The perfect car would be a Hummer (Humvee), or any four-wheel-drive vehicle.

Debbie: Sean has been at the Instilling Virtue School at the City of Ten Thousand Buddhas in Ukiah for four years now, and Lisa started junior high (7th grade) there this year. We are happy with the academic education they get there, the vegetarian meals, and the exposure to eastern culture and religion. The school lacks elective studies, athletics, and a music program. We commute an hour each way, and it works well only because I commute to work close by five days each week. The kids have managed to maintain their social life in Willits through soccer, swimming, and dance activities, so they don't lack for social stimulation away from their school friends in Ukiah.

Looking back, there are several things we would have done differently. We would have installed in-floor heating and a central vacuum system. We're still planning to add covered porches and walkways. Dirt, mud, and dust are a constant battle because of country life, pets, and building projects. We plan on giving more consideration to lighting and rock on pathways, and a covered car port.

Sean: I like the quiet, the peace, and that it isn't crowded where we live. I wouldn't like to live in town or in a big city because it is too crowded, noisy, and polluted, and I couldn't run around and play in open spaces. It seems like the only things to do in the city are to watch the boob tube or go shopping. I think living off-the-grid is cool, but sometimes there isn't enough power to use the computer, which kinda stinks. I don't like filling the generator with gas either. I like having trees around and clean air. It stinks that I have to travel one hour each way to go to school, and have to get up at 5:30 AM, but it is worth it for a good, challenging education.

Lisa: I like where we live because of the privacy we have. I also like to see the wildlife around us. I like making forts in the woods, and swimming in our pond in the summer. I wish there were more neighbors so I'd have friends to play with, but I don't want to live in a crowded neighborhood, so I guess the trade-off is worth it. I also like living here because I can have lots of pets. My dog is a golden retriever named Max, and I also have two rabbits, a cat, and a bird.

Wayne: Here's some advice: Install the best windows you can afford. We couldn't afford much when we started, so we are struggling with some windows that won't open, some that fog up between the panes, and some that leak air and rain. Someday we'll replace them, but it might be quite a while.

Debbie: Last year we had a "too close for comfort" fire where a neighbor's house burned to the ground! So we've installed sprinklers on our deck, and have done quite extensive clearing around our house. Also, I've

Looking back . . . We would have installed in-floor heating and a central vacuum system. We're still planning to add covered porches and walkways. Dirt, mud, and dust are a constant battle . . .

been dealing with Lyme Disease for about two years. Out in the country, you have to be careful about ticks! Check yourselves after being in the woods or handling your pets, and use tweezers to remove ticks, use tick collars on pets, and do everything you can not to get Lyme Disease!

Garbage

We are a nation of garbage specialists: we make garbage better than we do almost anything else. Taking proper care of what we discard is an inescapable part of taking care of ourselves. Garbage, like the dependent home, is a modern invention, and a phenomenon without parallel in nature. Only upstart humankind has perfected the art of making waste so disgusting that it cannot serve as food for other organisms. With wonder in our voices, we tell our children that the plains Indians used every part of the buffalo. Of course they did! Every part was useful. In one century, humans have evolved a garbage culture that has converted the Earth's plenty into a suffocating blanket that offers to bury the planet in pollution and waste. A few decades hence, when our grandchildren take us to task, independent homesteaders want to be able to say, "Not us; we were pioneers, trying to figure out how to do it right!" As each community's Mount Trashmore looms, market pressure—the rising cost of dumping—has made us more careful about our garbage. We pay better attention to unlumping our garbage: separating trash from materials that may be recycled. The more finely we practice unlumping, the less energy is required to return the refuse to its proper place in the natural order. As a people, we are slowly beginning to get control of our trash habits, and learning to reduce, reuse, and recycle. The

waning of the garbage culture proves that when motivated, we are an adaptable and thrifty people.

Anyone who has tried to find something mistakenly thrown in the garbage understands the problem; the motley accumulation in the bin shares only one quality: Someone judged them to be of no immediate use. Acculturated from birth to loathe our offal, we compound absent-minded reflex with a fetish: Once devalued by consignment to the trash, a discarded item is somehow sullied, and it is considered indecent to salvage it; polite people simply don't go through the trash, their own or anyone else's. So the reflex to trash something is perceived as irreversible.

Independent homesteaders regard "essential" garbage appliances with suspicion. Garbage disposals are a bad idea, adding energy to compost before sending it to the sewer or septic tank; this is like throwing good money after

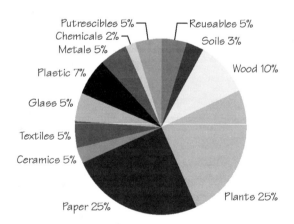

Most American garbage comes from industrial sources, but our household garbage stream is also outrageous. Putrescibles are items (such as meat) that putrefy. The biggest problem is that all these different kinds of garbage get mixed together. This pie is based on Berkeley, California's garbage, but is typical of other cities.

good. The garden needs the compost. Trash compactors were a marginally good idea until garbage trucks, dumps, and refuse transfer stations started applying much more powerful and efficient ones to the waste stream. Power mulchers accomplish immediate results, but patient composting in a garden mulch pile works as well and saves energy.

Our garbage waste problem and our energy waste problem are close kin; easily 75% of what most folks throw away can be turned to better use on the independent homestead.

— Start out by not buying trash or not bringing trash home.
— Next, for those of us who practice unlumping, it is a natural next step to separate and collect discards appropriately. A quick count of recycling sites around our homestead finds twenty-four caches, from "battery hell" where the nicads go when they die, to five distinct paper recycling bins.
— Finally, by reusing the materials we stockpile, or by passing our gleanings back to industries that can profitably reclaim their precious elements, we further the reform of attitudes about the reclamation and redemption of garbage.

The Unburn Pile

As a nation, the United States of America produces more yard waste than any other kind of garbage. We are fascinated by fire, and over the centuries have learned to use it wisely, for ceremony, warmth, and cooking. Around the construction site and the land, we have also learned to abuse fire. Laws forbidding agricultural fires in the U.S. and elsewhere are finally affecting the turf, sugarcane, tree, and rice farming industries, to the benefit of air-breathing creatures around the world. On the domestic scale, burning the backyard trash pile is one of autumn's pervasive rites . . . but it is wrong. Why burn? The rapid oxidation of all that organic material has no purpose beyond our cultural preference for bare earth. Of course, we cannot help it if a little plastic, a little painted wood, and who knows what else happens to have been dumped in the pile.

In Caspar, I have made the decision to stop burning the piles, but I keep making them. The traditional pile where we carry the cuttings and slash has not been lit for six years now, and, surprisingly, it is no larger than it has ever been. The slower oxidation fomented by the earth's microorganisms is less spectacular, but it keeps up with the garden. Unexpectedly, birds and other small creatures have developed a thriving community in our unburn pile, and fuchsia cuttings have colonized the pile's north side. Nature has begun to reclaim her basic building materials. We are more disciplined now about what goes on the pile: nothing human-made, nor even things shaped by humans, such as lumber and cardboard; these things are burned in the woodstove for heat, along with orange peels, the few putrescibles we produce, and other things that do not belong in a healthy compost pile. In time, a decade or two hence, we may abandon this pile and start another one elsewhere. A few months or years thereafter, we will have a rich mound where there might have been nothing more than ashes and dead earthworms.

Our fascination with fire has led to the industrial-scale employment of giant regional incinerators. In practice, these monsters seldom work as well as in the lab, and incinerators require large amounts of material to keep running. In Dade County, which includes Miami, and on Long Island, which incorporates the thickly settled suburbs of New York City, incinerator operators routinely purchase recyclable garbage from neighboring communities to keep their high-tech fires burning. In theory, incinerators can dispose safely of toxics, including chemical and biological weapons, but few communities would favor having these substances imported from afar and burned with questionable results, even if the cogeneration contribution of the incinerator to the local electrical utility's energy mix substantially reduced their electric rates. A parallel lesson has been learned since medical waste started washing ashore along the eastern seaboard in 1994: In the not too distant past, garbage was heedlessly flung into the ocean, which was assumed to be infinite. This idea being disproved, profiteers would now dump in the atmosphere, or even in outer space. Anyone with an elementary grasp on the meaning of sustainability would say this is a terrible (but typical) notion.

The Pendulum Returns

After half a century of intensive specialization, there seems to be a wide recognition that we need to take a systematically wider view of our vocation, academic disciplines, and fields of narrow expertise. Independent homesteaders, who have happily taken up the tools of dozens of trades, are pioneers in this move to broaden our understanding of the world we inhabit. Communities of these hardy pioneers and settlers around the world are repositories of polymaths and generalists who are ready to move the specialized world toward its new renaissance.

CHAPTER 13

SOLAR NEIGHBORHOODS

Dᴜʀɪɴɢ ᴛʜᴇ 1990ѕ ᴍᴀɴʏ Aᴍᴇʀɪᴄᴀɴѕ ʀᴇᴄᴏɢɴɪᴢᴇᴅ ᴛʜᴇʏ ʜᴀᴅ ʟᴏѕᴛ their communities, and started desperately trying to find them again. At the first EcoTech, an international conference on ecology and technology held in Monterey, California, in November 1991, questions about neighborhood and community unexpectedly engaged us all in fervent debate. Is a virtual community, such as a discussion group on the internet, really a community? Are there fences in real communities? Are ownership and community mutually exclusive? We could agree on little more than this: Neighborhoods and communities, at least in the United States, are seriously at risk, and something should be done by somebody, soon. By decade's end, this thread figures in the conversation whenever neighbors gather.

Like other social institutions in the late twentieth century (marriage, parenting, and family, to name an important few), community life has been obscured by economics and our ostensible need to survive and succeed. For most Americans, the call to neighborliness has been drowned out by the demands of career and the drone of TV. We chant "think globally, act locally" when we are overwhelmed by our insignificance in the face of regional, national, and global events. Yet we are so busy reacting we seldom have time to act at all. Traditional communities and neighborhoods have fallen into disarray because they require time, intention, and effort on the part of their members. The bunker mentality found in gated communities and neighborhood crime-watch programs serve a negative and fearful purpose far short of the friendliness and mutual support we want to find in small-town America.

Meanwhile, and not coincidentally, non-traditional communities, often incorporating global membership, are emerging and thriving. As you might expect, independent homesteaders, having stepped out of the conventional definitions of home life, are finding gainful ways to re-explore the meanings and functions of communities, as well.

Cultural landscape: Arnie and Maria Valdez's story

Arnold Valdez and Maria Mondragon-Valdez are seventh generation residents of Colorado's San Luis Valley. They are raising five children in the passive solar adobe home that they designed and built to demonstrate to their community new ways of using time-honored structures and techniques. Both Maria and Arnold are restless questers after new knowledge, but are also dedicated to bringing their new learning back to benefit their neighbors. They promote sustainable, traditional, energy-efficient design to help their neighbors thrive in Colorado's poorest county. The first part of their story was told in 1993:

Maria: Living in a solar house, you're like a cat—you track the sun. The kids follow the comfort: In the summer they spend more time downstairs, where it's cooler, and in winter, upstairs. You should have choices like that, not be trapped like in a house trailer.

I do my laundry according to the weather. If it's stormy, that's my break. And we take our baths at certain times. . . .

Arnie: That's because our solar hot water is hottest in the late afternoons.

Our children are very important to us. We have five. One has finished high school and is working and going part-time to college, then we have three in school, and an infant nineteen months.

Maria: Only one was born in the hospital. The rest were born here: It's a way of bonding to the house. We like the sense of place we have here. People don't have that when they live in the standardized housing in a city. People move too much, and when they do, they lose their culture. You get attached to land when you're young, and if you're always moving, you lose those values.

Our families have been in the San Luis Valley since 1852, and were in Arroyo Hondo in the late 1700s, so we have a cultural landscape and agricultural customs that are deeply rooted.

Arnie: We try to instill our values and sense of place in our children, but you can't force it. A couple of our children won't leave, but some of them will. Hopefully they'll all return, even if only to retire. We went out for schooling, and saw a bit of the world, and then we came back. There's not much here to support advanced education, and so you have to create your own work, become self-employed, or fall back to doing what your people did before.

I work in the adobe tradition; this house was the first new adobe in ten years when we built it, and it reawakened a traditional style of building. Both of us had ancestors involved in building Fort Massachusetts, and then Fort Garland, the northernmost outposts of the federal War Department in this part of the West. I'm coming around the circle, doing the same thing

Arnie and Maria Valdez work together in their greenhouse— the south wall of their home.

they did, but now I'm educating people and designing buildings: one evolution higher than my ancestors, but not different.

Maria: My great-great-grandfather—maybe I need to say one more great, I don't know for sure—was the first Hispanic to get a military contract for firewood. My dad has a propane business which my brother is taking over. So you see, my family has been in the energy business for a long time.

Our energy crisis started here in the late 1950s. Until then, we gathered firewood in the forests above the town, but the forest areas were closed to common use when a new owner purchased the land in the 1960s, and our traditional source was denied to us. Costilla County (*costilla* means rib) is considered the Appalachia of the West, one of the hundred poorest counties in the United States, and so the loss of that land for hunting, grazing, and wood gathering was very hard on our people.

Arnie: So now we turn to the sun for the energy we lost. We work with solar and environmental advocacy, working to develop regional self-reliance. And so our work continues to help us bond with the land.

Our resource center, the Peoples Alternative Energy Service, is a statement of energy self-sufficiency. From it, we do our community work, encouraging low-cost, energy-saving systems and advocating environmental issues. Because it's small and the overhead so low, and we're tenacious and committed, we've survived the Reagan-Bush years, during which so many solar organizations failed. Now we have momentum, and hope for a better future.

Maria: That was the building we lived in just after our marriage; then it was very cold, and we didn't know any better. When we built our house, we had better ideas about building solar,

and it was reasonable to use power from the powerlines for some appliances and light.

Arnie: Financing this house was a struggle, because it was adobe, and experimental. The Farmer's Home Administration stopped the funds once, and it was only because our senator intervened that we could continue. By doing so much of the work ourselves, keeping to the elemental concept that we were building shelter, and using found materials, we were able to keep the cost down to about ten dollars a square foot.

We originally had a 2-kilowatt wind generator and too few photovoltaic modules. Our system is connected to the grid so we can sell any extra electricity. The wind generator ran well for a few years, but the wind is so destructive, it sheered off the yaw column in one gust. Maintenance was tough because I had to climb the tower. We're in a marginal wind regime, and we got a wind machine meant for ten or more miles per hour; a unit meant for a slower regime might work better. We'd like to sell it, and enlarge our greenhouse instead.

In our community work, we teach from the resource center and our house, and use them as models and a laboratory. We make mistakes first here, on ourselves, and then we feel confident we can help people avoid similar pitfalls.

We plan to change the greenhouse to vertical glass, because it is more efficient, and will provide better space for us and the plants. We like the solar contribution and the diffuseness of the light, but the sloped glass must have the snow and dust swept off constantly.

State-of-the-art solutions are seldom suitable for low-cost applications, and so in the systems we design, the owner has to commit to maintenance. Because we work in the San Luis Valley with a heavy emphasis on small installations for low-income people in traditional communities, much of the best information comes from the Third World, where we work hard to keep up our network. The high-end information and applications are very interesting, but the best low-cost solutions are more common in Third World nations. It's difficult to connect, of course, because it takes money to get to the other communities and learn.

Maria: Because this is a poor county, outsiders think they can bring in their projects without asking and without opposition. Right now we are fighting a Texas-financed goldmine scheme which would spew cyanide and heavy metals into our water. We're citizen enforcers; we helped catch four noncompliant events, resulting in a permit violation fine and increased environmental monitoring. We fear the buildup of an industry like mining, because it would bring with it quick infrastructure, and spoil the local culture which has been building itself for so long.

Arnie: Of course, the community is split. They can all see that it might mean the end of the dream, but some see jobs, and think that's more important.

Maria: This issue, of holding the community together in the face of economic stress, is not so easy as solar.

Arnie: As soon as jobs are involved, the issues get cloudy.

Maria: This is the oldest community in Colorado. We're trying to keep the community intact, because it has a rare and endangered cultural landscape.

Arnie: Which needs clean water, clean air, and clean economic solutions to survive.

Maria: Who would want to build with adobe if it had poison in it?

Arnie: We believe that building security through a sustainable lifestyle is much surer than social security, which builds dependence. In San Luis, the Church is an important part of that life, and I'm looking for ways to work with the Church. We have tried to work with universities and government, but they send us specialists who are insensitive to the community's needs. That's why our "town square" is a checkerboard of paths to nowhere. HUD introduced energy-inefficient substandard housing in the 1970s: two-by-four frame constructions with quarter-inch plywood roofs, no insulation. Early in the 1980s, DOE put an active solar system on our museum, which is too complicated for us to fix, and so it's broken. The Church is much more a part of our community.

We have built an environmental center, and we plan to build a chapel on the hill above town, but this is a very fragile environment, and looking out over the whole San Luis area. . . .

Maria: We're afraid we're shooting ourselves in the foot by creating such

Because we work in the San Luis Valley with a heavy emphasis on small installations for low-income people in traditional communities, much of the best information comes from the Third World.

a spiritual center. Will people come, fall in love with the valley, and come back to buy, build, and gentrify our community? We don't want to market our cultural landscape. We want to maintain our cultural enclaves, because they are our treasures.

Architecture and buildings have an important role, because people respond to symbols. We are trying to come to grips with architecturally motivated tourism. There should be sacred space around cultural enclaves, and the cultures respected as national treasures; we can't tolerate resort development, subdivisions, mini-malls, and fast food chains.

Arnie: There's no building or land use planning code in Costilla County, and the only limit is the Farmer's Home Administration, which is the banker for low-income community homebuilders.

Maria: We'll be okay for a while, but regulation is coming. We hope for something enlightened, like Davis, California's alternative building code. As the oldest community in Colorado, we have to work to protect what we have, and the ambiance is working against us. We are inundated by people fleeing their cities. Aspen has already become a pseudo-culture of second, third, fifth homes. In Santa Fe, inexpensive homes are now celotex boxes covered to look like adobe, and adobe is only for the wealthy. We are too proud to become the residential sacrifice area for the whole United States.

Arnie: Malcolm "Capitalist Tool" Forbes's development, Trinchera Ranch, is a sample of this kind of profiteering: a huge site subdivided with no sensitivity to the land, eight hundred miles of roads leading nowhere, a jetport and conference center for the elite. When urban subdivisions are transplanted to a mountain setting, you cause plenty of trouble. Now erosion and wildlife problems make it clear what a bad idea it was.

Maria: Without a sense of vernacular architecture, for which purpose, and for whom the subdivision was made. . . . It's a crime against nature and the community to develop like this. Will buyers, if they are lucky enough to buy a buildable lot, ask for help from people who know the land and weather?

Arnie: That's why our parish built the environmental center first even before the chapel, on the hill just above town: a simple, stand-alone structure where we'll offer an environmental view of the community. We hope it will merge ecology and spiritualism, and reinstill the environmental aspect of living with spiritualism. We hope visitors will get a sense of the environment, and maybe they'll leave with a sense of respect.

1999 Update from Arnie and Maria:

Maria: The gold mine did happen, but they had nothing but trouble, and we were vigilant. They had a permit saying they'd limit cyanide in their holding pond to about 4 parts per million, and we caught them with a large migratory bird kill and about 700 parts per million in the pond, and we busted them. They destroyed the evidence, we think by shooting the dead birds so full of lead they sank to the bottom. But they got a hefty fine and had to install a big detox plant. They left because the gold was contaminated with copper, which caused their equipment to fail. They've gone, but it's not a done deal yet, because it's hard to restore land and trees at high altitudes.

Up at Trinchera Ranch, Forbes was clever: By selling lots to small absentee owners, he made it so that there were too many people to challenge. He's even got a mausoleum up there now.

We have a new environmental crisis—I guess we're always in the right place at the wrong time. We now have high-impact commercial logging on Taylor Ranch, which is the whole center of Costilla County, up on the Crystal Range. Zachary Taylor was one of their earlier expansionist forebearers; I believe he was important in taking the U.S. to war against Mexico.* This one family with a history of abuse monopolizes the fragile uplands, 77,000 acres. The ownership of the land has been disputed since 1960. Their rhetoric is "sustained use," but their actions show they're here to take everything they can as quickly as possible. Colorado is very backward about land use and environmental protection, and it's difficult to keep property owners from despoiling their property. The Taylors have nothing to do with the land; they live in North Carolina. They don't care about the land, only the money they can grub out of it. Most of the wood goes to pulp; over the years the good trees have gone into local homes, so the loggers are taking high-altitude trees which really aren't good enough for timber. The logging operation tears up massive amounts of landscape for small amounts of profit. Where the mining company was careful to employ locals, logging brings in workers from the South and Northwest. Many of us say "Thank God" because we don't want our people involved in what they're doing. The contractors aren't unionized, and they use independents who don't have the right kind of

equipment. Working conditions for the loggers are terrible.

Our watershed is at risk, and the community is angry. The downed firewood our people used to gather sustainably is now getting ripped off the land while it's still alive by these absentee profiteers. The impacts on erosion, watershed, and environment are stunning. Since there are no pulp mills nearby, huge trucks rumble through day and night, and even though the county has asked them to follow certain routes, they drive right through neighborhoods, and some of the older historic adobes are being destabilized. It's hard for the community to respond meaningfully, because it's working class against working class, and the real perpetrators are hiding in their enclave in North Carolina counting their money while the underclasses fight among themselves.

Arnie has taken the job of Costilla County land use administrator. Before that, we were one of three Colorado counties without a land use policy. On his second try, Arnie has won a Loeb fellowship to Harvard. Talk about the hoops you have to jump through! He had to do seven interviews! They asked him what kind of people really make him angry, and he told them "Eastern professors who know everything and have clean hands."

We graduated our son Iein from Williams College, and he's working in Georgia now. Our other boy, Airam, graduated from high school last year and is in Adams State College in Alamosa. Soleste just turned seventeen, Atsa is thirteen, and little Maya is almost eight. That might seem like too many children, but our relatives live right down the road, and we're an extended-family kind of community.

Arnie: The last two years as land use and zoning director have been hard. The enforce-

---

*Maria is being gentle. General Zachary Taylor led the U.S. Army into Mexico and provoked a war in 1845, then rode his "war hero" status into the White House. The bullying gene apparently continues to run true in the Taylor family. —mp

ment part is a real challenge—most people don't volunteer or come in. We've sued the Taylor Ranch to stop the logging, and that will be the ultimate test of the land use code. Overall it's working, but the one big crack is enforcement against the Taylors. Our experience with Trinchera has been more open and cooperative. It's been important to me to be part of the formative process, but now I'm looking forward to finding someone to continue the county work while I go to Harvard for the Loeb fellowship. After two semesters of advanced environmental studies at the Graduate School of Design, I think I'll want to return to architecture.

Maria: We're remodeling our house. We will keep the old adobe framework, but there will be more glazing. We are trying to get away from the use of wood and fossil fuels, trying to make a "greenhouse house." We want to show that you can live even more comfortably by being more solar-reliant, with more solar growing space. There will be an indoor fountain to help keep the humidity up in the house. Right now the house is torn apart, and you can see that it's getting to be a megastructure. We like the idea of preserving tradition in a modern context. We're using recycled metal instead of wood, giving the house a post-modern front while making it more solar. No way we're going to use wood with the work we're doing against the Taylor ranch. Every time we see a logging truck, we get a clear message: We need to be even more independent and self-reliant. It makes no sense to us to build with lumber, but how much less sense does it make to take these trees for cardboard boxes?

Arnie: The house is all framed out and roofed. I have to put the 16'x32' wall of glass on the south, which will give us a two story sunspace with an overlooking balcony from the second floor into the greenhouse. The roof integrates with the greenhouse, and the glass wall is vertical not slanted the way it was before. In the end walls, I'm integrating cement block, adobe, and recycled glass block, and it's made a nice merger.

Maria: Here in San Luis we're doing preservation architecture with the Colorado Historical Society, finishing up an inventory of the entire village of San Luis, 150 properties, describing the historic context of historical vernacular Hispanic architecture in Costilla County. By defining the cultural landscape we take the first step toward protecting important traditions.

I'm still working on my doctorate, slow but sure, and helping Arnie with the county stuff—they're getting two for the price of one. I came up with a dissertation topic that addresses the way traditions arise in a small community. I'm interested in the way organic scholars emerge in the community because it needs to come up with its own solutions, not be more marginalized by being told what it should do by outside experts. My subject is hegemony and cultural dominance, and their effect on organic scholars. It may take a long time for a village to produce its own intellects. . . . we come out in all different ways. Every village has their organic intellects, but where do they go? You can't live in a subsistence base in the U.S. anymore unless you're subsidized, and if you're subsidized you are forced to give up your values. I know it sounds corny, but our work is all about fighting to maintain the spirit of place. Preserving the values and people of little communities like San Luis is important because that's where the country's diversity comes from. ⸺

Bonding with the Land

The North American continent's first human inhabitants enjoyed an elaborate and caring relationship with every aspect of their environment, even the rocks and shadows, as well as the animals and each other. A deer was taken only when the food was needed by the clan and could be spared by the environment, and only as a gift; the hunter and those who ate the meat made a conscious effort to explain their need to the spirit of the land, and begged for continuing tolerance. This heightened sense of humble interdependence fell before the rifles and plows of pioneering Europeans whose culture encouraged a taker's view of the environment. Believing themselves to be God's chosen, and that the land had been created to provide for them, they wrested a living as long as the land could provide.

The American land is powerful and still untamed, and human claims to dominion are shallow. As we mature, many of us experience the onset of a deep, earth-oriented conservatism, which turns some of us into animists and worshippers of nature. And yet, the greediest and most denatured among us continue to spend entire lifetimes unaware of these restorative influences.

My children and their generation are the first to have lived their whole lives with awareness that the host planet has a finite carrying capacity. While some of these children imitate the insouciant greed of previous generations, others have preserved the child's sensitivity to other life, and a resistance to the short-lived benefits of avarice. Are they the beginning of a new race of American natives? These young ones struggle not to be angry at those who have gone before and who have raped and pillaged their heritage, bequeathing a darkened prospect of hard work with little luxury. Sometimes they fail in this struggle, and their anger

My children and their generation are the first to have lived their whole lives with awareness that the host planet has a finite carrying capacity.

⸺

shows in their music, mores, and mistrust of their elders. Children born of off-the-grid parents, who have suckled energy consciousness with their mother's milk, start with an advantage in addition to their generally happy upbringings: For them, conservation is an ingrained habit. We are lucky, we older independent homesteaders, that these sensitized young ones will be caring for us and the planet in our dotage. It is important that we do all in our power to help them pull the rabbit out of the hat.

Land for the cost of a car: Nancy Hensley's story

Nancy Hensley lives with her son in an off-the-grid home she built with her brother's help in the wooded hills of Greenfield Ranch, an "unintentional community" of pioneers and settlers who found affordable land and built off-the-grid homes in the lightly oak-forested grassy uplands above Ukiah, California.

This story starts in about 1969 when my cousin was working on a dream he had of making it possible for good people to get good land in the country for the price of a new Volkswagen. His research turned up the Greenfield Ranch, a beautiful 5,300-acre sheep and cattle ranch that wasn't making enough money to pay its taxes. With lots of help from friends who wanted to be a part of his dream, he was able to purchase the ranch. He divided it into smaller parcels, most of which have the essentials for country life: water, access, and a building site.

The Ranch has a history of being a thorn in the side of the county planning and building departments. About the time the county figured out we intended to subdivide the Ranch to the size parcels allowed by the ranch's original zoning, the county changed our zone. At that time, Mendocino County was already in trouble with the state, which was finding problems with the county's lack of a general plan. The state's pressure forced the county to impose a moratorium on all subdivisions from December of 1978 to September of 1981.

Building went on even though we couldn't subdivide. The county just didn't know how to deal with all the non-code homes that were popping up all over the Ranch. They couldn't just bulldoze them, even though they may have wanted to. After many years and long hours in the county supervisors' chambers, most of these problems were solved. But now we are faced with the steeper fees of the 1990s, and the tougher standards for subdividing and building.

Back to 1971. My brother John, my parents, and I bought in as partners on two hundred acres. John and I moved onto the Ranch, set up a tent, built an outdoor dining table, pumped up the camp stove, filled kerosene lamps, hauled water from the spring, and started to build a house.

Getting outside power was out of the question. This was a pristine environment sparsely populated except for coyotes, raccoons, red-tailed hawks, magnificent madronas, and trees with names like Grandfather Fir. There was a powerline running to the old Greenfield Ranch House at the center of the ranch, but when it was blown down in a storm we voted not to put it back up. The Ranch has an active board of directors, and everything is done democratically. The theme of our lives was simplicity and self-sufficiency. Years later, when my parents and I began plans for a new house a hillside away, we looked again at the prospect of grid power. We could get outside power for about six thousand dollars. We all agreed that we could buy a lot of solar power for that much money, and were relieved not to get hooked into the same companies that create nuclear power.

John, along with my father, a neighbor, and two close friends, built our new home with passive solar in mind, but we didn't take the intensity of the summer sun into account, so we added a grape arbor to the south side of the house. It has made a big difference by shading the entire south side of the house in the summer and allowing full sun exposure in the winter. On cold winter days a good oak log keeps us warm for twenty-four hours. We burn less than a cord of wood a year unless we get an unusual amount of snow or icy weather.

Water is a big issue here, and fire is our biggest threat. The land divisions were made so almost every parcel has a water source. I pump my water from a 140-foot well up to two 1,350-gallon tanks, then gravity-feed the house through more than half a mile of pipe. I use a gas generator and submersible pump, but I want to convert that to solar, soon.

When the Ranch does have a fire you really see what the word "community" can mean: Everyone from miles around shows up to help.

Occasionally we have had problems with poachers hunting deer. We have the rare white deer here and we are very protective of them. The phone network kicks in as soon as there is a problem, and neighbors start setting up roadblocks. We wait for the fish and game officers to take care of "intruders." We're working on a solar entry gate, but it's a complicated problem because there are a hundred families living on the Ranch now.

At home, we use as much as a thousand gallons of water in one day when we are irrigating our large vegetable garden, boysenberry patch, flower garden, and fruit and nut tree orchard. When we aren't irrigating we can use as little as seventy-five gallons a day for household use. We use an inefficient propane batch water heater because we have too much calcium

Nancy Hensley's garden seldom fails to make her smile.

in our well water, which causes scale to build up too rapidly to allow us to use an instantaneous heater.

Our battery bank is 1,600 amp-hours, retired from the telephone company. We have six 2-volt batteries weighing two hundred pounds each. They give us a good reserve of power in cloudy weather. The newest addition to our house is a 12-volt energy-efficient Sun Frost refrigerator that really needs two more solar panels added to our tracker to properly support its power needs. Without those extra modules we are finding ourselves low on power after three or four days of cloudy weather. Then we run the generator for a few hours to boost our battery bank back up.

We use 12-volt power upstairs so we aren't kicking on the inverter for lighting in the bedrooms. The downstairs is 110-volt house current for all our appliances in the kitchen and for entertainment in the living room.

It isn't obvious to guests that this house is independently powered, and they will often leave lights on. We need to educate them about the discipline of being your own power company. Living in a house like this takes getting used to. I like to be able to get a house sitter when we go away, but it can't be just anyone. We usually try to get someone from the community.

Even though I go to town almost every day to work, I find time for the garden. I like to think of it as my retirement plan, knowing how to grow my own food and chop my own wood.

1999 Update from Nancy:

Years later I'm still explaining to people that I do have electricity, I just get it in a different way than they do.

About fifteen years ago, when my family first bought this beautiful piece of property, we considered the options for getting electricity to our building site and the choice was pretty clear. We could pay an outrageous amount of money to bring in powerlines and poles, or go solar. With our limited budget we opted to set up a few modules in the front yard propped up with two-by-fours.

We picked up some retired telephone company batteries at an auction. They were a steal at $400. There are six of them, 2 volts each, and they provide a mega-storage capacity of 1,600 amp-hours. We poured a small cement pad and built a small shed to cover them. They resided just outside the living room on the east side of the house. We wired the house for both 110-volt and 12-volt to keep our options open. Then came the inverter, a slick Trace 2012SB, which needed to be installed close to the batteries because the cables were massive and expensive.

Those are the highlights of the first years; I suppose there were other

I am glad that this life pushes me to be conscious of what I am taking from this Earth and to know that I am doing as little harm as I know how.

details of the system I'm forgetting, but I never was all that involved with it. It gave us the power that we needed to have lights, music, and television. That's all I really cared to know about it.

Things were going along just fine, then just for the fun of it one day I borrowed an electromagnetic field meter and tested the fields around the inverter—whoa! As I said, the inverter needed to be by the batteries, so it ended up on the book shelves in the living room by the couch. We noticed that if you were lying on the couch watching television your feet were in the red danger zone, your stomach in the yellow, and only the tip of your head was in the safe green zone. "Hmm, what to do? We'll just move the inverter." Then we realized that we would need to move the batteries as well. Not a small job. These batteries not only had their cozy cement pad and shed, they weighed 200 pounds each! What a hassle.

Everything was calm for a while but we had been noticing a faint propane smell off and on for a few years. After reading that propane contains benzene, which is a carcinogen, it seemed it was time to look into this. Our great little propane refrigerator wasn't vented to the outside and in fact wasn't on a exterior wall, and the kitchen layout didn't allow us to move it. A 12-volt Sun Frost refrigerator seemed to be the answer. After getting the Sun Frost installed we could no longer satisfy our energy needs with four old 35-watt Solec modules and we decided it was time to upgrade with a Wattsun tracker and the addition of two Solarex 60-watters. We squeaked by with these for a few more years and just recently added two more 80-watt modules to the tracker. The batteries recover much more quickly now after a cloudy spell, but we still have to run the gas generator for three or four hours every few weeks in the winter to keep them healthy.

That's the brief history of our lives with solar power, but the real question here is why do we still live "without electricity" when the powerlines have gotten so very close? Believe me, I've been tempted. I've even had some friendly arguments with a couple of neighbors who are further from the lines than I am, since they thought that I was standing in the way of them getting the lines to their homes. I've sworn at the generator when I couldn't get it started, and ignored the batteries for months at a time. And I've wondered what it might be like to be living here alone when my nineteen-year-old son is off on his own. Wouldn't it be easier, maybe even wiser to "give in"?

Frankly, I don't like looking at the tracker out in the garden. The system can be somewhat of a nuisance during those long dark stretches of January. I have even taken to turning off the freezer section of the refrigerator to conserve in the winter. And if you have ever cleaned the gunk off the battery in your car you can imagine what it's like to clean our monster batteries. But really worst of all, it makes me dependent on other people. I can't just write a monthly check and have it all taken care of. When something goes wrong I have to call someone to fix it. And I've never really taken to mechanical, electrical things.

Is it all worthwhile? Occasionally, I have had to search my soul, and when I do there really is again no logical option for me but to live this way. It may seem a little harsh sometimes, actually going out into the rain to replace the fuse in the tracker, but when I do I am always glad to be outside. Being my own power company helps me to stay aware of my surroundings, the seasons, and our energy consumption. I am glad that this life pushes me to be conscious of what I am taking from this Earth and to know that I am doing as little harm as I know how. I am thankful to my friends and neighbors for helping me with the system—especially Ross. I used to strive for independence in every way possible; now I have learned the joys of community interdependence. Ross is just as happy with a pot of soup and a pie for his work as he is with a check. In the end it is all really worth the peace of mind that comes from knowing that my dependence is on my friends and not on a faceless power corporation. ⌐

Independent Children

Nothing ties us to our home and neighborhood more tightly than children. Nonetheless, independent homesteaders are not immune from the forces that necessitate two-income families, latchkey programs, and television-as-childcare.

I looked for, and expected to find, ways in which off-the-grid children differed from their on-the-grid peers. Instead I found that children living on independent homesteads or in families striving to control their energy use are no different from children who live in any supportive family environment. People who work with children will quickly understand that this is not to say that children raised off-the-grid are "normal" children; a supportive family environment is not at all common in our workaholic culture, where family life is submerged in the daily struggle to make ends meet in our relentlessly consumerist civilization, and the TV set is the chief pacifier and source of intelligence.

Gender stereotypes can be markedly less

rigid off-the-grid. As I already noted, system-management functions are performed by whichever family member is present when the system needs attention, even if the original system was designed and installed by the men. There seems to be excellent technology transfer, in the sense that all family members soon understand the basics of the system and are willing to tune it and their activities to fit the circumstances. What appears to be a remarkable mastery of amps, watts, volts, and other energy topics is nothing more than technological and environmental awareness; most of these children have spent their lives on "voodoo power," and its qualities seem quite natural to them.

Many children reported to me that their friends like to visit them because they love the freedom and openness of a natural setting, and because the household's focus on energy management is intriguing. I was also informed, in surprisingly strong terms, that it was at times awkward to return the visit, because on-the-grid friends live in wasteful ways that make energy-conscious children uncomfortable. When asked what is missing from their lives, most children were stumped. "Maybe I could play video games more . . . but no, I like to go outdoors, even when it's raining," one boy told me. This sentiment is revolutionary in a nation where children increasingly keep indoors in all weather.

Making the home into a rich educational environment benefits both children and their parents, whether children are schooled at home or in town. Liberated from the tyranny of generic housing, thoughtful parents can install child-oriented features such as light switches and kitchen work spaces low enough for short people to reach, so children can take care of their own needs and participate in family chores. Small doors on children's rooms and child-sized furnishings help children understand that the home was built to provide for their needs. When adults work at home and their activities become part of the daily routine, children are encouraged to perform their own work, the work of learning. In many cases, these children benefit from an environment enriched by the work of the parents, which is very different than the lonely households that the children of working parents return to after a day away at school. All this is a sharp contrast to a conventional world where children are sent away to learn, and return to be treated as noisy inconveniences incapable of affecting their own world and consigned to their own rumpus rooms and dinner times.

Home schooling is an attractive alternative for many off-the-grid families, and is enthusiastically embraced by the younger children, aged twelve and under. Their favorite part is being able to stay home, living in the learning environment instead of visiting it for a short, intense period after enduring, for most rural students, a crushingly long commute. With some of the time pressure removed from learning, lessons come about naturally and in context rather than being dictated by standardized curriculum, schedules, and the added complexity of too many students and too few teachers. Among the families I interviewed, home schooling has been enjoyable and productive for the younger children and adults who participated. One benefit of home schooling in the early years is that it often gathers together a convivial group of children, and creates work in the home for one or more adults. This results in what might be called tribe-building, which carries over into the lives of the

families involved no matter what educational paths they follow as the children grow up. Small groups of like-minded souls who congregate to educate their children together can provide a less rigidly programmatic, more serendipitous way of learning than is typical in a regular classroom. Since lay teachers come from differing educational backgrounds and seldom continue teaching the same age group year in and year out, they solve educational problems spontaneously, intuitively, often naively, but always with the child at the center of the solution. Life and work are sharply separated in the conventional setting, but can mingle freely in the independent-home learning environment, and this integration provides unexpected bonuses. In public schools, gardening may be one more distraction from the mandated programs, but in an independent home, some of a child's most important lessons can be among the flowers and legumes.

Having served ten years in the oxymoronic role of teacher to teenagers, I believe that curriculum becomes less important than social interaction as the children pass through puberty. Almost every off-the-grid-raised teenager I encountered expressed, and was granted, a chance to go to a conventional school. Children who enjoy the tribe-building of early home schooling seem to make a good academic transition to conventional school even as they often find their new classmates distractable, disruptive, and unfocused. My home-schooled informants insisted that for them school's most important lessons were learned before, between, and after classes. They admitted that formal class sessions were occasionally interesting but mostly irrelevant to the world they expected to inhabit after graduation. These children expressed a strong appreciation for the truth, and conceded grudging respect for adults who told it, but were quick to express contempt for institutional education's slowness in awakening to their own and the planet's real future. I took dark comfort from my sense that children raised in the freer and more natural off-the-grid environment were specifically angry, while many of their on-the-grid peers seemed to drift sullenly in a river of helpless negativity. College students from off-the-grid backgrounds told me, in 1998, that much of what they were taught appeared to come from a time capsule buried in the 1960s, when the teachers were themselves in school. Several teenagers predicted that they would be entering a world much different than the one their parents had found at the end of their schooling, and wondered how they might ever be able to achieve the qualities of life their parents now enjoy. When they asked me, "Did it seem that way to you?" I was forced to confess it did not. The few children I met who stayed in a home-schooling environment well into adolescence were in some cases brilliant, but narrowly informed and somewhat maladroit socially. Peer exchange is so important for children of this age that any benefits of home schooling, unless an unreasonably large cohort of children can be assembled, are overshadowed.

At the same time, my decade of teaching leads me to conclude that much of the important learning going on during the teenage years comes from caring adults in a richly educational home and community environment rather than from a traditional schoolroom. Independent parents tend to raise independent children who are immune to much of the emptiness and waste that plagues their peers. Lessons about the work ethic and acquisition of employable skills were apprehended more easily in homes where adults worked at home. When the adults involved in early childhood home schooling continue to involve themselves actively in secondary education and their children's teenage lives, the family relationships, although inevitably strained, held.

As much as we would like for them to remain safe on the homestead, at some point most children leave their independent homes to seek the bright lights and consumerist temptations of the city. These adventurers are strongly motivated, of course, by their wish to meet people, to experience the world, and to define their own identities. This causes temporary alarm among parents who may find their only solace in reclaiming the space in their childrens' abandoned rooms. Our alarm is often temporary, because most independent children soon report home about the joylessness of urban life, and begin their own movement toward a more peaceful lifestyle, often near their parents, or toward a more purposeful and conservative life in the city. Amazingly, they grow up, just as we did! It will be interesting to watch this process as it completes a second generation and enters its third.

A problem universally acknowledged by parents and children alike, and one that is obviously not a circumstance limited to independent-home-raised children, is the difficult passage our boy-children must make through an adolescence infested with beer and internal combustion engines. In southern Humboldt County, California, where off-the-grid homes are more common than elsewhere, the too-much-testosterone (TMT) problem is compounded by bad weather, difficult roads, distance between home, friends' homes, and school, and the paradoxical values of an illicit, marijuana-based economy. Neither permissiveness nor strictness seem to work well to counteract testosterone poisoning, and young men literally bounce off the walls, carving tire tracks up the steep roadsides between Ettersburg and Redway. It is no comfort to off-the-grid parents that TMT also plagues on-the-grid and urban communities.

I can recommend only one stratagem for helping a young man survive the awkward years between the onset of personal awareness and the time when he can take up some real personal power. Cautiously and with suitable adult aloofness, encourage him to build, occupy, and maintain his own separate shelter—an independent homestead of his own.

Make friends with what you need: Michael Reynolds's story

Michael Reynolds is an architect working in Taos, New Mexico, and the originator of the earthship, an innovative building design employing retired tires filled with tamped earth. Earthships are a modern application of the age-old southwestern adobe tradition. Because of their simple materials and great thermal mass, earthships and their rammed-earth, adobe, and cob kin are appropriate homes for the southwestern climate and for inexperienced owner-builders. Reynolds and his associates in northern New Mexico also pioneered a controversial way of owning land, which is still evolving in 1999. Cooperatively owned communities at Reach, which I visited in 1993, Star, farther out onto the remote plateau, and Greater World, near Taos, allow builders to buy land for about the cost of a car. By making land easily affordable to young and innovative builders, these cooperatives make possible experimental, sweat-equity homes and communities. More of the Star tale can be found in Ken Anderson's story, in chapter 8.

It's simple observation: If you look up and see an avalanche coming at you, you get out of the way. That's common sense. For example, it's insane the way we build nuclear power plants and ship the electricity all over the country. Not only is there a threat from the plants, we have to deal with the crooked businessmen who sell and regulate the energy.

I came out of school interested in doing low-cost housing, and everywhere I turned, I ran into obstacles.

One night, probably more than twenty-five years ago, watching TV, I saw two specials back to back: Walter Cronkite's piece on clear-cutting, in which he predicted a housing crisis in the near future (and he was obviously right about that), and Charles Kuralt predicting a waste crisis in which we would be buried by cans. It just fell into place for me: Take the source of the problem and turn it into a building material. It didn't take me long after that to patent my beer-can building block.

Once I've got an idea, my mind never stops. I started building houses out of cans. Then I saw piles of tires at the dump, and I thought, "energy crunch . . . thermal mass. . . ." Nothing matches a three-foot rammed-earth-mass wall for energy efficiency. I could immediately see that, with simple skills and available, resilient materials, it would be easy to build cheap housing.

If you get into the right frame of mind to look at byproducts, the answer is obvious. It's an unarguable phenomenon: We're running out of trees, but we've got big piles of used-up tires. If you were a spaceman, it's one of the first things you'd see. Here's another obvious one: When we build a house, we're building a box to keep ourselves warm, right? Then we build a little box inside the bigger box, a refrigerator, and we bring in nuclear power to keep it cold. The spaceman would just shake his head. I'm working on a thermal-mass refrigerator.

What you have to understand is, everything you need is out there. An earthship requires just a subtle amount of heat to stay comfortable through the winter. We've been putting in little three-hundred-dollar gas heaters that burn fossil fuel, but that doesn't make much sense. I started trying to think, where is heat? And how can I make friends with it, and get it? So we've come up with a little heater to provide just a little heat through the winter. It's a steel box; in September you put a bale of alfalfa in it, and wet it down and close it up. Pretty soon the hay is decaying, and it gives off just the amount of subtle heat an earthship needs through the winter. In spring you open up the box and turn the compost out into your garden. These little passive heaters can be built into benches, furniture, anything. Right now, we're working on finding out how many bales you need for how much heat.

We figure that building codes are an existing phenomenon, and we believe we can work with them. For example, building officials don't like composting toilets, so we asked, why not? Well, they told us, people use them wrong, there's at least a dozen things you have to do to make them work right, and if you forget one thing, you end up with raw shit in the compost. That's dangerous. So we came up with the idea of our shit fryer, a special-purpose solar oven that cooks shit to powder at about 300 degrees in a couple of days. We took some of the powder and mixed it with drinking water, and sent it off to the health department, and the response came back: Lots of particulates, but sterile. We showed those results, and the fryer, to the inspector, and it made him happy. Working with the inspectors, we hear their fear, and that becomes our design determinant. We solve it, and give them the solution; we simply disappear the problem

Michael Reynolds.

they have. We get the solution to work, we don't care what it costs or how it looks at first, then we refine the solution, and build cottage industry and low-tech jobs for the people around here.

My whole life, I've seen inflation getting worse, and politicians chasing it across the sky. I aim to make it so that people don't need a job to survive. They need a job if they want a VCR or a Corvette, but if they want to reduce their needs to subsistence level, they shouldn't need jobs. My overall objective isn't to create some architected environment; I'm trying a new approach, living past architecture into culture and economics. We talk about that in the *Earthship* books, how to get a house without the mortgage companies. We've made it happen here, up the mountain at Reach and out on the mesa at Star. It's just not right for people to go into hock for land, to pay big dollars for a teeny piece of land.

My idea is to get the profit out of land. For our first project, at Reach, we bought fifty-five acres of steep hillside, no power or water, and we let people buy in, a thousand square feet for a thousand dollars. They get free run of the rest of the land, and a support group of other like-minded people. Well, more people wanted in than there was space, so we started a much bigger project down the road at Star.

As soon as we took infrastructure and profit out, took that weight off, the whole project rose to the surface like a cork, a buoyant success beyond our belief. It's not a subdivision, but more like a country club, which works okay in New Mexico because it doesn't trigger any of the zoning difficulties. Usually, a subdivision's big problems are infrastructure, sewage contamination, water table. We're not doing any of that; we're resting lightly on the land. All our water

comes from roof catchments, so we're not using groundwater. There are no septic tanks, and all our waste materials are recycled inside the shelter, so there's nothing that leaves the house. We're recycling greywater and not making any blackwater. This approach slips through the traditional boundaries and concerns. If we tried to build close in to a town, we might run into density restrictions, but not here in New Mexico or Montana. Our associations just haven't been challenged—it's like we're eighth-inch gravel passing through a quarter-inch screen.

Let's say we want to grow flowers in the garden; there isn't an architect or scientist alive who can go out and put together the chemistry and mechanics of a living flower or a growing tree, but through our knowledge of science, we can make good soil. Well, the community is the plant, and it's made up of human nature and other things we can't make or change, and we know that certain kinds of soils grow certain kinds of plants. Like in a city, you can see why people are the way they are, because they're growing in spiky, dry, urban soil, and they need a lot of supervision, and are always looking for someone to take care of them. Now if we have luxurious soil, no stress because there's no debt for land and shelter, I trust the heart of the human beings. The community becomes the vessel that lets us transcend, and advance into our human potential. So you see, I'm not trying to design a community, I'm just building the right kind of soil. . . .

1999 Update from Mike Reynolds:

Our solar toilet has evolved; we got it working fine, but discovered there's too much resistance to the idea here in the U.S. We needed to try another approach. But the shit fryer works! We took a couple of them to India, to serve as toilets at base camps at high altitudes where nothing composts. Here we are donating two $1,800 toilets . . . and the Indian customs officials won't let them through! So we built new ones with local materials, which was great because our hosts got involved and learned how to make more. We installed two shit fryers at the base of Katzenjunga . . . and they work perfectly.

Our work on the shit fryer put us in the right frame of mind to take another look at sewage. We'd been working to extend the ideas of our indoor greywater system. Besides producing happy greens and flowers, at the end of the system the water is clear, but it isn't potable. We started building systems using that water for flushing toilets. Flush with that water and you're using the water a second or third time, not flushing with fresh, potable water, so you can use as much of it as you have. In our new system, when you flush, the effluent goes through standard plumbing to a conventional septic tank, which makes the health guys happy. Of course, we're already taking most of the water that conventional systems put into the septic tank—shower, bath, and sink water—through our indoor greywater system, which reduces the load on the septic system and makes it possible to size the system smaller. We wondered, what would happen if we solar-heated the septic tank? We discovered that the anaerobic process can be accelerated by about ten times. This "solar-powered incubator" makes the septic system run about ten times faster . . . so it can be even smaller.

Normally, the effluent from a septic tank flows to a leach field, which we know creates all kinds of problems. Leach fields pollute soil and groundwater, seep into wells, waste land, and generally don't work as well as they should. We've isolated our leach field from the soil,

using pits about three feet deep, strategically located to enhance the landscaping of the house. Here in the Southwest, where officials require "zero landscaping" because of the water scarcity problems, this is already a major improvement. Unlike a conventional leach field, our treatment cells are lined with rubber, so that nothing escapes into the environment. Plants and microbes in the treatment cells have the opportunity to transpire and clean up the effluent. We originally used pumice in the bottoms of these cells as a sort of condo for the bacteria that digest the waste and purify the water. Pumice provided more open spaces at the bottom, which makes the bacteria work better. But pumice is expensive, doesn't come from here, and will get used up some day. We noticed that plastic jugs are turning into a major recycling hassle, and figured we could crush those jugs up by jumping on them in the back of the truck or driving over them, then they'd be a perfect matrix for the bottom foot or so of the solar incubator septic cells, replacing the pumice.

As for the plants in the incubator cells— you just can't believe how well they grow! Even before it was outlawed, we'd given up on landscaping around Taos, because the soil is so poor. Now we noticed that anything planted in the moist, fertile soil of the treatment cells turned into a jungle without any care at all.

Everything we've done here extends and improves the earthship's circulatory system: Rainfall from the roof collects in the cistern, is pumped and filtered for interior water uses, the greywater is cleared up in the interior planting beds, stored again for flushing, flushed to the solar-heated septic tank, and finally waters the landscape. Every drop gets used four times, without ever getting dumped into the environment! Recently every earthship we've completed

Nautilus earthship, south wall. See more pictures of Nautilus House in the color section.

has incorporated this whole water treatment system. Health officials in the state of New Mexico and Taos County have given us a clean bill of health: They want to see more, because here in the high desert we're all having water problems, wells and aquifers running dry and getting polluted. This is exactly the kind of solution these guys want to see. We learned to give it to them in a way they can understand, with valves throughout so if someone wants to, she can shut down the solar subsystems and run the house's plumbing like a conventional system. Our sewage system is set up using conventional parts, and we built all the alternative systems with valves that can be set to show the officials, including realtors and bankers, exactly what they expect to see. If we don't threaten them, they're happy to let the new inventions ride along. It's like giving a dog a pill wrapped in a lump of hamburger.

Right now our biggest project returns us to our original goal: Earthships were supposed to be easy to build and affordable, in the hands and within the reach of the regular people. But in practice, unless you're a good builder, the cost was getting up to $100 per square foot, which is not much

With very little care, just making homes for folks, we turned a wasteland into paradise.

better than other forms of shelter in the Southwest. Even if it's reasonable, it's out of reach of the single mom. So we designed a simpler model, a sort of Volkswagen of earthships, which we call the Nest. It's a simple, one-room, 700-square-foot building with rammed-tire footings and a vertical south wall. We can assemble all the parts in the back of a truck, drive it in, and throw it up in a week. Pretty soon, naturally, people were asking, "Can we make it a *little* bit bigger?" and so we changed the design so it could be modular. Now we're calling this "the vertically faced earthship" or VFE. The original earthship is our Buick, and the VFE is our Chevy Nova. The U is a little shallower in the VFE, and the roof beams run from front to back, but it's just a newer, refined, packaged, assembly-line solution—like automobiles. And we have the cost back down to $75 a square foot—much closer to everybody's grasp.

Our affordable community experiments are thriving. Reach is about to be rolled over to the people who live there. Like any community, we've seen a 40% to 50% dropout rate, because some folks who think they can live that way find they cannot. Some homes are rented, some have been sold. Star has a ways to go yet, mostly because of its bad long road, but the people and building going on out there are in good shape, too.

Our new project, Greater World, is right off the highway, so it's easy for passersby to find us. We built several demonstration earthships which folks can rent for a few nights to find out how it feels and how the systems work. Greater World turned into a school right away, with people coming from all over the world for a few nights or up to six months. Due to our recent war with the building and planning establishment, the school is presently on hold, and we take our teaching to the students whenever we can. I'm also working with a group of Maoris now—they are a people of the earth, thrilled by the ideas we're proposing as compared to the European lifestyle that drives them to the highest suicide rate in the world. I'm going to New Zealand to set up an earthship school with them.

Out at Greater World, there was a gravel pit that no one wanted, devastated land, but oriented east-to-west with a big south-facing cliff. We started building split-level earthships into that cliff, using water catchment and the solar incubator and treatment cells, proving that damaged land like that can be reclaimed. With very little care, just making homes for folks, we turned a wasteland into paradise. We're thinking of calling it Lemuria.

I mentioned our war with the county planners. There's been a lot of illegal subdivision going on around Taos, substandard structures, mobile homes, and the worst sewage arrangements imaginable, so the county hired a new guy to crack down. Unfortunately, because Star and Greater World are high-profile projects, the guy decided to start with us, even though we weren't what the county wanted cleaned up. The result: We waste time in court, even though the judge has already ruled we are not running a subdivision scheme, and we'll probably win. But the planning guy is too far into it now to back down.

All the fuss attracted the attention of our local state senator, who told us he'd help us get the subdivision law changed. As soon as he saw what we were doing, he pointed out they were giving away money, begging for people to solve the problems we had under control, and here all we wanted was permission to keep working. With his help, we've introduced a new section to the subdivision law that allows for sustainable alternative developments. It's a strict criterion: You've got to satisfy local health codes, and meet or exceed Unified Building Code provisions; you've got to have your ducks in a row.

This part of our work touches so many issues, from recycling and housing for low-income people to better use of materials and resources. The law's ready to be considered as soon as the legislature opens, and other states have already expressed interest.

At Greater World, where we've got about twenty homes finished enough to be lived in, I'm beginning to see my gardening project pay off. Because folks don't have mortgages and utility payments and all that hanging over their heads, they spend their spare time helping newcomers build their houses or making general improvements to the land. ⟜

In the sere New Mexico high desert, the plantings in a sensibly designed gray water system add a needed bit of greenery. This garden is inside Nautilus House.

Caveat emptor: J. P. Townsend's story

After she read the first edition of this book, J. P. Townsend sent me an email alerting me to the other side of the earthship story. After checking out her sources and reading the archives, I asked her to tell her story. Her experience and the advice growing out of it applies to anyone who wants to find land, design and build a house, and launch an independent lifestyle. J. P. and I share the hope that readers of this edition will not have to make all of the same mistakes, no matter where they buy land or what kind of dwelling they decide to build.

I'm an old hippie deluxe. I spent the Nixon years in Europe knocking around the boulevards with the motto *"Soyons débiles!"* My memories of Paris and London in the early 1970s are heady: salons and suppers with revolutionaries, writers, filmmakers, and feminists, the northern light of London winters glinting off red buses and whitewashed buildings low enough to allow a sense of horizon and big sky. I arrived in the U.S. in time for Nixon's resignation speech. When my camera bag was stolen I secretly sighed with relief, gave up being a photographer, and started graduate school in anthropology. I thought I had done my preliminary research well, and I chose the Ozark area in order to be uplifted, I thought, by hardy pioneers from the voluntary simplicity "movement." Fieldwork in the Ozarks proved more stressful than I could have imagined. It was a hard life these hardy pioneers were leading, with odd results for the children I observed. It seemed there were no "happy families" in my research group. I came up against my own fears, finding myself unable to listen to the painful stories of incest and spousal abuse I was hearing. The stories affected me in ways I could not shrug off. My subjects depressed me utterly!

I escaped to the mountains for the Reagan years, and in Boulder, Colorado, I began to meditate. As my interest in Buddhism deepened, I wanted to learn the language of literary Tibetan to get closer to the teachings and maybe help translate some of the sacred texts into English. So, during the Bush years I found myself, hippie, scholar, meditator, and female, in despair at the trashing we humans are responsible for on this planet. While I was studying in Kathmandu, some friends there took a permaculture design course offered by visiting Australians. I caught on, by osmosis I guess, that here might be an antidote to despair: building soil, growing food and trees. In 1994 I studied permaculture myself—too late to save me from the disaster I am about to relate!

Let me get it over with: I lost a small fortune by hiring a huckster to build me an "off-the-grid self-sufficient" home. I relied on the design and construction services of a "visionary"/architect/contractor/author/high-

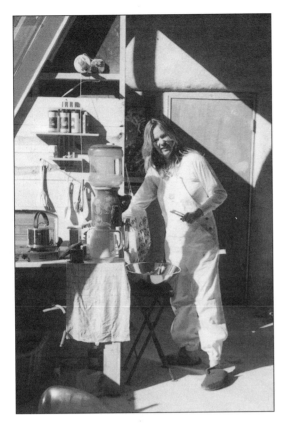

J. P. Townsend.

profile home-dealer. As I type this, the man is still pushing his services and products on the internet, in print, and in person to anyone who comes within earshot. Using the hippest new-age notions, his pitch describes seamlessly integrated earth-, user-, and pocketbook-friendly building styles embedded in utopian communities, unbelievably cheap! He promotes his wares with the unstoppable fervor of a high-pressure car dealer on latenight TV. Meanwhile, his state's attorney general, county officials, and other disgruntled victims are howling for his head.

How Did I Get Here? This Is Not My Beautiful House! Looking back, I see I made three beginner's mistakes:

Mistake #1

I fell for a "visionary" sales pitch. I bought into a vision that was presented as: (1) the answer to everything, (2) fully realized, and (3) cheap. I failed to engage and trust my critical faculties, but went for wishfulness. The result of my folly is that I am still in possession of a legal and financial liability I can neither use nor sell. If something seems too good to be true, it probably is.

Mistake #2

I learned this the hard way: A sustainable lifestyle cannot be packaged and sold like a consumer product. Advertising and commerce would like us to believe otherwise; this is the nature and process of "greenwashing." Everything "alternative" is not always "greener." Be skeptical. We grow into sustainability slowly, by adopting thoughtful strategies and principles that fit the stages of our journey. For us in the West, this strategy seems to involve rolling up our sleeves and doing battle with our habits of consumption. In our civilization this is not easy; a supportive community helps us know when we have *enough*. If, in spite of my sad story you are still emotionally attached to the purchase of a nifty self-sufficient home-deal package, I urge you to get a bank loan; do not pay out of pocket. The documentary and legal protections that a bank officer requires might be *worth the interest charges.*

Mistake #3

On the first day of work, if the architect/construction boss doesn't arrive on site with a clear schedule mapped out on paper, cut your losses then and there! Fire 'em, do not be shy! After my experience, I realize I was intimidated and disempowered right from the start by my belief in the builder's self-proclaimed expertise and my self-proclaimed ignorance. I lost control when I tried to exchange cash for a dream. I was well intentioned, I wanted to be kind to the environment, but I didn't want to take responsibility for the hard parts: I didn't know about picking a decent site, I knew too little about photovoltaics and catch-water systems and how they are wired or plumbed. And sadly, I didn't think I needed to know these things in order to live sustainably, nor was I encouraged to learn by the builder. I thought I had made a deal with him to provide me with an item I could own the same way I maintain my car without knowing much about the internal combustion engine. I have learned that designing and building isn't the hard part, it is the *necessary* part. When we can do it with friends we trust, brainstorming using design ethics like those of permaculture, it can be an ecstatic and healing journey.

Buddhists have "four reminders" of the importance of meditation practice. The second reminds us about "impermanence": Nothing stays the same, and everything that is put together comes apart. Houses dissolve. I want to build keeping that in mind, constructing a "biodegradable" home, not something fit only for the landfill. I have become an extremist, an aspiring neo-Luddite, dreaming of a low-tech life. I take my inspiration from a set of "Ecotarian Guidelines" put forward by the Tyaya People of Bocare, County Kerry, Eire:

— Do not buy anything you cannot or will not ultimately produce yourself.
— If you cannot produce an essential item yourself, make sure it will last your lifetime, without further input.
— Buy only foods that are organic, bioregionally grown (start big and shrink it), unprocessed (uncooked, unshelled, unhusked, etc.) and unpackaged (bulk or refillable containers).
— Materials: Reuse things, use scavenged or found things, and buy only secondhand. Use only natural materials.
— Do not buy new timber or wood products.
— Do not purchase electricity.
— Do not purchase animal products.
— Do not store money in anything other than an eco-institution.
— Learn to be strong willed inside shops. The pressure to buy what you want is fantastic, but can be overcome. . . .
— **Start the change next time you go shopping.**

Although I depend on technology in my work, I daydream about drastic changes and meditate on the third of the four reminders, which is the law of karma: Good and bad deeds follow you as surely as a shadow.*

Natural Communities

At the end of the twentieth century, ecologists are just beginning to appreciate the subtle and pervasive interconnections that sustain natural communities. In their own search for missing community and connection with Nature, two generations of social pioneers have precipitated a re-evaluation of many assumptions that shape civilization in a much more complex and often destructive way. For many who settled at Greenfield Ranch, Star, and other innovative communities, the immediate desire was for an affordable place to build a home, plant a garden, and raise a family, but the long-term benefit has been renewed focus on the matrix that supports home and family, a rediscovered sense of community. Clusters of self-reliant homesteading families living under their own power, actively working with land, sun, wind, and water, have shown that it is possible to create a restorative environment wherein human habitation is tightly woven into the fabric of nature. Such neighborhoods raise strong children, conserve land values, consume resources sparingly, take care of their needy, and withstand the stresses of change well.

The long-term benefit has been renewed focus on the matrix that supports home and family, a rediscovered sense of community.

---

*In case you need to know all four "reminders" or literally, in Tibetan, "the four thoughts that turn the mind toward the teachings," J.P. has provided the other two: "The first 'reminder' is to contemplate that death comes without warning . . . this, my body, will be a corpse. This reminder lends an urgency to practicing meditation. The fourth reminder is that samsara has defects. Samsara is a Sanskrit word that encompasses the cycle of birth and death, the wheel of life, our shared existence. This reminder is helpful for folks who have a fortunate time of it within samsara. . . . say, Bill Gates. It's a reminder that even if you have a lawn and a swimming pool in this life, still there is suffering: war, famine, destruction of our planet and many species. Remember this from time to time as you enjoy the pleasant circumstances of this lifetime, and allow your heart to open to the suffering. Turning your mind toward the teachings means just this: Be heartbroken, and very, very kind."

CHAPTER 14

LIVING WITH CONSTANT CHANGE

PEOPLE SEEK EQUILIBRIUM, BUT LIFE IS CHANGE. MANAGING AN independent home requires almost continual tuning to the cycles that abound in nature. Long-time off-the-gridders welcome the invitation to synchronize with their environment, and make these adjustments instinctively.

In chapter 4, I discussed the need for daily reconciliation of energy budget expectations with actual production and consumption. Each self-sufficient system—water, space conditioning (heating and cooling), hot water, waste, as well as electricity—requires such attention, always with an eye to anticipating the next cycle, and managing it better. As this accounting becomes intuitive and habitual, and we become better managers of our living machine's adjustment to its surroundings, we might expect the process to take less time; instead, we may find ourselves taking more time, but also more pleasure, in maintenance tasks. Independent homesteaders report that the mainspring of their days is wound by their home's interaction with the natural world.

The passage of day and night, the shortest cycle, dominates our understanding of the way a home performs. Photovoltaics gather energy during the day. In most places, wind also has diurnal rhythms. In an independent home, certain tasks (including bathing, laundry, and use of power tools) are arranged to fit the daily energy-gathering cycle.

The next shortest cycle is defined by weather. In my temperate coastal climate, as in many others, weather patterns are never as sharply delineated as day and night, and have a seasonal texture. In Caspar, in summertime, two or three pleasant days are often followed by two or three of fog. In winter, storms seldom last more than three or four days. The satellite photo on the internet or evening news helps us see and plan for the times when a storm train, a line of two or three successive storms, will whistle past us from the Gulf of Alaska or the tropical Pacific; relying on the sun for hot water, we bathe and wash dishes before the first storm hits.

For workers and students, the pattern of week and weekend changes the perception of home: This interesting and artificial construct may yield clues for off-the-gridders about how their lives might change if they took control of their own time. For work-at-homers and those truly immersed in the daily cycles of maintaining a self-contained home, the weekly cycle is barely noticeable.

The seasons impose another order on our weather. Longer-period seasonal patterns, driven by the sun's place in its annual precession, dominate our shorter-term weather forecasts, and modulate the amount of light that reaches us. Growing things respond to the seasons, as does the wind, outdoor temperature, and the likelihood of falling water. Interspersed among these perennial rhythms, the annual round of holidays, birthdays, and anniversaries punctuates the life of home and family.

Almost everything that exists in time is cyclical, and the house itself has a life cycle, from inception to decay. First, the home is planned and envisioned, then it gradually materializes. New houses, and any new structure, require a shakedown period during which we encounter and resolve unforeseen problems. Once the worst of the bugs are worked out, and the space becomes livable, we enter a stable phase. Independent homeowners are famous for the way they attenuate this stability; start-up bugs are barely controlled when we initiate new projects to enlarge, improve, refine, and rebuild the space we have wrought. Inevitably, before the last nail is driven, decay begins, and parts of the house become worn and abraded by the friction of life and seasons. Maintenance and repair are crucial, or the cycle begins to wind inexorably and prematurely down to the last phase, abandonment. Many independent homesteaders proudly insist that we will never finish building, and so we find ourselves planning for new construction while simultaneously working to stave off decay. Within this century-long life span of a house, we

The passage of day and night, the shortest cycle, dominates our understanding of the way a home performs.

encounter the many overlapping cycles of family—our children's birth, growth, graduation, and departure, our own vigorous youth giving way to age—as well as shorter cycles of equipment life and the complex choices associated with repair or replacement.

Solar Cycles

As emphasized in earlier chapters, responsibility for our own energy requires us to anticipate our needs and opportunities every day, and re-adjust our systems accordingly. Here is a simple example: If I know that a few hours of sun remain before a storm, I might check water and power supplies; if they contain more than needed to weather the anticipated storm, it may be wash day, since I know that water consumed will soon be replaced and the batteries can be recharged before night falls or the storm closes in. If I hesitate, the only thing conserved will be dirty laundry.

Living under our own power, we learn to seize opportunity quickly, to revise schedules flexibly in the face of nature's greater determinants, such as seasons and weather. In our independent home, energy systems and resources on which we rely are practically family members, and we consult them, checking their capabilities before making our final plans. Considerations of this kind were once shared by everyone, but such responsibilities are nowadays delegated by most people to invisible others, the water company and the electric utility. What has been lost?

For independent homesteaders, the annual modulation of daylight, as winter's brevity of days and lengthy nights reverses itself in spring and then reasserts itself in autumn, affects the way we dress, our daytime activities, even the hour and circumstances of our rising, dining, and going to bed. Each season has its tasks. In spring, we check and repair the ravages of winter, remove storm battenings, prepare the soil and put in our garden, and plan the summertime renovations that will make the coming winter easier. In summer we cherish our shade, tend our plantings, take care that our water is not overdrawn and our batteries do not overcharge, make the changes necessary to survive the next winter, and plan the changes we will put in place next spring to make summer more enjoyable. In fall we harvest and prepare for the onset of winter; we get in the firewood for the winter after next, check the chimney, confirm the health of batteries and equipment that will be called upon to give us comfort through winter's long nights, and begin to dream of spring. Wintertime's arduous totality, for all but those who live in the deserts and tropics, is the true test of an independent home: If home stays habitable through this trial, the rest of the year is easy. We watch the sky and the satellite photos, fill the woodbox when we sense an oncoming storm, batten down and hope as the storm batters past, then mop up and make urgent repairs that cannot be postponed until spring. Those of us who choose to live at the end of the road are subject to the vagaries and blessings of sun and storm, and we come to love their endless pageant.

Daily and seasonal source/demand curves provide an interesting way of looking at our energy cycles. Like the patterns of weather and seasons, they vary within ranges dictated by site, structure, equipment capacity, and family disposition. Using this knowledge of our patterns, we can optimize our energy.

Gradualism

I find myself, at times, sitting on a step, a stool, a rock, contemplating a part of my house. When I built it, I knew that elements of it were right, absolutely; the foundation and the anchoring of frame to foundation had to be right, for instance, because I planned a house taller than its longest ground-level dimension, a sort of tree house on a site where tree-levelling winds are a wintertime commonplace. The kitchen, a favorite place of mine, came right into focus very soon, and every year fewer of its components remain provisional, awaiting better replacements or their designed obsolescence that curses our consumer society. From the beginning, the overall vision was clear, but only with time and the freeing up of cash flow by the completion of more critical projects has it been possible to make some necessary alterations. I doubt if such progress can be hurried, but my partner Rochelle disagrees. "Why can't we just do it now?" she wonders. I fall back on my chronic lament, "Because I don't know how to make it exactly right . . . yet." Luckily for me, she is patient, and has seen enough completed phases to know the process may be glacial, but continues.

From the start I have known that parts of this house were temporary; funny, how some of those temporary elements persist!—one day the entry will have its water garden, and the foundation work for the stairway and tower in the northwest corner will begin before this book is published. Living within an assemblage of permanent and temporary parts, new visions come to me, more or less grandiose or practical. Occasionally an idea will be so right that a flurry of construction will make the vi-

sion real. So it was with the bathroom door— only twenty years in coming, but then in three days the idea, the wood, and the hardware all materialized. Talking to others, I found that this is the pattern of stubborn, naive builders who are unwilling to be limited by conventional wisdom, but wish instead to invent their own functional spaces. Thus our homes are living, growing sculptures, and as surely as salty wind peels paint, temporary sections will decay and require patching while just-finished additions are readied to face their winter shakedown cruise.

Some people may choose to immerse themselves in other pursuits, and let experts envision and construct their living space for them. The most visionary experts understand that homebuilding is a participatory activity that relies for its impetus and personality on the inspiration and insight provided by the homeowners. Pioneering spirits will insist on going toe-to-toe with the forces and frustrations of living in space they have built, assiduously tackling every new project and repair. Settlers, and others for whom the intrigue of living off-the-grid is not a way of life, may find ways to have others handle some of the more technical details of homebuilding. Many of the original pioneers now report that they will hire done what they once did themselves. All who participate actively in building their own homes agree that the process is one of their life's most gratifying and exciting experiences.

Generations

Families have cycles. All too quickly we go from the year when we toddler-proof the cabinet to the year we yield up the family car's keys to our new-wheeled teenager. Residences are

Homes that grow to match the needs of their inhabiting families may not be as architecturally gracious, but they become treasured because they respond well to the family's changing needs.

often planned and built at the time when family life centers around young children; at the time, it seems these inquisitive small ones will always be with us. Suddenly, especially in the slow-developing timeframe of an owner-built house, children grow up and begin to leave. Small, dormitory-style accommodations work well for young children, but as they grow up, no space within our home suffices. Frank Lloyd Wright found one solution to this conundrum for a university professor's family: Rooms for transient children were enclosed in a semi-permanent way adjoining the living room; as children left and the professor's growing eminence required more space for entertaining, the bedroom walls were removed to enlarge the living room. As noted in the preceding chapter, another way to improve a family's home life during the brief period between the time when children begin bouncing off the walls and the time they leave, is to build, or help the children to build, their own separate cabins. These stand-alone shelters make fine guest cabins, special-purpose outbuildings (Linda Edwards packages her mustard in her daughter's casita), and habitations for the children when they return on visits. Such outbuildings allow inveterate builders to experiment with new structural and energy technologies.

Haste in designing a house, and failure to plan for the future, creates space that will not take good care of us as our needs evolve. Architects and house designers may have masterful ideas about stairway management and bathroom configuration, but their interest generally ends at the moment the project is finished. A proper ancestral home must serve a family for decades and generations. Architects serve the housing industry, not families, and their turnkey houses often lack the personal touches and flexibility that preserve a family through decades. Homes that grow to match the needs of their inhabiting families may not be as architecturally gracious, but they become treasured because they respond well to the family's changing needs. In the throes of inspiration about redesigning or adding to a house, remember the long-cycle needs of family.

For example, how will our homes nurture us as we age? My house, with its three levels and lots of windows, will work fairly well for us as long as we can climb steps; I am gambling that we will be able to do that as long as we live. Catwalks and belaying points around the windows have made it possible for me to wash and maintain them until now, but soon I will need to enlist someone more agile and daring to work on the third-floor windows and the roof. My electrical, plumbing, and heating systems are left in the open to the extent that the building codes allow, for ease of service, and I have tried to map and label their invisible segments.

I would like to say that all circuits are labeled and their routing known; only in researching this book did it become clear to me how important this home documentation can be. Already, in a mere twenty years, I have completely lost two wire runs. I know where one end is, but I cannot find their other ends without ripping out walls.

Aging is an inescapable condition of any family's cycle in a home. Many independent homes are situated in remote and difficult terrain, and are plagued with uneven floors, narrow doorways, and many other features that make them young people's houses. When I revise this book again in 2005, I may have sad stories to recount about once-proud owner-builders forced from their ancestral homes by such imperfections.

Public Consciousness and Cyclical Change

When I started compiling a comprehensive list of cycles that homes live through, I missed a very important one, but Mac Rood reminded me. "Don't make long-term business decisions based on short-term government policies," he admonished. In 1993, after a little Ice Age in the political fortunes of solar energy, the U.S.'s second longest political regime of the twentieth century appeared to come to an end. With a new president from the alternative energy generation and an allegedly pro-environment vice-president, energy-conscious people hoped for a return to a more sympathetic hearing in Washington. In Hawaii, an island nation held in thrall by fossil-fuel suppliers despite the most clement solar situation on the planet, the opportunity for evolutionary energy production goes begging. Cully Judd, who lives in a grid-connected solar-powered demonstration home in the middle of Honolulu, says "Hawaiian Electric Company should get down on their knees and beg to pay me a dollar for every one of my nonpolluting, absolutely renewable kilowatts!" But state and federal regulators stubbornly navigate into the future by looking in their rear-view mirrors.

We have learned not to hold our breath while awaiting governmental enlightenment. Slow progress is evident. In 1993, only two states had enacted net-metering, whereby small energy producers can return their excess energy to the grid for the same price they buy it; in the next five years, ten more states joined the net-metering parade. But the development of an energy policy for the next millennium is neither rational nor consistent. Scientists widely agree that we have extracted half of the planet's available oil, and so should now begin to experience the inevitable run-up of prices unless some dramatic reduction of demand is achieved. Arab oil, since 1979 a reduced fraction of the source for U.S. domestic consumption, is again inching its way back past 30% of our supply, the point at which the sheiks have pulled the noose tight twice before. Meanwhile, El Niño and mild winters left short-term domestic oil supplies at such high levels in 1998 and 1999 that gasoline was cheaper than it had been for decades. The presidency for which so many had such high hopes has been stymied by entrenched interests; its showcase "Million Solar Roofs" program of tax credits is neatly contradicted by deregulation of electrical supply, which has providers promising lower rates and longer guarantees of unsustainable pricing. With so many contradictory signals, we should not base any kind of decision on government energy policy.

The solar industry survived its nuclear winter, and is stronger for the experience. With so many cycles at play—economic, generational, the quick-time quadrennial political cycle and its weird perturbations for two-year congressmen and six-year senators—makers of solar energy equipment have kept their eyes on the sun. "The solution," they say, "comes up every morning." As an undercurrent to all else, the environment's cycle endures. Some people are so deafened by the foreground noise of the jobs vs. owls debate that signals from the natural world are inaudible. For others these signals— topsoil loss, species extinctions, endocrine disruption—are an imperative heartbeat. Fickle fashion and government programs ride the crests of waves of public awareness, and like all short cycles, the change can be brutally fast, thrashing adherents against the rocks. In the twentieth century, Los Angeles saw solar hot water go in and out of fashion twice; each time the

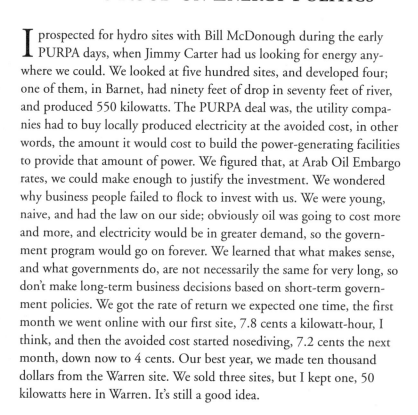

Mac Rood on Energy Politics

I prospected for hydro sites with Bill McDonough during the early PURPA days, when Jimmy Carter had us looking for energy anywhere we could. We looked at five hundred sites, and developed four; one of them, in Barnet, had ninety feet of drop in seventy feet of river, and produced 550 kilowatts. The PURPA deal was, the utility companies had to buy locally produced electricity at the avoided cost, in other words, the amount it would cost to build the power-generating facilities to provide that amount of power. We figured that, at Arab Oil Embargo rates, we could make enough to justify the investment. We wondered why business people failed to flock to invest with us. We were young, naive, and had the law on our side; obviously oil was going to cost more and more, and electricity would be in greater demand, so the government program would go on forever. We learned that what makes sense, and what governments do, are not necessarily the same for very long, so don't make long-term business decisions based on short-term government policies. We got the rate of return we expected one time, the first month we went online with our first site, 7.8 cents a kilowatt-hour, I think, and then the avoided cost started nosediving, 7.2 cents the next month, down now to 4 cents. Our best year, we made ten thousand dollars from the Warren site. We sold three sites, but I kept one, 50 kilowatts here in Warren. It's still a good idea.

Mac Rood in his Mad River powerhouse.

promise of cheap energy left sensible technology stranded. In the last solar binge, following the second Arab Oil crisis, millions of active solar hot water systems were installed with tax credit money (an enlightened but poorly implemented program) and cobbled together by fly-by-night installers (a horrible outcome). How many of these early-1980s systems are still working? When they broke, how many owners could find reputable service people and replacement parts? No one really knows, for this was a program without follow-through, a dream scam for exploiters, and a black eye for solar technologies.

One relevant lesson was suggested by Amory Lovins: "Nixon made a wonderful speech about population, perhaps one of the best ever made by an American president, but nobody ever thanked him, and so he never mentioned it again." We can affect political cycles only by exerting economic pressure, and seeking out influential people, insisting they take the right actions, and letting them know that we appreciate their work when they do so.

"The solution comes up every morning."

The Political Weather

We may attempt to shift our lives to take advantage of political cycles, but we may also try to guide those cycles. Politics blows hot and cold, first welcoming, then rejecting, then again welcoming alternative ideas about energy. At the state and municipal level, the agencies and commissions that regulate the energy monopolies should be public guardians, insuring that the best possible local policies are instituted for consumers. In practice, these institutions are packed with industry shills and apologists, and attempt to maintain equilibrium by denying or resisting change.

Depletion of conventional fossil energy sources will not be denied for much longer. As Dennis Weaver, actor and vocal advocate of energy sanity, says, "If we continue to stick our heads in the sand and ignore it, live in denial, and say we don't have a problem, it's only going to get worse." The work of regulating power monopolies is complicated by energy technicalities and the fact that regulatory agencies are part of the political apparatus, administered by appointees, and infested by interested parties. Public utility commissions are—or can be, because of their statutory obligation to serve the interests of the people, and their genuine power over the utilities—a powerful tool for reforming energy practices in a meaningful way. Increasingly, energy activists are learning the language, joining the club, and putting this tool to work.

Making cycles serve us:
Leigh Seddon's story

Leigh Seddon designs and installs renewable energy systems at Solar Works, Inc. in Montpelier, the capital city of Vermont. He lives on-the-grid in an energy-efficient house with a solar and wood heating system. He is a solar internationalist, having taught courses in applied solar energy in South Africa and the Middle East, and to foreign engineers visiting this country. After twenty years of experience with renewable energy, Leigh is at the forefront of entrepreneurs installing grid-connected photovoltaic (PV) systems. He continues to be an energy activist, a key member of the Vermont Public Interest Research Group, and a solar fox in the bureaucratic chicken coop of public utility regulation.

Our goal is a solar chicken in every pot, renewable energy systems everybody can afford. Over the years we have relied on angels to pioneer the way.

Noel Perrin (whose story is told in chapter 2) is one such pioneer. He installed the first utility-interconnect PV system in Vermont. The story goes, he was pestered to do something about pollution by his environmental studies students. He was already connected to powerlines, so it made sense to him to put PV modules on the roof, and sell his excess output to Central Vermont Public Service (CVPS). At first CVPS wanted a million-dollar insurance policy and a full-time operator for a plant generating at best 200 kilowatt-hours a month! To shorten a seven-month story, we got them to be a bit more realistic. Noel was able to satisfy them with a good homeowners insurance policy and an inverter that shuts down if the grid fails. We weren't quite as successful on the buy-back rate. He gets four cents per kilowatt-hour and buys power back at about three times that. I guess it was a matter of principle, not economics, for Noel. He's an example of the crusading angel, who establishes a principle at some personal cost, so others may follow. Now it takes half an hour to set up another utility buy-back contract, and Vermont has net-metering so the buy-back rate is equal to the customer's retail rate.

Because of pioneers like Noel, we are now poised to roll out the biggest rooftop PV program on the east coast in Rhode Island, a state that has electric retail competition and net-metering. In our pilot phase, we'll be installing systems on the roofs of about five hundred homes, small businesses, and schools. We have a range of PV system sizes starting as small as a 250-watt AC module on your roof. We expect this to cost about $12 a month. Of course there's no payback—you'll get about $3 a month of solar power for $12—but it allows people who are interested in supporting renewable power to vote with their pocketbooks.

What we're looking for is leverage: We are trying to find out who will most value systems like this. Schools are a natural. One of our programs will be to get citizens to contribute toward the installation of PV systems and monitoring equipment on their local elementary and high schools. In Rhode Island, we're hoping to involve forty or fifty schools. The school kids will help install our systems, which will then become part of a new school energy curriculum.

We will also be working with the new retail energy service companies that will be competing for a customer's attention. So for the first time, we'll be seeing prime-time advertising letting people know they can have their own PV systems. This way we can tap the small percentage of customers who will be glad to pay a premium for clean, sustainable energy.

Utility restructuring offers some real positive aspects as well as some negatives. Among the positives: Eventually customer choice will lead to better decisions than those made by government and business. I firmly believe the market will work when people are empowered to choose and make educated choices. One drawback is that significant choices may not be available.

Across the river in New Hampshire, we saw a pilot competitive market flooded with companies advertising "green power," but no real choices, offering mostly existing fossil fuel and nuclear generation. To do restructuring properly, we have to make sure that real choices and real information are available. California seems to have done a good job with disclosure standards, while the New Hampshire pilot had no viable standards at all, and no way to enforce them. Big questions remain about the will and ability of government to regulate the energy marketplace.

The major drawback of restructuring is the bailout of the big bad guys for their bad investments. Most utility executives see restructuring as a chance to redeem their past misjudgments and make their badly damaged companies whole, but somebody is going to pay for that. In some states, notably Pennsylvania, restructuring is nothing more than a utility bailout. In some states, the bailout is balanced with good new programs. Massachusetts, Rhode Island, Vermont, and others are moving toward a regional portfolio standard that requires new energy providers to supply a mandated percentage in renewables (which includes large hydroelectrical projects), with a further percentage committed to emerging renewables—like PV and wind energy.

In California, Massachusetts, and Rhode Island they've instituted a "wires charge" to pay for some of the social benefits we used to see paid for by the regulated utilities, such as energy efficiency and development of renewable resources. In Massachusetts, this charge will produce hundreds of millions of dollars a year, of which at least thirty to forty million dollars will be put directly into renewable technologies. So I predict we'll see much more use of renewables under restructuring in progressive states.

Financing efficiency involves honest valuation of externalities, but the utilities are fighting tooth and nail to keep from adding externalities to "avoided cost," the cost at which an efficiency is said to be worthwhile. This makes no sense. In Vermont, we will probably get a statewide "Efficiency Utility," but the argument is about what size budget we have to work with. The difference between an unadjusted average cost of six cents without externalities, and eight cents including a 30% adder for externalities, makes it much more reasonable to finance efficiency. In

Leigh Seddon explains the State of Vermont's electric vehicle charging station.

the new dockets we're working on it looks like the cost of power must be raised at least 30% just to account for the health aspects of air pollution. To account for the comparable costs of nuclear waste disposal, water impacts from hydro projects, lack of job creation (renewables create jobs), and other costs associated with conventional power plants, it's not inconceivable that you'd be looking at an adder of 50% to 100%. I don't think we'll ever get there (because we won't stand for it politically; we prefer to see it hidden in our taxes), but 30% is a whole lot more than the 15% we've been able to work with in the past. In two years, we've seen the externality adder doubled, and in another couple of years, it may double again.

Honestly, I think programs like this are the icing on top of the cake. A million solar roofs sounds great but doesn't do enough to address global warming. Yet that doesn't mean we shouldn't do it. We must get started putting the technologies out there. What will finally make a difference will be customer choice, consumers voting with their own power systems. I think as a country we lack the political will to stop the next energy war in the Middle East, but as individuals we can make a difference. And there are a lot of individuals—30% in New England—who say they'd pay more for renewable power.

The key thing holding the renewable energy market back, I believe, is a dysfunctional market, where we willingly spend billions for a war to keep oil prices low. What is the real cost of our fossil-fueled economy? We don't have even a vague clue, but we know the subsidy is immense. In Vermont, and a few other states, there's a move to quantify the external costs of generating electricity. For example, our utilities figure new coal capacity will cost eight cents a kilowatt-hour, but by the time you figure in acid rain, air pollution, and other immediately identifiable environmental concerns, you can add seventeen cents more. Who knows what other costs will have to be paid down the road?

It is also important to remember that efficiency is the bridge to a renewable energy economy. We can't power the U.S. economy sustainably until the infrastructure, from homes to factories, is much more efficient than it is now. We could have the same economic output using half the energy, as Amory Lovins has shown. Vermont is a good example; during the last ten years we've met all our load growth through efficiency and renewables. If we wanted to, we could continue this trend indefinitely.

We consider a compact fluorescent in a house to be a conservation power plant. In Vermont, the average electric customer uses 700 kilowatt-hours per month. We have worked with families and designed houses to get by on 100 kilowatt-hours quite comfortably. This turns out to be a very powerful tool for promoting renewable energy. And as we demonstrate super-insulation, super-refrigeration, solar water heating, PV systems, and other new technologies, these technologies are being picked up by mainstream builders. ⬌

What will finally make a difference will be customer choice, consumers voting with their own power systems.

⬌

Shift Happens

In the context of geologic time, where a millennium is short, perturbations in the flow of energy are common. Pink dolphins and rays inhabiting the Amazon, which flows to the Atlantic, are more closely related to Pacific populations of their relations, indicating that the ancestral Amazon drained into the other ocean, and then the whole tectonic plate tilted. It is not at all unusual to find places in the wild where a watercourse is dry because of the way land above it has changed, for instance, when a river's course has been changed by a rockfall, mudslide, or earthquake. In the geologically active West (we like that expression because it does not make our hearts beat fast as it does when we say "on the earthquake-troubled Pacific Rim") changes can be sudden. In 1992 a tremor knocked down buildings and changed river courses in Humboldt County. Planners and insurance companies are questioning the wisdom of lowland development in Florida, along the Gulf Coast, along the Mississippi's banks, and on the shores of California and Hawaii after devastating losses due to big weather in recent years.

History is filled with lacunae—the Anasazi who dominated the southwest United States, the Mayans of Central America, the Moche and Tiwanaku of Peru and Bolivia, the Mesopotamian, Harappan, and Mohenjo-Daro along the Indus: All these civilizations fell, but the reasons for their fall are unclear. Increasingly, scientists are looking to sudden and dramatic climatic change for explanation. The current warm era, called the Holocene, was originally thought to be stable and climatically benign. Are we due for a shift?

In what we like to call modern times, humankind has exploited an unusually long period of climatic stability to produce an unprecedented level of civilization. We are pleased to presume this accident reflects our mastery and management of the planet's energy flows. As self-appointed masters of the planet, we decree

JON STOUMEN ON SECONDARY PERFORMANCE

We met Jon Stoumen in chapter 1, as O'Malley's stalwart advisor and spouse. Jon practices the arts and crafts of solar architecture in Healdsburg, along California's Russian River, where urban forest fires, floods, earthquakes, and storms determine which homes stand and which do not.

When we look at the way a building goes through its normal response to its climate, regional environment as well as the way the residents live, their uses on a day-to-day, normal basis, we are evaluating the house's primary performance.

We must never forget secondary performance, which is what we see when we look at the way buildings fail. Even in benign climates, buildings finally fall apart if left long enough (especially unmaintained). By looking at how they fail, we can learn to make better buildings. An example: When there's a big fire, study the houses that survive, to see what their builders did right. Monocropping is as much a problem with modern architecture as it is with agriculture: Too many buildings of a certain genre get built without getting tested, like Maybeck's houses in Berkeley, with nice, pretty, combustible wood shingles. The ones in the hills all burned up in the Berkeley fire in the 1930s. Another example is the devastation that Hurricane Andrew did to minimal-code-standard buildings in Florida.

Secondary performance is how a building responds to severe environmental stress. A building has to perform in its at-rest state, and also withstand encounters with severe environmental stress, much the way most organisms have to perform. If we look very carefully at successful regional buildings after one, two, three centuries, we can understand why some buildings are still standing when others have fallen. In many cases we cannot build them in exactly the same way as they were built before, so we knowingly substitute a modern response to the same stress. When we look at the old buildings in New Mexico, built with thick adobe walls, rein-

lakes, islands, rivers, and are resentful when natural processes disturb our doings, delay our plans, or interrupt our swift movement. In recent decades, government has spent enormous sums to resist the forces of geologic change, to clean up after they strike, and to indemnify those who have run athwart nature's intentions. Is it wise or affordable to continue this policy? Present-day humans of a certain social status insist on their right to go anywhere and do anything at any time, willingly paying large fees and endorsing earth-changing works to do so. Born of the same stock, people who live in independent homes find ourselves indulging in such follies with the same insouciance, even as we wish to participate in the transformation of human consciousness to a higher level of planetary awareness.

---

forced by interlocking vigas running all the way through to the outside and beyond, we see a simple, appropriate response to primary and secondary challenges. We who work in California are concerned that adobe structures aren't good in earthquakes because we have seen some collapse. But some old adobe structures survive many earthquakes while others fail; the difference may be in the quality of construction, the wall-to-opening ratio, for example. The old churches and buildings that have survived have very few, small openings. More modern adobe structures may have giant openings, thinner and taller walls relative to their thickness, and these are the ones that fail.

This idea struck me outside a discount store in Marin. We had just been in Maine, where we had ridden kayaks on the ocean, and it was totally cool. O'Malley was inside getting some stuff, and there were two kayaks on the roof of the next car. It turned out the kayaks were owned by kayak instructors, so I asked them, what's a good kind of kayak to get? One instructor said, the bewildering thing about buying kayaks is, if you buy a kayak that's stable in the kind of water that you learn on, benign conditions, and you want to take that kayak in rough conditions, it won't perform well. And if you buy a kayak that will perform well in rough conditions, then it won't perform well for beginners in easy conditions.

That started me thinking about those two kinds of performance. A house doesn't always exist in rough conditions, and it doesn't always exist in mellow conditions, so the trick is to make one that will exist in both conditions and will exhibit characteristics that will be cool all the time. One of the great things about the old indigenous architecture is, they didn't have access to a great number of materials and techniques to build with. Take an Eskimo kayak for example: You've got these people living on the top of the world—super-, unbelievably cold conditions—and they were able to build with few resources and a lot of knowledge, totally sophisticated, streamlined, lightweight, stronger-than-shit boats that they trusted their butts to, in the most severe conditions in the world all the time, a boat-building tradition passed down from generation to generation. Kayaks are almost an architecture. They have to perform, they have to be warm, dry, flexible, light.

So the aim is to build a house that, like the kayak, has good performance characteristics. We want a house that can perform well in still water and also perform well in white water.

A synthetic town: Laughlin languishes in the southern Nevada sun, consuming coal and belching particulates while it harvests tourist gambling dollars.

The cataclysms of our planet's continuing formation, its greatest and most unpredictable cycle—great storms and droughts, fires and pestilence, volcanos, earthquakes, ice ages and thaws, comets, meteors, and all the other manifestations of Gaia's restlessness—are all evidences that the earth is still inventing itself. There may be a periodicity to these catastrophes, but it is too long and intricate for humans to comprehend. When we choose a home, we hope that our site is blessed with immunity from the more dramatic upheavals of an unfinished planet, but we would be fools not to give some thought to the characteristic disasters of our region in hopes of learning from earlier failures and successes. A thoughtful blending of traditional designs and new means may help us guide our homes safely through the chaos. Just as we try to understand the unpredictable shifts in political weather, we must attempt to attune ourselves to natural processes great and small, befriend them, or, in the worst case, anticipate nature's tantrums and head for higher, sounder ground.

Jon Stoumen suggests we build our homes "to perform well in still water as well as white water," and his advice makes sense, especially to independent homesteaders who mean to survive the coming changes. Whether the white water ahead is of our own making, or merely represents our planet's return to its characteristic instability, we must strive to be prepared. And then we must do the best we can to preserve the comfort and grace of our lives.

CHAPTER 15

THE WILL TO
SURVIVE

"The Future ain't what it used to be."
—*Casey Stengel*

WE KNOW WE CAN LIVE WELL IN ENERGY INDEPENDENT HOMES WITH-
out imported energy. With increasing urgency, the question must be
asked: Do we have the will to survive? Are we willing to make the hard
changes required to rejoin the planet's natural economy soon enough to
preserve ourselves and the many other highly evolved species that will
perish unless we alter our filthy, greedy overconsumption of nature's limited resources?

In the preceding chapters, my fellow pioneers and I have sketched
ways to improve energy use in existing homes or by building anew,
thereby reducing the load we place on our planet's strained resources.
Our good news, that the new energy frontier is open, safe, and a rewarding place to explore and settle, is not without a continuing and increasingly pre-emptive challenge.

We the people of the United States comprise a twentieth of the global
population yet we consume more than a third of the planet's energy and
generate more than half its waste. Can we summon the will to reduce this
inequitable consumption and profligate waste? Being human and American, in the absence of a compelling imperative like a war or a depression,
we willingly adopt only answers that insure continuation of our treasured
standard of living. In this book we have seen that by eliminating waste
from our fossil-fuel-powered economy we can make a good start toward
sustainability . . . but that strategy has problems. Waste and pollution are
inherent in all fossil-fuel-burning technologies, and eliminating this

energy source completely is impractical. Furthermore, having squandered half the fossil-fuel resource in less than a century, *any* continuing use of heritage fuels is ultimately unjustifiable, not just because it fouls the planet, but because it unfairly deprives our heirs of potentially important resources. We must improve the efficiency of all energy uses, even those powered by energy falling to us freely from sun, wind, and water, and actively seek new sources. This call for sweeping change extends beyond our energy usage to every resource-oriented activity: As a people we are being called on to change ourselves utterly, and most futurists fear that we are too soft and too weak to answer the call.

Bunkers will not save us: Leandre Poisson's story

Lea Poisson lives in a futuristic cement-and-glass house he and his family built on a clement southern slope in southern New Hampshire. An inveterate designer, Lea is constantly seeking better ways to dry fruit, move heat within a building, and keep a community strong. In Solar Gardening, *Lea explains his ideas for extending the growing season and increasing food production even in the harshest climate using simple gardening appliances of his design. I asked him to talk about the theories that took him out into the garden from architecture.*

In the mid-1960s I was teaching a class at the University of New Hampshire called Environmentics, based on the study of everything that had to do with how we, as a species, had gotten to this point in history, and how these factors might project us into the future. While researching for the course I read the Petroleum Institute's report on how much oil there was left on the planet. The other force to reckon with was the population bomb. With so many people on the planet, the present solutions could not keep up with the problem. My conclusion: The most valuable gift we can give our children is an intact ecosystem. Since then, the world's population has doubled, and the biosphere is on the ropes. Seems our civilization has a cancer: It's built on growth, and borrows from the planet's future to survive.

It doesn't take much looking to see that we're not doing very well. How many monarch butterflies did you see last summer? I heard the Mexicans cut down the forest where they winter. There used to be so many orange newts around here that children would collect them and hold races, but last summer I don't think I saw more than two. I've been nurturing the bluebirds, and so there are a few in the woods; visitors wonder what those birds are, because they have never seen a bluebird. The problem is right here.

By 1970 I had enough information to realize that Western civilization was on a collision course with destiny. When population explosion, resource depletion, pollution, and species extinction became apparent to us, my family decided to move to the country and try to live in a totally different way, more gently on the planet, to become more self-reliant for our own basic needs. The first thing we did was to grow as much of our own food as possible. There are immediate economic benefits to living a self-reliant lifestyle. Income tax is figured assuming you have to buy everything you need to live. Provide for your own basic needs, and you increase your non-taxable earning potential.

When we bought the land, it was so overgrown that we didn't know there was a view. We cleared land for the garden, and discovered lots of fallen stone walls from the time a century ago when this was a farm. Fortunately, I love to rebuild stone walls, creating order out of chaos. Illustrates my Garbage Principle: You don't really have to clean anything, you just arrange it in straight lines and it looks orderly.

People ask if I'm living some kind of alternative lifestyle and I tell them, "No, just trying to save our collective asses." I remind them that Jeffersonian democracy is based on the self-sufficient landowner who cannot be tyrannized. The Founding Fathers wanted everybody to have a piece of land and democratic access to the sun. Too bad the Jeffersonian dream has evaporated.

After a few years of working toward self-reliance I went back to my own college, Rhode Island School of Design, and talked a lot about the need for doing for ourselves, unplugging, canning and storing food in the root cellar. One of my students said, "Well, if I ever get hungry in the city, I'll come out with a gun and take your food away." Hot-headedly I responded with "You do and I'll blow your expletive-deleted head off!" Afterwards I realized that he was right, that if we don't all make it, nothing really matters. We have to share our skills and knowledge and work with our community. The bunker mentality won't work: It's just a matter of time before someone needy rips off our root cellar.

While looking for creative solutions to the problems of growing, processing, and storing food I came up with the idea for my solar dehydrator. Since I didn't want to manufacture it, I developed some plans showing everything you'd need to know to make your own, and started selling the plans. I guess there was a ready market of like minds, because I've sold 35,000 sets of plans, which helped put my children through college.

You see, people have to work toward their own integrity, doing as much as possible for themselves. For example, it doesn't take special skills to build the Siberian stove I designed, just hard work and cement. The one heating

Jeffersonian democracy is based on the self-sufficient land-owner who cannot be tyrannized.

this house has been working for nine years. It burns slabs of waste I buy from the local lumber mill, four hours a day when it's very cold. I had to design special doors for the firebox because there was nothing on the market that could withstand the 1,800-degree combustion temperature. The heat is stored in 10,000 pounds of thermal mass, and gradually radiates into the house. We bake bread and pizza in the firebox when it cools to the right temperature; we call it "catching the wave." ⌐

Shelter without pollution or compromise: Alan Crabtree's story

Aside from eliminating waste from our new energy budgets, we must satisfy our-selves with reusing land that has been damaged by abusive extraction, and salvag-ing materials and energy that were discarded without being completely utilized. Before going "dumpster diving", many homesteaders go prospecting for plots of regenerative land that have been scalped of their vegetation for immediate profit, and direct their new growth into patterns more suitable to the needs of settlers.

Another immediate saving can be made by reducing the grandiose but useless exaggerations of the space and material needed to make a home. Sharply resisting the inflated preferences of Architectural Digest-intoxicated architecture, the greed of lending institutions, and the slavish housing market, many independent-home builders have found that no more than 450 to 600 square feet per inhabitant provide sufficient space, depending on the degree of outdoor living permitted by the climate.

When Alan Crabtree attended one of my week-long independent living work-shops he had already bought his land in northwestern Washington state, in the "blue zone" of the old Insolation Regimes map. His site, in the Cascade Mountains north of Seattle, is a mixture of cliff-like north slope and flat topland high above the Stillaguamish River. His site is far from the nearest powerlines, even though a large high-voltage transmission line runs along the foot of his property, and so he came to the workshop seeking to learn about renewable energy systems, and was not bashful about asking for my recommendations for designing his house. Alan lis-tened carefully to me and to anyone else with expertise to share, and then built a wonderful new off-the-grid home.

In 1997, after a year of research, planning, and jawboning with would-be suppliers, Alan and his wife Becky Teagarden invited us up to help wire his renew-able energy system, and we shared a grand campout on their spectacular site. A year later, I asked him to tell me again about finding the land and building the home.

My parents had a house out in the country where we couldn't see any neighbors, just trees and fields. Because of that experience I wanted to build a house where I could look out and see nothing but my own trees, fields, and unspoiled nature. Unfortunately, because this is the Northwest, where logging is a way of life, you'd have to be George Weyerhaeuser to buy enough land so you'd never see a clear cut. We have a nice vista to National Forest lands that are steep enough they'll probably never be logged, and we're close enough to the mountains that we can see them, but far enough away that we don't catch their constant rain. We average forty inches of rain a year; twenty miles east of us, they get twice that.

You can never find a perfect piece of property; you have to make a list of your priorities, and then do your best to find the land that fits most of them. We were looking for a large enough property so we'd have five or six acres of flat, buildable land surrounded by state and national forest land. We sought a place where Becky could have pretty much any size animal she wanted to have. Becky and I had several conversations about the relative benefits of starting small or moving into a finished house, with Becky voting strongly for the latter.

My work isn't too far—35 miles—which is an average commute for most people in the Seattle area, but my plan is to work part-time, going in fewer days, and, ideally, with improved technology, I'll soon be able to do a lot of my work from my home. I'd like to make the drive once a week. Becky has the same strategy, although for now her 75-mile drive to work makes it worthwhile to keep a town house.

The first thing we built on the land was a wood-pole metal-sheathed barn. When we started building the house, a friend loaned me a generator, but we never even took the plastic dust cover off it. The two smartest things I did were to build the barn first for dry storage, and then to get the power system installed temporarily and use it for building the house. Both made building easier, and by the time we moved in to the house, I had a good working knowledge of my energy system, how to use it, and how much it could power. Everyone was leery at first about whether the system would work, but after it became clear they had abundant power without feeding a generator or listening to it sputter, they really got into the beauty and silence of the natural setting. At any given time during construction there might have been two power saws and a miter saw going at once. When the crew got used to the idea that the inverter needed a second or two to wake up, they stopped thinking about the power system and worked as they would have on any job site. No one ever said "I wish there was a generator up here"!

We thought about this house for a long time before settling on a laminated cedar log house. The walls are made of one-by-seven cedar inside and out with three inches of polystyrene insulation in the center, so their

*Becky Teagarden and
Alan Crabtree's
mountain home.*

R-value is right around 22. We put in an R-50 ceiling. Once a year we can expect a week with daytime maximums of 100 degrees, and maybe a week with minimums in the low twenties. The elevation here is 800 feet, so we don't get a lot of snow—it snows more in Seattle! We only see snow three or four times a year, and it stays on the ground for about a week each time. Generally we don't get snow until mid-December, but we can usually count on one late spring storm. Our heaviest winter rain and snow is in January and February, when we might go for two weeks without seeing any real sun.

I'm not sure we got it completely right the first time, but we spent three years researching different parts of the house, talking to people and learning from their experience. Everything we did, given the money and the technology at the time, we did very well. I would have liked to build without the construction loan, which constrained us to the bank's timetable, uncertainty about financing, and extra inspections. The bank was a lot stricter because I was an owner-builder. They only gave us nine months to complete the house. With this kind of loan, you only get paid after things are installed and inspected. At the beginning of every month, their inspector would come out, certify what I'd done, and then I'd get a check.

One of our house's unique features is the steep solar roof. We decided to use a homemade rack and invest in more modules to maximize winter energy gain. Once we thought about it, a steep south-facing roof made good sense. In the long run we won't have to worry about adjusting the racks or maintaining the modules. They're clear up out of harm's way. When we were first planning the house I didn't expect to mount the modules on the roof, but after our workshop I was eager to redesign everything to take the best advantage of the site's solar window. No one had any problem with the roof, except they thought the angle strange. The angle we chose is 63%, our latitude (48°) plus 15 degrees to get into the meaty part of the winter solar window.

When we designed the system, we compensated for our cloudy, rainy climate, low winter sun angle and narrow solar window by sizing our battery bank for five or six days of moderate use without any charging going on. Cutting back to just the essentials, we could make it for ten days. We never even considered buying a generator. After a couple of winters, we're ready to develop our micro-hydro site, which flows well from January through June. Developing it will require a 1,000-foot wire run, and we hope to be able to take advantage of some breakthroughs in micro-hydro technology.

We designed a passive sunroom on the front of the house—since we don't live anywhere near a street, I always think of the southern exposure as the front—which works well when we get winter sun, and heats the house passively most of the time. There's abundant biomass on the land, so wood is our natural back-up heat source. The bankers wanted us to spend $2,000 for a nice wood heater we'd have to stoke every few hours, and were unwilling to finance, for a lot more, around $10,000 installed, the masonry

fireplace we wanted based on your advice. We got a Finnish soapstone stove called a Tulikivi from a local mason who does nothing but install masonary fireplaces, which is big enough for our small space. The Tulikivi comes as a kit from Finland, although they also have a quarry in Virginia.

The Tulikivi is a large mass—4,000 pounds, a footprint about two feet by four feet up to countertop height—with a number of baffles inside for the smoke to travel through, so all the heat transfers to the masonry before the smoke goes out the chimney. We build a fire once a day, or twice on really cold days, filling up the small firebox—under three cubic feet—with dry, small wood which burns hot and fast for about thirty minutes. In a masonry stove, you don't want big logs that smolder; you want it all to burn up really fast. The stove's exterior never gets so hot that you can't touch it, but it's warm enough that on a cold day it feels good to sit leaning against it. One good fire early in the morning keeps the house warm for most of the day, when the outdoor ambient temperature is typically in the mid-thirties.

Our finished twenty by twenty-four-foot house and eight-by-twenty sunroom has a big open upstairs room, so our total square footage is about 1,120. It feels big because we haven't divided the house up; the only interior door is the bathroom's, which is there for guests. Our loan was for $160,000, which gave us enough to pay off the land and also paid for the barn. According to the bank, our square-foot cost was extremely high and they needed me to convince them we weren't wasting their money. The reason: We were building something about half the size of a typical house (which around here is about 2,200 square feet), and small is costly, and the energy features (super-insulation, electrical systems, and masonry stove) increased up-front building costs but will save us a lot in maintenance and operating costs. They wouldn't give us money for the stove at all, just enough for a typical heating system. Because I did a lot of the work myself, we were able to pay for everything within the loaned amount. It's a conventional thirty-year mortgage, but the plan is to pay it off within 10 years. We took this route because the alternative—saving a lot of money, building small, and adding on as you go—didn't make as much sense as building what we wanted in one burst. The costly renewable features weren't so much a hard sell as a challenge to find the right lender. We found a local bank whose mission is to finance loans that larger commercial banks won't touch. The biggest problem was getting a fair appraisal with comparable buildings that the bank could accept. Except for the San Juan Islands, there's virtually no one off-the-grid in this part of Washington. The bank ended up accepting the size and type of the house to establish their comparable value figures.

Our road has been the only bad part of the experience. When we were looking for property we found a number of nice parcels we couldn't afford, or unbuildable land that we could afford. This 31-acre parcel was being sold by someone who had just logged it. Access was over a crude logging road. As we were building the house, a representative of the state Department of Natural Resources walked in off their neighboring parcel and told us we couldn't use his road. Not only was I fooled about this problem, but so was the title company. They admit it's their problem, and now they're on the hook for solving it for us, but they were moving so slowly that we finally filed a lawsuit to get their attention. Within the next month or two it should all be resolved. We know we'll eventually get proper access. The lesson, as far as I'm concerned, is that whenever you're buying remote land, the record-keeping isn't as careful as on a city lot, and so you have to be more prudent in checking things out. Title insurance was a good thing for us, and stands to save us a lot of money. They'll probably have to pay someone $5,000 or $10,000 to get us an easement, and I'll probably only have to pay for improving a quarter mile of road. ⌐

Mass-producing Mistakes

Appropriate new homes are unlikely to repeat the mass production errors of the twentieth century's housing boom, which continues to erect unfit shelters on cookie-cutter plots of once-productive crop land. Because rational architecture requires that each site and solar regime be carefully considered and each home deliberately suited to its prospective tenants, we may expect a reduction in the velocity of the race to replicate. I have often heard people say, "But there just isn't enough wilderness to go around. Not all of us will be able to build on forty acres at the edge of the woods. . . ." And unless we change our behavior, this is undeniably true. One of the most pressing tests of our will to survive calls for an immediate and dramatic change in our apparent urge to breed ourselves into extinction. Overpopulation is the root cause of *every* threat to our species. Many of the children in this book, like my daughters, clearly and insistently express their intention not to have children . . . and while I may grieve, because my sensitive and sensible offspring would make good parents and I would love to be a grandfather, the responsibility for that decision has passed to them, and I trust their vision of the future. In my view, it is supremely arrogant for old ones to urge war or breeding on their children.

Without reducing our human rates of reproduction to one natural child for each woman, our species has little or no chance of surviving without terrible dislocations in the very near future. Ever an optimist, I assume that our children will be able to adopt this standard for themselves and help peoples everywhere on the planet do the same; there is not time enough for a slower revolution. Assume this, and work with my daughters to make it so . . . and look what changes come about, almost by magic, within a single generation:

Current "housing stock," the sum of mostly over-large, energy-unfit houses, becomes an energy mine containing more than enough housing to eliminate homelessness, if we can bring an end to energy inefficiency. By applying the energy wisdom known to this book's energy independent homesteaders and others who have already discovered happiness within reduced energy requirements, we can uncover and refine the "energy treasures" buried among the energy spoil: houses with large south-facing roofs for harvesting sunlight, well-built houses, ancestral homes. By assiduously mining and recycling materials while the unfittest homes are demolished, building materials for new homes and energy retrofits can be provided without taking more than the continent's forests and the earth's crust can provide without further damage—importing virgin building materials from other continents is foolish and, if you will permit me the coinage of a word, fuelish. There may even be a surplus of materials that can be shared with less-developed peoples on this continent, to help them bring their housing stock up to a level we can all agree is worthy of human habitation.

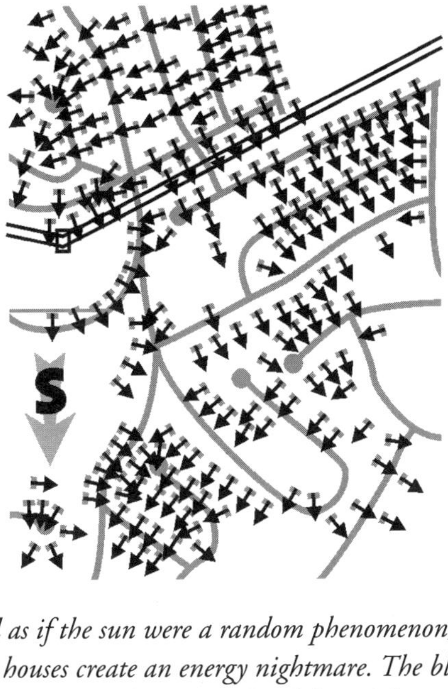

Sited as if the sun were a random phenomenon, tract houses create an energy nightmare. The black arrows indicate the center of each house's best solar exposure, which reveals how few of these houses on New York's Long Island are correctly oriented to the south.

Mechanical Crucifixion

We are too quick to favor iron over people. Evidence of this is everywhere, from the elaborate automated phone systems that greet us when we call our banks to the devices we use to play with snow and water. Addicted to speed, infatuated by power, we hurtle out of control through our lives. Nowhere is this more obvious than in our automotive fixation. Amory Lovins tells us that improving fleet mileage, one measure of the efficiency of our automobiles, by seven miles per gallon, would completely eliminate our dependence on Persian Gulf oil. My own informal studies show that this same effect could be achieved if just half the well-meaning but hypocritical owners of four-wheel drive Urban Assault Vehicles bearing Greenpeace and Nature Conservancy emblems replaced them with high-mileage low-pollution vehicles. Fuelishness!

The paving of America is a scandal. Pavement is the poorest possible replacement for productive land. Hastened drainage decreases the "turns" of the water cycle: In a healthy primary forest, water typically falls from the sky, is taken up by plants, and transpired back to the clouds to fall again three or four times before returning to the sea. When the forest is clear-cut for slash-and-burn agriculture, this recycling is reduced to between two and three turns, but when precipitation falls on hardscape, roof, or pavement, the local cycling factor falls to one. In America, this runoff usually bears the telltale toxic sheen of petroleum pollutants from toxic-petroleum-waste roofs, vehicles, and other pervasive fuelishness.

We might easily wonder at the precedence humans give to our always-hurrying automobiles, and looking at a modern neighborhood or shopping district, we find another major clue: While humans still think of themselves as the dominant species on the planet, it is clear that corporations, which benefit from the manufacture and ownership of automobiles, are really the top predators. It has not been many years since "What's good for G.M. is good for America" was a succinct representation of our economic wisdom. We now watch sorrowfully as our cars eat our princesses and our children. Yet only after a century do a few questions emerge about the turn so-called civilization has taken.

Because there are so many of us, does it not make sense to question every instance where a machine takes over the meaningful work of a human?

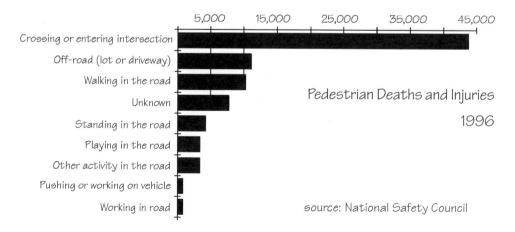

Cars eat people, and it gets worse every year. Better stay in bed.

Who will continue the work?

Scott Suddreth's story

Humanity relies absolutely on its children, on successions of younger generations, to solve the problems that baffle us. Convention too often paints the present crop, Gen-X and younger, as a materialistic, self-indulgent lot—one wonders: Who taught them? Those of us who espouse and actively teach responsible living and sustainable design often encounter many young and bright-eyed students who are intrigued by the challenges ahead and eager to start solving them. Unfortunately, the institutions that they inherit are still much rooted in the past, still professing extractive, rapacious, and short-sighted disciplines and subjects. A single bright light shines in an unlikely place: There is one college-level program in sustainable design in the United States. I asked Scott Suddreth, until very recently a graduate student in that program, to tell that story.

We founded the Appalachian State University Solar Energy Society (ASUSES) "to educate, enlighten, and inform students and the community about solar applications as well as other renewable energy sources"—goals we share with our faculty sponsor, Dennis Scanlin, head of the Appropriate Technology program at ASU. We want to design good systems using affordable equipment so that average people can participate in a solar revolution.

Our first public event was the First Renewable Energy Christmas Tree Lighting, for which we put together a small renewable energy system consisting of three photovoltaic panels, a small wind generator, and a small micro-hydro. We held a raffle to pay for everything, and it worked so well the club had a little money left over after the event.

We got a big shock when the 1996 American Solar Energy Society's conference came to Asheville. For some of my classmates, a national conference with fewer than four hundred attendees was their cue to seek a more active field, but I was excited by the opportunity to help pioneer the field. I learned desktop publishing to bring the best technological tools to my work. Right after returning from Asheville, we set up a demonstration booth at the student mall for Earth Day: solar ovens, a solar concentrator, a solar food dehydrator, and tons of solar literature from the ASES conference. The overall response from students was promising.

In the fall of 1996, club enrollment was up to fifteen people. To improve our educational outreach, I decided to learn about websites and designed a webpage for the internet. The internet has proved to be invaluable in my work to promote solar energy. (You can find our website's address in the resource list.) For the Second Renewable Energy Christmas Tree Lighting, we invited two local elementary school classes to make ornaments, and despite a snow storm, more people came than the year before, and the raffle was successful. The snow made our tree simply beautiful and put the evening in a holiday spirit. The ceremony went very smoothly with everyone cheering when the lights came on. What made it seem so magical to me was the way it all just seemed to fall into place.

A class assignment launched my study of the current state of renewable energy technology in Watauga County, a northwestern county in the mountains of North Carolina, to clarify Dr. Scanlin's theory that our region is full of solar technology. In the 1970s, a locally published magazine, *Mother Earth News*, printed dozens of articles about renewable energy and sustainable development. Because most of these ideas were easily replicated on rural farms and tax credits made them affordable, I had a hunch that Dr. Scanlin was right. The best method for finding renewable technology turned out to be driving around the county looking for anything that looked solar. I found over half my participants this way. It certainly helped that I grew up on a farm in neighboring Caldwell County and could present myself with quality southern Appalachian manners. I was inspired by the enthusiasm people expressed when I asked them what they thought about their technologies. To my delight, 95% of the system owners I found by driving around looking were thrilled to brag about their systems along similar lines: "I do not understand why everyone is not doing it, too!"

Only two homes were completely off-the-grid, one using a micro-hydro system and photovoltaics to supply energy, the other depending solely on photovoltaics. These two pioneers had years of experience making their own power but wanted to remain anonymous because of their independent life styles. In other words, the people who had the most to offer were reluctant to help teach others.

I concluded that in Watauga County the two main barriers holding back mainstream acceptance of renewables are the lack of experienced technicians who understand how to install and maintain the equipment, and the need for further education and acceptance of modern renewables in general. This finding turned a compact fluorescent lightbulb on in my head: Because the purpose of the Appropriate Technology program at ASU is to teach people about renewable energy technologies, I began to feel an obligation to use ASUSES to play a more active role in continuing the county's technology-transfer process.

Looking for help, I found Matthew Willford, who had been living without electricity for two years when I met him in 1995. I was delighted to discover that Mateo was already one of the most successful grassroots organizers in the Watauga County area. He was already organizing an environmental awareness festival appropriately named EarthFest, and invited the club to set up a demonstration to provide renewable power to all the other booths. To pull off the event we needed the help of Rod Baird, who had just started his own renewable energy business named Rock Castle Solar. I recruited Rod and designated him club technician. EarthFest, which took place at Turkey Tom's Music Park, gave us a great inside look at the amount of work that goes into bringing a grassroots function together.

We went right from EarthFest to the 1997 American Solar Energy Society's national conference in Washington, D.C. The overall mood at the conference was of tremendous excitement sparked by talks about global warming and deregulating electric utilities, a

Scott Suddreth shows Maura Scanlin how a solar concentrator works.

concept that seemed to spell a bright future for renewables, because they offer environmentally friendly energy. I left D.C. with a fierce fire burning in my soul. Riding home from Washington, I began to realize the significance of our Appropriate Technology program at ASU: We have the only established university-level Appropriate Technology program in the country! I spoke with several architecture and engineering students who expressed jealousy of our program and its hands-on approach.

When fall semester 1997 started I faced the harsh task of recruiting a new club, because eight of the ten club members graduating with me in June didn't continue on to graduate school. I was happy to be joined by two new graduate students, Scott Taylor and Heath Moody, who accepted leadership roles in ASUSES. Also at the first meeting, we developed a program of one-hour workshops to be taught following our meetings. To improve outreach and technology transfer we made our workshops free for the public, and asked the ASU News Bureau to notify local newspapers and radio stations. Growth was slow at first, but at the third meeting, when I was scheduled to speak about passive solar design, I was shocked to find seven community people waiting to hear me speak. I started out nervous but relaxed after a few minutes. This was the first time in my life that I felt like a teacher.

For several years, a National Tour of Solar Homes has taken place in October, and in 1997 the North Carolina Solar Energy Association (NCSEA) asked ASUSES to organize the tour in our area. Using my

Every year, Appalachian State University students trim a Renewable Energy Christmas Tree.

directory and Dr. Scanlin's contacts, we found six homes, and a local reporter wrote an article about me and the tour which wound up on the front page of the *Democrat.* Twenty people showed up for the home tour—our best achievement ever. Scott Taylor compiled a nice booklet, ASU let us chauffeur the tour participants around the county in school vans, and during lunch we demonstrated a blower door test to impress our guests. Topping this tour next year will be a challenge!

Mark Estepp, the chair of the Technology Department, was switchmaster for the Third Renewable Energy Christmas Tree Lighting, which was the club's last project during my year-and-a-half term as president. We rebuilt the cart for our electrical equipment, and I commissioned my wife Anne-Marie to paint a beautiful scene showing windmills, waterwheels, and solar panels with our name, ASUSES, on the front cover. Again the raffle was a success. About fifty people showed up for the ceremony, including many familiar faces from the community and university.

I look forward to the challenges that lie before me even though I know the road ahead may be rocky. I feel a driving force inside which is fueled by teaching others how to harness energy in an environmentally friendly manner. I do not know what obstacles lie before me, but I do know that the sun will rise tomorrow!

The Tough Keep Going

After the longest period of peace and prosperity in American history, most all of us alive in America today share an unusual history: lives singularly without epic stress or challenge. Granted, terrible armed conflicts have occurred and we have endured harsh community and personal tragedies, but during the last half century the most formative challenges we have faced as a people have been gentle, compared to the frightful times our ancestors faced. Our grandparents' lives were punctuated by World Wars and the Great Depression, but we only have the Arab Oil Embargo, AIDS, Vietnam, and a few memorable earthquakes and big storms to complain about. Living with such unprecedented peace, and at such a clement geologic time—no ice ages or serious volcanoes anywhere nearby—we have had time to invent luxury and perfect the home.

We can be certain, based on the history of the human race and the obvious changeability of the planet we inhabit, that our good fortune will not go on forever. The paradox in this is that most of us have never been called upon to act superbly, heroically, selflessly . . . and we appear almost to have forgotten how. We have certainly forgotten that our species is the planet's steward and designer precisely because we are the most adaptable species ever to appear, and much of our daily energy is devoted to the suppression of change, the very quality we are most fitted to endure and thrive on. Have we also forgotten how to adapt?

This book, and the people in it, are students and lovers of change and challenge, seeking and finding it under their beds and in their kitchen gardens, making change where none previously existed. The unseen and unacknowledged changes that threaten the species are suspected and anticipated, even eagerly by some. We have willingly placed ourselves at the edge of civilization, testing our ability to weather a closer relationship with nature, to sustain our pleasures and comforts using no more than the energy nature freely sends us with the wind, water, and sun. Very few things ring with sureness, but we can all subscribe to Scott Suddreth's certainty: The solution comes up every morning!

CHAPTER 16

FORWARD!

AT THE FOUNDING OF OUR NATION, OUR ANCESTORS HUDDLING ON the strands of a generous but wild land had no choice but to honor the spirits of the woods and mountains. The bravest among the settlers ventured into the woods and across thickly forested mountains alive with wildlife—a virgin ecosystem of unexploited natural riches. The original peoples of those woods and mountains, together with their kin on the plains and deserts, were devotees of the sun, and walked in grateful awe amidst wonders: The land provided them with plenty, and by their habit and understanding they easily lived within the sustainable limits of the land's wealth. To the European eye, such wilderness begged to be conquered and put to use.

Growing in pride as well as in population, the offspring of these explorers, our ancestors, believed that this great land was ordained by their god for the taking, and began their epic westward predation, clearing the woods, slaughtering the game, and poisoning the streams, secure in their belief that the land would be infinite. If not infinite, America proved to be immense and abundantly regenerative, and so for centuries the effects of their assault and battery was only locally and temporarily evident. As long as the prospect of virgin lands stretched westward to the horizon from the top of almost any frontier hill, our ancestors knew that profit could be taken without worrying about mistakes, which could be left behind. The land and climate were harsh at times, and settlers kept alive the pretense of man against nature, the heroic wresting of wealth out of wilderness. Who could criticize their heroism, especially when it provided riches and comfort to all willing to risk the struggle?

Those who had tracked the land before, and who took a longer, subtler view of the world, retreated or were overwhelmed.

In recent decades, we humans have raised our voices and extended our hands to clamor and grasp for more and more energy. Our reach, which is global, and our greed, for which I have no words, have led us temporarily and spectacularly astray, costing other species their lives and rightful homes, and causing death, disease, and disarray among ourselves. The Indians who had lived on the land for millennia were removed from the places that sustained them, and confined to ever poorer reservations, or assimilated, by force and by law. The land itself was subjugated, its timber taken, topsoil eroded, and mineral wealth exhumed, and the remaining lands were left to be ravaged by the unabated elements. The land's greatest wealth, the fertility of its soil, now washes at unprecedented rates down the great rivers and into the sea. The Oglalla aquifer, a great reservoir of water underlying the western plains, has been pumped nearly dry. Neither the ancestral forest, with its ancient but relatively fragile web of cooperating species, nor even the mighty rivers, were safe against the onslaught, because the Euro-American, although puny in singularity, had learned to hunt as an army, in numbers great enough to defeat Nature. Having taken the easiest pickings on this continent, acquisitive developers now turn to work their evil magic on other continents, and other peoples. A culture founded on quick seizure and rapid growth now finds itself hurtling toward an inelastic wall: the end of natural abundance. We have used up Earth's easy resources and stressed her ability to recover and restore.

Along the peripheries of our cultural reach, where the roads and powerlines play out, and where we can still find a hilltop from which the vista seems limitless, the land retains its vigor, and nature's wisdom can still be sensed. We know, even as we resist such knowing with the atavistic enthusiasm of the damned, that the limits are before us. Some who settle along the fringes maintain an eschatological view: "Humanity's headlong rush to devour nature has gone too far, and the end is near." These prophets remind us that nine of every ten species ever to walk the planet have gone extinct. Because extinction is so common, the passing of the human species, like that of the dinosaurs before us, would be as ephemeral an occurrence as the wind in the trees, having no permanent importance in the larger scheme of things.

The voices in this book tell a different, more hopeful story. Explicitly, or between the lines, I hear them describing a turning of the tide. Sending that message inward from the periphery, we seek to pass back toward the center news of a regenerative partnership with the forces of nature, an appreciation for the wisdom of the land, and a plan for salvaging the future for our children. We have evolved; we have adapted ourselves and our tools to sustain an honest, joyous, and fulfilling life without damaging the fabric of nature or destroying the geological and biological wealth of our host planet. Our approach to energy cooperates with earthly processes rather than exploiting them. We thrive under a conservative approach to matter and energy, and a shared sense of the value of all life.

In chapter 5 Stephen Heckeroth compared Earth's geological history to a four-and-a-half mile walk, with each mile representing a billion years. In the last few hundredths of an inch of this pilgrimage, with this sense of partnership, this vision, and these tools, we hold out a chance to transcend and survive. The vista we cherish before us is the most exciting ever seen: the possibility that we may do as California governor Jerry Brown suggested: Serve the people of the world, preserve the planet and all its life forms, and go on to explore the universe.

What We Know

We know that to survive we must find the solutions to puzzles that have until now defeated our species. We must stop overpopulating. We must make war no more. We must use energy sensibly. If we continue to hurtle down the path of overpopulation, warfare, and prodigal use of energy, we will crash. If, as individuals, we can each slow our pace, embrace our neighbors, develop an extraordinary regard for re-

sources, and commit personally to living reasonably within the budget allotted to us by the sun, we may succeed.

We know that the tools are available to us. A solar-regulated life may be lived anywhere on this planet in comfort and plenty. We need not wait for technological miracles; the miracles we already have, if competently applied, together with the inevitable refinements and breakthroughs, will provide enough for all of us. By learning to live with the energy supplied freely by the sun, we elaborate a model than can, with few limitations, be refined and applied everywhere on the globe. We will find sufficient to our needs the sunlight that falls on us and on the plot of land we care for, along with the winds and the waters that are moved by sun and seasons (as well as certain exotic and beloved necessities, including our chocolate and oranges).

But now we know that we face real limitations, which, though few enough, may be brutally unyielding. Here are our challenges:

— *Overpopulation:* We must accept that quality of life and the carrying capacity of an environment bear a constant relationship to each other. If we wish to improve individual life, we must learn to control the way we breed. This requires a great deal of us and of our institutions: We must evolve our notions of family very quickly, because what was right and proper half a century ago is now absolutely wrong. One natural child is enough for each mother. Every other species on the planet regulates its population, or is regulated by natural controls, to stay within the carrying capacity of its range. Having disabled the natural controls within the last century, we must learn to regulate ourselves or expect more powerful controls to assert themselves.

— *Greed:* We must learn that amassing wealth steals from others while spiritually impoverishing the wealthy. The model of perpetual growth is impossible in a closed system such as that of our planet; growth and consumption can exceed the fertility and regeneration fueled by momentary sunlight only at great cost and inevitable, eventual failure. The colonial pattern of depleting local stocks and then stealing the resources of weaker peoples by war or trickery cannot be allowed to continue.

— *Impatience:* As we innovate, we must learn to discern and live within the earth's prevailing cycles, treading gently at first, harvesting only gradually and in the fullness of time, being always careful to heed early signs that tell us if we are on course or gone astray. We must learn to resist the temptation to rush quickly to market without first understanding the whole cost and effect of new products and endeavors.

— *Conformity:* We must learn to tolerate and celebrate our own particularities, and the diversity in humankind and every other kind of life. We never know when we will need to appeal to the hidden strengths that variation carries within its designs. At the same time, we must honor the human and the living, and shun the deadened, and dead-end values, of lifeless corporations and commercial constructs that destroy life and dishonor diversity.

We know that the time available to us is limited. Accidents encountered along our present course have in the recent past frightened us into periods of terrorized immobility and reluctant sanity. Chernobyl, the Gulf War, AIDS, the ozone holes, global warming, El Niño, the flooding of Midwestern rivers, urban forest fires, the harshness of hurricanes and cyclones: These are signposts that we ourselves helped to erect, and all bear the same small print: "Turn back; go no further; this way lies danger."

Quietly now, from the hills, down the muddy roads, the pioneers are coming back, reporting new-found discoveries and earthy delights. Neither gold nor timber this time, nor peoples to enslave, nor fortunes to be made, but for an even brighter hope: We have within ourselves, and falling from the sky, everything we need to live sustainably and well. This bounty from the energy frontier can be used to guide and enable a planetary renaissance.

What We Must Do

First, we must be rational in our own lives and all our works. As we live and when we build, we must be attentive to the offerings of our environment, and live within the boundaries they suggest, right down to the smallest details. As we live with and within nature, and as there is no waste in nature, we must waste nothing. The voices in this book give good counsel. Wes Edwards recommends a small house. David Katz and Robert Sardinsky agree with Einstein's precept that "Everything should be as simple as possible, but no simpler." Felicia Cowden tells us that "Solar power's biggest gift to the environment is showing people that it is possible to live well without being wasteful." Hearing these words is not sufficient; we must put them into use.

Next, we must work to move ourselves, our neighbors, and our institutions along a course leading toward sustainable practice powered by renewable energy. This will demand all our patience, our intelligence, and our powers of persuasion. As opportunities for environmental exploitation and quick profit collapse, we must find ways to pay the whole cost of everything we buy—to make the marketplace responsive to long-term needs, as expressed by our children and grandchildren "unto the seventh generation." But we must take an even longer view, beyond our familiar horizons, to the widening future we wish for our planet and all those living on it. By bringing the global vision home, Leigh Seddon, Steven Strong, Maria and Arnie Valdez, and, as you will soon read, Amory and Hunter Lovins, urge us to participate, to help our leaders and representatives—who are more lost than we are — find the right way.

Finally, we must be lucky and brave. We can hope for serendipity. We can hope to find, by accident, the few tools and clues we need but do not yet have — better energy storage technologies, more efficient ways to collect the sun's rays, a more compassionate and far-sighted attitude toward the less powerful—but we must not count on technology or government to bail us out. We must be vigilant, perpetually ready to take the better path when it is offered, and just as quick to backtrack when we see we have strayed down a wrong path. For example, we are sure now that the internal combustion engine, though familiar and addictive, is infernal. Are we willing to give it up?

The soft path: Hunter and Amory Lovins's story

From their mountain fastness beside Capitol Creek in Old Snowmass village, Hunter and Amory Lovins have been quietly working for many years to visualize and then realize a soft path for human survival through the hard realities of economic and nationalistic bullishness. By making their own home, and their Rocky Mountain Institute (RMI), into working models of the solutions they propose, they have become authoritative and experienced prophets. Their advice is sought by governments and corporations around the world. We talked in 1993, first about the house in which they and the Institute live, and then about the future. Several of the items Amory predicted for the future at that interview were: "a photovoltaic panel with a micro-inverter built right into the frame, so you just connect it to the building's wiring" and "a variable selectivity window with photovoltaic sensors and drivers, also in the frame, which distributes intelligence throughout a building, so you point a gizmo at it and push buttons to change the setpoints." These items are already available or very near the marketplace. Since that time, RMI has published several important papers, shaped the dialogue about hyper-cars, and continues to thrive as the world's foremost energy think tank.

Amory: In building the Institute, we tried to anticipate every threat. One nice catastrophe we're unprepared for would be an earth-sheltered elk on the flat roof falling through the greenhouse glass. We learned some important things in the uninsulated cardboard sieve of a house we were in before we moved here. A power failure, for example, caused a cascade of nasty events: With electricity out, we had no water, and the furnace stopped, and without water pressure or electric heat-tape, the pipes and radiant heating tubing soon froze.

In contrast, we tried to build our own place so there were three ways to do everything.

Hunter: It works well the way it is. We fix things that break, like the inverter, but we're not thinking of major changes.

Amory: There's no battery on the photovoltaic system, because we have a utility interconnect. Our electric co-op insisted on a grid-excited inverter, to guard against our energizing the line when they think it's not energized. Interestingly, our power has turned out to be of slightly better quality than theirs. If you regard this as a household, our 2-kilowatt array is two or three times what a very efficient household's requirement should be. The average use of the whole building is 1,200 watts, but only a tenth of that is for the household; nearly all the rest is the office equipment in the Institute's headquarters at the other end of the building.

Amory Lovins.

Hunter: This system is more hands-off than a normal house system, which takes quite a lot more ongoing maintenance.

Amory: But the maintenance required here is of a higher order, because the equipment is more intelligence-intensive.

Hunter: A house ought to have stand-alone integrity. This system is intelligence-intensive only because it was cobbled together—for example, the hot tub system. When it was built, the equipment was less common and more specialized, like house-trailer appliances, which, when they break, have to be sent out to be fixed. We now have local experts who come to work on our systems. For example, there are lots of people who can properly set the seasonal angle on the PV array or check the chemistry on the active solar water-heating loop.

One of my pet peeves is that so many people don't think about how passive systems work, and do stupid things. Even so, this house is pretty forgiving: If you leave a window open, the thermal mass reheats the overcooled air so the house temperature recovers.

Amory: The problems are often simple, like when the lower fishpond crashed, it turned out that algae had grown over the bubble hose. We lost the fish for a few days, then fry washed down the recirculating stream from the unaffected upper pond and re-colonized the lower one.

Hunter: And we have annual attacks of cotton scale.

Amory: Living here, we've learned that some of our original ideas can be further refined. We don't run the main air-to-air heat exchanger at night in midwinter, because letting the sun regulate the air is simpler and better. The greenhouse fan that heats up the CO_2-depleted boundary layer on the leaves (so the plants grow better) is usually connected to a solar module, and runs only when the sun warrants (although I notice it's disconnected now, which is one of the puzzles of having so many people in one house). We plan to run the waterfall (recirculating system) the same way, with a 12-volt pump and dedicated PV, which would better match the pond's oxygen requirement. Things work quite well, even where right now it's partly manual and someone from the Institute has to occasionally turn a few things on and off.

Hunter: It is much busier here than it was ever meant to be. The building was planned for twelve people, but there are now thirty-six at the Institute, installed in six buildings.

Amory: We only have administration and outreach in this building now. Oh yes, and there's seventeen more at E-source in Boulder. . . . That's a lot of corks to keep underwater!

This house is pretty forgiving: If you leave a window open, the thermal mass reheats the overcooled air so the house temperature recovers.

ROCKY MOUNTAIN INSTITUTE

The Rocky Mountain Institute (RMI) is a non-profit resource policy group whose executive director, Hunter Lovins, and director of research, Amory Lovins, are authorities on energy strategy. The Institute's main offices are in a comfortable, super-insulated four-thousand-square-foot rock, timber, and glass building beside Capitol Creek in Old Snowmass, Colorado. The building's twenty-one-desk office is on one side of the central bio-shelter, and the kitchen, entryway, and Amory's work pod are on the other.

It gets cold at 7,100 feet, down to forty below every winter, but the Institute building has no conventional heating and derives 99% of its space heating from the sun shining into the bio-shelter, a large greenhouse built right into the building's center. Energy-saving features are used throughout the building. Its slip-form rock walls are sixteen inches thick and rated at R-40, and the roof is R-60. Nearly all the windows are argon-filled double glazing with a thin-coated polyester film within, which are five times as efficient as normal single-pane windows at keeping heat on the desired side of the window. The glass greenhouse roof glazing and windows in the curved east, south, and west walls allow winter sunlight to reach the massive north wall that serves as thermal storage to help keep the temperature uniform within the building. The glazed greenhouse roof continues above the roof line so that it can vent excess heat. Unusual and experimental features abound, including a solar clothes-drying closet and air-to-air heat exchangers. Even the copying machine and the teapot were chosen for their energy efficiency.

The building feels like home to a large, busy family. Staff and visitors gather at the dining table between the kitchen and the greenhouse. The multiple open levels and curved walls give a sense of openness and unified space. On the north wall a recirculating waterfall, purportedly tuned by a master waterfall-tuner, feeds a small stream and two ponds that runs through the greenhouse, which is filled with tropical plants and home to a cranky iguana; everyone else appears to be happy and productive.

Hunter Lovins moves RMI's autonomous rat catching system to a new location.

JON STOUMEN

We've upgraded the other buildings that RMI owns, putting in high-tech glass. They all have efficient lighting, of course.

Hunter: We worry that governmental officials, who should know better, thoroughly confuse conservation and efficiency, and call for all Americans to sacrifice . . . . Exactly the wrong thing to say; back to the days of Carter's cardigan sweater.

I wish I thought that anybody in the Administration really understood the fundamental problems outlined in, say, the book *Beyond the Limits.* I don't see anyone acting like they understand. The new bunch may be a lot better than what has been in there before. It seems to me that most good people are going to say, "These new guys are so good, we don't have to do anything anymore." That is a risk that could get us all in a lot of trouble. There are enormous interest groups that want the business of government to carry on as it has for so long. [Hunter is called away.]

Amory: I got a call today from a guy in the power generation business, who says that his industry has decided to lobby for the BTU tax to be levied after generation at the power plant, so that all the conversion losses go untaxed. Instead of taxing the fuel for 1,000 BTUs, only 300 BTUs or so would end up getting taxed because the rest is upstream. Their precedent for this is that the refiners think they've got permission to pay on their output, not their input. The gas pipeline people want to pay at the retail, not the wholesale end. This is all breathtakingly dumb. The whole point is to tax depletable fuels as they come out of the ground or into the country, so you give an incentive to reduce not only end-use losses but also distribution and conversion losses.

I second Hunter's concern. Nixon made a wonderful speech about population, perhaps one of the best ever made by an American president, but nobody ever thanked him, and so he never mentioned it again. If a president tries to do something good, but gets opposition and not support, he'll back off. This also happened in the early Carter years, where we figured our people were in there now, we didn't need to work. We really need to work harder to make sure that the things we are trying to do actually happen. Efficiency and self-sufficiency issues now need public support, vigorously expressed.

I'm very interested in making changes around here that allow me to transmit electrons rather than protoplasm, using Picture Tel. You see, this protoplasm really doesn't like to travel. I've done Picture Tel for things like addressing a bunch of CEOs in Germany without having to go there, and it worked fine. There is a learning curve, and so the available equipment and software is decreasing rapidly in cost. The big price is still the equipment. We're having a terrible time with our telephone company, because we need upgraded service, and this is a rural system. I guess the local switches just can't handle it, so it may take a while.

About a seventh of all companies now accommodate telecommuting. At least seven million American workers already do their work at home, and their employers see a 15 to 20% gain in productivity when they do. Not to mention a lot of private and public costs in commuting avoided.

Electronic presence seems to be very well accepted as long as it is technically good. You need to be able to send graphics, and see what is going on. We are working on upgrading our systems, but it works very well to communicate with the Boulder office using fax and modem.

You're interested in the effects that a band of self-sufficient homesteaders might have on the global energy picture, yes? Small effects, I think, but material, and growing, and heading in the right direction.

It's realistic to plan on self-sufficiency if the desire is there, even if you include all the energies that go into a home. We certainly had it in mind when we built the Institute building. We used indigenous rock. The core of foam insulation and the Portland cement from Salt Lake made this an energy-intensive structure to build, but when amortized over a long period, that cost is acceptable. The oak timbers from Arkansas were sustainably harvested, in the sense that replacement oaks were planted when these great beams were harvested.

I think a much broader question is, how will the utility system evolve? Ultimately, it will look like the telephone company looked (when we had a telephone company): basically bookkeeping, dispatch, and operations mediating between a lot of distributed sources. Carl Weinberg, who managed research at Pacific Gas & Electric (PG&E) before his retirement, proposed the distributed utility, in which you don't baseload the power plant, you baseload the grid — an excellent concept. In the past, utilities would focus mainly on generation, less on transmission, still less on distribution, even less on customer service. In the future, it is going to be exactly the other way around. Generation has already become a commodity; PG&E just dissolved its engineering and construction division because they're not planning to build any more big plants. They are reshaping the company to match the new priorities of customer service and distribution emphasis. Even in more built-up areas where there is less of an economic or cultural incentive to go independent, you'll still end up with technically similar systems. When we ultimately do have cheap photovoltaics, they will be all over the place, and utilities will simply write off a lot of power plants and not build more.

You're interested in the effects that a band of self-sufficient homesteaders might have on the global energy picture, yes? Small effects, I think, but material, and growing, and heading in the right direction.

Many utilities offer skid-mounted PV systems of from one to three kilowatts to drop in wherever line extension would be uneconomic. Utilities in western and south Australia and even just outside the metro Sydney area are in a bind, because they are supposed to sell rural electricity within 10% of the urban price, even if it means extremely long line extensions. So they're starting to put in a lot of PVs. I think adding PVs to the mixture, along with end-use efficiency, makes a lot of sense.

[Hunter returns triumphantly, the long-missing organic rodent control mechanism—a four-foot rat snake—around her shoulders.]

The way we compare electric resources is all screwed up. There are about a dozen important advantages of dispersed, renewable sources (and usually of end-use efficiency, too), which are not being taken into account. The removal of fuel-price risk means that a lower discount rate should be used, which makes PVs cost-effective today. The lead-time difference by itself makes them worthwhile: You can add capacity with PVs much faster than by building a plant of any fuel type. And accounting for externalities makes PVs worthwhile, as does avoiding distribution.

You can add capacity with PVs much faster than by building a plant of any fuel type.

An old friend of mine from Holland, Dr. Caes Daey Ouwens, built a 50-average-watt house, similar to ours except with a gas refrigerator. He then did a stand-alone PV system, instead of taking a drop off the pole a few meters away and installing a service entrance panel. So he was able to avoid a small part of the distribution cost. He couldn't avoid the cost of running wires down his street, but still he got a ten- to twelve-year payback from capital cost plus energy savings, which kind of surprised him. Then he went to a settlement in Indonesia that had subtransmission running right by the village. There he was able to avoid the costs of drop, step-down, switch gear, transmission into the village, and wiring the village. Instead he gave everybody some efficient end-use devices and stand-alone PVs on amortization at the utility discount rate, without subsidy. The result: They got a positive cash flow immediately, because the debt service was less than the villagers were already paying for radio batteries and lamp kerosene.

If that's true for a Third World village, it ought to be true for a bunch of us, too. The Indonesian grid cost-per-kilowatt-hour is pretty low, and so the basic point is quite revealing. In December 1993, I offered a paper on this whole subject, "Better Ways to Compare Electric Resources." There is a whole lot we are leaving out when we compare bus bar cost at a flat discount rate.

How long will it take for us to recognize this and put it into practice? Probably five to ten years. Some jurisdictions will probably do it earlier, which is why I want to write this paper this year, to put some pressure on.

The slowest-moving parts of the energy-delivery apparatus are the regulatory agencies and the bulk of the utility industry. They are pretty much hand-in-hand at how fast they come along. The most progressive utilities, PG&E being the outstanding case, will be leading the rest. Florida Power is coming along; Tampa will take longer. There's been a surprising movement lately in Commonwealth Edison, American Electric Power, Southern Company. Some of the systems that used to be real stick-in-the-muds are now showing a lot of signs of intellectual life. We are seeing, I think, a generational change, as the people who put the old policies in place gradually die or retire. But some of the change is surely due to an incurable attack of market forces. It costs more, now, to back up and clean up grid power than it does to build a small, self-sufficient source. The Federal Aeronatics Administration (FAA) is using this approach to switch ground-control avionics over to PVs even when there is already grid power to the site.

The biggest revolution will have to come in the building trades, because stick-building is so appallingly primitive. Good structural engineering dictates that a stud wall should be 15% wood, but generally it's more like 35%, because carpenters are so sloppy with corners and blocking. The institutional barriers to such reforms are staggering. We've identified about twenty-five parties—in conception, finance, planning, building, maintenance, and so on, *ad nauseum*—who speak different languages (there's no Rosetta Stone), and who have perfectly perverse incentives. The system penalizes efficiency and rewards inefficiency and obstruction, so we've got to find ways to turn that around, just as we did in the utility industry. The problem is, there's no customer feedback from end-users in housing design. We already know that lack of improvement is the death of any manufacturing business, and if we manufactured any product with as little customer feedback as when we make buildings, we'd have gone broke long ago.

We have the technical knowledge to yield buildings that are ten times as energy efficient, and cost less to build, but we've seen hardly any real improvement in fifty years. When an architect is studying, does he ever talk to a practicing mechanical engineer (M.E.)? He's too busy studying how to build something that is, as they say, all glass and no windows. When he's done, he pitches it over the wall to the M.E.s, and says, "Here, cool this!" Designing a building nowadays isn't team play, it's a relay race! We've got to find a way to make the rewards proportional to the performance efficiencies of the building. As it is now, fees and profits keep going up every time the baton is passed, and nobody's working or thinking too hard. That's obvious: If the building's equipment is simpler, there'll be less fee and less profit.

We think we've found a way to disconnect this loop: We're offering a two-part fee, first for design, then a bonus based on x% of the energy savings. Ontario Hydro has a similar program where the design team gets a performance-based rebate from the utility equal to three years' worth of energy savings.

We've just signed off on the basic design for a new tract house in Davis, California, for PG&E, in which we were able to push the designers to see what it would take to get rid of the cooling hardware entirely. In the process of planning a modestly air-conditioned home, we challenged them to find, and set aside in a basket of cooling package measures, every technique, no matter how small, that might be used to reduce the cooling load further: double drywall, white roof with radiant barrier, super-windows, refinements on that order. When they'd finished the first design pass, they'd come up with a house better than Title 24 standards with one and a half tons of refrigeration instead of the three tons they expected. Then they shook out the cooling package basket, and found that it added up to just enough to make even the ton and a half go away. But the best surprise was the overall cost: Assuming that 36% of the initial electric use was user plugloads that couldn't be analyzed until the occupants showed up—hair dryers, TVs, computers — they had realized a savings of 62% of the site energy, or 89% of the analyzed load, below the 1993 Title 24 standards. The marginal capital cost of the measures (compared with the base case) was minus two thousand dollars; and the marginal maintenance cost was minus seventeen hundred dollars. Win, win, win. Never mind that the house will be quieter, more comfortable, and more healthful.

Let me ask you something. How do you feel, meeting here in the Institute? Buildings are supposed to make you feel good, but too often the questions that get asked have to do with equipment, not with application. We made an effort to introduce curves into our interior space; if God meant us to live in boxes, wouldn't she have given us corners? The greenhouse's plants produce plenty of oxygen and ions and nice smells. And we tried to design out the mechanical noise and 60-Hertz electrical smog you find in almost every building. Hear that? That's pure silence (except for the waterfall, which is more or less tuned to alpha rhythm). If you knew you could live and work in space like this, would you be willing to work anywhere else?

To most of us, such things seem so obvious, and yet . . . my friends, we are in for difficult and dangerous times, but the consciousness is rising. We might yet get out of this alive . . . who knows? ✒

The Future's Gifts

Throughout this book I have tried to keep an elemental, particle-sized focus on the work we are trying to do—making independent homes for ourselves. As each minuscule detail is perfected and brought into a comfortable symbiosis with its neighboring details, a nearly organic web of life develops. Continuing to travel, my perspective keeps widening, and I find growing centers of awareness everywhere I look. Each is slowly, ever so slowly, extending its reach and influence; where the edges touch, a strong, fibrous community begins to take hold, and a broader local awareness is born. This network shimmers on the brink of consciousness—the hundredth monkey may not have joined hands yet, but she is standing in line, and her eyes are sparkling with prescience.

In the next decade, we will see consciousness dawn as surely as the sun rises. Possibilities that were unthinkable a quarter century ago, but which become practical as I write and you read, will become routine and habitual everywhere. For two centuries, since the time when we learned to harness energy and use it to drive our tools, we have been assailed by restlessness and hunger. We found power so intoxicating that we have taken two centuries to recover our more ancient knowledge that consuming energy is not an imperative, but only a means. Hung-over by the lingering effects of earlier overconsumption, we must now come to rely on the diminished resources remaining to us. Those who have already taken this path declare that the result is not impoverishment or hopelessness, but wealth, health, and an invigorating sense of self-determination and independence.

Deep-rooted human values, natural values, will eventually bring our focus home. There we will find ways to restore the authentic humaneness of our society, to empower and enfranchise, and to broaden our regard for every form of life. We will put our restlessness and greed aside, along with its illusory trappings of national pride, race pride, sex pride, species pride, and at last we will join the intergalactic family.

We need not look to heroic technology to bail us out of our dire energy predicament; the time of the quick fix is finished, and we cannot count on miracles. Our history is no Greek tragedy, to be resolved by gods descending from machines. Awakening from a two-century-long bad dream, where we made ourselves sick and nearly insane with the belief that we could conquer Nature with technology, we now know that technology grants tiny boons—an image of the way a photon behaves, the splendor of buckminsterfullerene. We have almost all the necessary tools, and must now dedicate ourselves to using them well.

The independent home is a masterpiece of appropriateness and hope. Over the next two decades, as the spirit of independence comes down from the mountains, pervades our culture, and is welcomed home by every family, we will all at last attain our energy independence.

A great weight will lift from the planet, and from our souls, wherein we feel our oneness with her most powerfully—and we will be able to get on with the sustainable business of living.

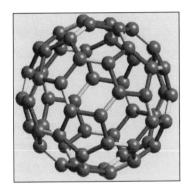

A recently-discovered form of carbon, called Bucky Balls because of its reliance on the same geometry as Fuller's geodesic dome, comes in a variety of other useful shapes. This is C-50, one of the commoner forms.

HOME ENERGY AUDIT FORM

This worksheet helps you perform a quick and approximate energy audit of your home. The information is very valuable for making sensible decisions about energy features. Please remember that this is supposed to be quick, and need not be complete (although the more complete, the better). If you don't know answers, estimate. If you haven't the foggiest, leave blank.

For electricity, the number of kilowatt-hours is more important than the dollars spent, but for other energy sources, the dollar value is more relevant. Gasoline means the amount of vehicular fuel your family purchased.

trends from previous years are often instructive...

Energy Consumption	last four quarters:				something happens here	next four:			
	Jan-Feb-Mar	Apr-May-Jun	Jul-Aug-Sep	Oct-Nov-Dec		J-F-M	A-M-J	J-A-S	O-N-D
Electricity	kwh $	kwh $	kwh $	kwh $					
Heating fuel type:	therms $	therms $	therms $	therms $					
Firewood source:	cords $	cords $	cords $	cords $					
Gasoline	gallons $	gallons $	gallons $	gallons $					
Water source:	gallons $	gallons $	gallons $	gallons $					
Sewage destination:	gallons $	gallons $	gallons $	gallons $					
Garbage destination:	barrels $	barrels $	barrels $	barrels $					
Repair & Maintenance	$	$	$	$					
Improvements	$	$	$	$					
			TOTAL:	$		TOTAL:			
						SAVINGS:			

House	material	insulation
Foundation & floors		
Walls		
Roof		
Additions		

Windows	type	percent of wall	year installed
South-facing			
West-facing			
North-facing			
East-facing			
Skylights			

Appliances	year purchased	fuel	hours per day	months per year
☐ Furnace				
☐ Air conditioner				
☐ Refrigerator				
☐ Other refrigerators?				
☐ Freezer				
☐ Hot Water Heater				
☐ Hot Water Circulator				
☐ Water Pump				
☐ Other big loads				

Site Map Form

South Elevation — the house from the sunny side

family name: _____

location: _____

number of members: _____

house - square feet: _____

number of rooms: _____ bedrooms: _____

land - elevation: _____

area: _____

Plot Plan

Your house in the center, showing roof slopes, outbuildings, roads, well, powerlines, neighbors' houses, trees...

Show, if you know, where storms and winds come from, the directions of the best views, and other influences

100 paces

Please remember: this is *not* an art contest. Just do the best you can...

South is down

GLOSSARY

afterbay: the pool below the hydroelectric generation plant where the water "borrowed" to spin the turbine is reunited with the stream.

airfoil: a shaped blade or wing designed to provide lift. When the blade splits a moving fluid such as air into two streams, and the shape forces one stream to travel farther, the pressure on that side is reduced by the Bernoulli effect, and the blade tends to "lift" into that stream.

allelopathic: an interaction between two species in which one or both interfere with the processes of the other. In plants (such as the eucalypts or bolting lettuce), this is often accomplished by production of a material tolerable to the producer but poisonous or unpleasant to other life forms.

alpha rhythm: brain waves in the 8 to 13 cycles per second frequency range, usually typical of a normal awake person in a quiet resting state.

alternating current (AC): electricity that changes voltage periodically, sixty times a second (or fifty in Europe). This kind of electricity moves more easily than direct current (DC).

alternator: a form of generator that produces alternating current; common in modern automobiles.

ambient: the prevailing temperature.

amorphous: "without form"—often used to describe the chaotic silicon lattice of thin-film PV modules.

amortize: paying a debt over time in periodic installments that typically include interest.

ampere: the unit used to measure the instantaneous flow of electrons, theoretically, 6.02×10^{23} electrons. Abbreviated as amp.

amp-hour: measure of a battery's ability to sustain a flow of energy over time; 60 amp-hours indicates a battery can deliver one amp for sixty hours.

animist: one who believes that everything, animal, vegetable, or mineral, possesses a soul or spirit.

anode: the electrode (in a cell) toward which negative charged ions are drawn.

appliance: "a thing used as a means to an end," a machine to achieve some useful purpose, such as to keep beer cold or make music.

architect: from the Greek "chief builder"; "A person skilled in the art of building; one who understands architecture, or makes it his occupation to form plans and designs of buildings, and superintend the artificers employed." (*Webster*, 1828)

architecture of dominion: building as if humankind ruled everything.

Ark, the: Noah's boat, or by extension, any isolated repository of rare and endangered life forms, for example, the Hawaiian island of Maui.

average-watt: the wattage consumption of an appliance or operation averaged over a long period. Appliances often use more energy when starting.

avoided cost: the money saved by the utility because it is able to use a cheaper source; the minimum amount set by PURPA legislation that utilities were obligated to pay independent power producers. When this idea originated during the Carter era it was meant to define the whole amount utilities might avoid paying by buying surplus electricity from small power producers. If, for example, the utility could use clean renewables instead of buying OPEC oil, it might also be able to avoid paying for transporting and storing the oil, and for cleaning up the combustion pollution. If the replacement source was deemed sufficiently abundant and stable, the utility might even be able to avoid paying for building a new oil-fired power plant, in which case the avoided costs might be substantial. During the Reagan era, utility interests weaseled the definition to exclude all factors but the cost of the oil. Ironically, many in the renewable energy business believe that their industry is stronger because it grew without government subsidy and intervention.

back-to-the-land: a social movement in reaction to urbanization, in which people leave their "civilized" jobs and homes in the city and suburbs to return to a simpler, more pastoral life.

bagasse: the sugar cane after the sugar is squeezed out, a fibrous material that can be used to fire boilers to produce steam to spin turbines to generate electricity.

balance-of-system: equipment that controls and measures the flow of electricity.

barostat: a device that opens or closes a switch depending on the barometric pressure.

baseload electricity: the smallest amount of electricity consumed by a utility's customers. Baseload is provided by slow-to-start, relatively inexpensive-to-operate generators, while peak-load is provided by quickly dispatchable sources.

battery: a collection or grouping of similar entities, hence a battery of artillery, or of tests. In electrical usage, a collection of cells (often misused to mean a package containing one or more cells).

battery charger: a device to apply electrical current to a battery or collection of cells in order to store electricity.

belt-and-suspenders: redundant systems.

berm: earth mounded in an artificial hill.

bioengineering: using engineering techniques to "improve" life (as in designer crops or genetic engineering), or applying biological techniques to accomplish manufacturing work previously done by fabrication (heat, beat, treat).

biomass: wood, bagasse, indeed any mass (usually combustible) created by biological processes.

biosphere: the thin layer of air and water surrounding the Earth that supports life,

blackwater: used water out of a toilet or other grossly unsanitary source.

blower door: an adjustable panel that blocks a doorway in a house, fitted with a fan controlled by a barostat.

blower test: a test of the leakiness of a house conducted with a blower door.

boule: a barrel-shaped single crystal of purified silicon or other semi-conductor metal. The thin wafers of silicon used in photovoltaic cells and integrated circuits are cut from the boule using a thin diamond-bladed saw.

bridging fuel: a fuel available in limited quantity that can be used to "bridge" the time required by technology to discover a better source. Petroleum is considered by experts to be a bridging fuel, and atomic power was supposed to be an unlimited source.

BTU tax: an often-proposed means for rewarding conservation by taxing consumers for every BTU of fossil energy consumed.

bus-bar cost: the price of electricity at the point of delivery; for most of us, the retail cost.

carcinogenic: cancer-causing.

carcinogens: substances known to cause cancer; benzene, a byproduct of combustion of natural gas and propane, is a carcinogen.

cathode: the positive plate in a battery.

caulking: goo pumped between ill-fitting parts of a house to discourage water from infiltrating. The goo comes in a number of colors, compositions, and abilities to adhere to glass, wet wood, plastic, paint, and metal. Caulking in most cases cracks or shrinks, and so the process must generally be repeated periodically.

cell: the smallest self-sufficient working mechanism. A single-celled animal may have been the first life form. A solar cell is a single slice of doped semiconductor and supporting hardware that produces a voltage determined by the chemistry of the semiconductor; the larger the cell, the greater the amperage, but the voltage remains constant. A storage cell in a battery is usually a container full of electrolyte plus a cathode and an anode that also produces a voltage determined by the chemistry of the cathode, anode, and electrolyte; as with the solar cell, a larger storage cell will produce more current, but the voltage remains roughly constant.

centrifugal pump: a mechanism for moving fluid by centrifugal force. The fluid enters near the axle of a rapidly spinning impeller, which impels it away from the center. Since the fluid dynamics around the spinning impeller are chaotic, centrifugal pumps are inefficient.

charge controller: the device in a stand-alone energy system that feeds electricity from the source, typically a PV array, to the battery bank. The charge controller protects the batteries from overcharging.

charge gradient: an electrical "slope" created by mingling ("doping") a semiconductor with atoms that give up or take up electrons more readily than the base material. To a free electron, such a gradient looks like a hill to roll down, away from the negative and toward the positive.

circuit breaker: an automatic switch that senses too much current and turns off. Fuses perform the same function, melting when too much current passes through the fusible element; unlike fuses, a circuit breaker can be reset after it cools off and the overcurrent condition is corrected.

CNC: Computerized Numeric Control, in reference to milling equipment that can be programmed to form complicated curves.

collector: in the solar world, a device that harvests solar energy, sometimes by heating water or any transfer fluid, sometimes by heating air. Properly speaking, photovoltaics aren't collectors, they are modules. See also **photovoltaic modules.**

compact fluorescent: a fluorescent light about the same size as an incandescent lightbulb, created by folding a small diameter tube; a compact fluorescent usually also includes the electronics that convert house current to the pulsed high voltage electricity that excites the phosphors in the tube. In a typical fluorescent fixture, the electronics (called the ballast) are separate from the tube.

comparables: in real estate, similar properties that have been sold recently. Because an off-the-grid home's energy system changes many of the fundamental costs of ownership, and because independent homes are seldom sold, finding comparables for such a home may be difficult. Since bank loans are often based on comparables, this makes financing off-the-grid homes more difficult.

compost toilet: any toilet designed to make the unmentionables go away by composting them, usually along with grass clippings, sawdust, or moss.

conduction: one of the means whereby a quality is carried through a material (convection and radiation are two others). Heat conduction takes place when energetic (hot) molecules bounce off lower-energy molecules, transferring some of their energy. Heat can be carried by conduction through solids—that's how a stove heats the water in a teapot. A material's resistance to conduction is measured as its R-value. Electrons can also be conducted through a wire or other conductor by moving from atom to atom; in this form of conductance, the resistance is measured in ohms.

conserve: to use as little of something as possible, so that the balance may be preserved unused, or so that a finite supply can last as long as possible.

consumer: one who consumes. In this book, this term is used in two ways: We are all consumers, in that we consume oxygen, food, and so forth; but "consumer" is also used in a pejorative sense to denote someone who consumes for the sheer sport of consuming.

convection: one of the means whereby a quality is carried through a material (conduction and radiation are two others. Convection can be observed (by adding a marker such as incense) in the way heat moves through the air of a room—generally, colder air falls, causing warmer air to rise because it is lighter, its molecules being more energetic and farther apart.

conversion: any process in which one form of energy is converted into another; for example, a generator converts spin (mechanical, rotational energy) into a flow of electrons (electrical energy).

conversion loss: the energy lost in converting one form of energy into another. The sum of the conversion losses between source and final desired effect defines system efficiency. An internal-combustion driven automobile is an extremely inefficient system: Conversion losses in an automobile may be as much 95% of the original heritage fuel consumed.

cost-effective: an accounting term that describes a process wherein the overall value of its benefits are greater than the sum of all its costs.

CRI: Color Rendering Index—a measure of the faithfulness of color perception as compared to sunlight. Under certain conditions, *un*faithful color rendering is desirable, as for example over the salad bar, where lights are chosen to accentuate the greens.

current: flow. In electrical terms, current means the electrons flowing through a conductor, measured in amperes.

curtain-wall: a building with glass on the outside and the structure within; to survive a hot afternoon in such a people-cooker, curtains will be handy.

cutting windspeed: the speed at which a wind-spinner's blades start making a noise loud enough that it carries upwind and annoys the neighbors. The noise is due to chaotic, non-laminar flow around the blades, caused by the airspeed of the blade tips breaking the sound barrier.

dead dinosaurs: a flippant term for fossil fuels. In fact, most fossil fuels were laid down hundreds of million years before the dinosaurs.

deciduous plantings: trees and shrubs that are leafy in summer when shade is welcome, but lose their leaves in winter, allowing more light to pass through their bare branches. Deciduous plantings on the western side of a house reduce heat build-up that makes some houses unbearably hot during long summer afternoons.

degree-days: a measure of the need for heating and cooling; one heating degree-day calculated at a base temperature of 68°F would mean that the outdoor temperature averaged 67°F; ten heating degree-days would mean that the outdoor temperature averaged 58°F. Degree-days are usually accumulated over months and years to describe a climatic regimen. For example, Fairbanks, Alaska, "enjoys" 13,940 heating degree-days a year, but in the warmest month, July, only about 160 degree-days of heating are required. Only 200 heating degree-days are needed to maintain the base in Miami, Florida, but 4,198 degree-days of cooling are required.

direct current (DC): electricity that presents one unvarying current to a load. See also **alternating current** (AC).

doping: a chemical process essential to the fabrication of semiconductors, in which "impurities" in the form of atoms of electrically active elements are introduced into the pure semiconductor matrix. Phosphorus, gallium, and selenium are common dopants.

drip irrigation: a technique for delivering irrigation water precisely at the point of need, thereby reducing the amount of water (and hand-watering) required.

(service) drop: generally, the wires that swoop in from the powerline to the service head above the electric meter; any connection between utility distribution and consumer metering.

Edison cells: an early commercial nickel-and-iron battery pioneered by Thomas Edison.

efficiency: a concept based on the question, "How much of what went in came back out?" In an efficient system, you get back as much as you put in. In physical systems, a tangible cost, or a conversion loss, usually reduces efficiency.

electrical co-ops: Beginning as cooperative organizations sponsoring electrification, particularly in the 1930s, electrical co-ops have matured into a rare (in the U.S.) but successful and efficient means for providing electricity.

electricity: a flow of electrons providing usable energy. Because no one has ever seen an electron, the theories attributed to electricity have a mythical quality, but anyone who has grabbed a hot house-current wire can attest to the reality of this kind of energy.

electrification: bringing electricity to the users—the term is mostly used in the context of Rural Electrification programs popular during the 1930s.

electrolysis: a chemical process in which the hydrogen and oxygen atoms in water are separated by the application of electricity.

electrolyte: "battery juice"; in lead-acid batteries, more-or-less dilute sulfuric acid. Consumer batteries use a solid electrolyte made up of filler impregnated with chemicals. The electrolyte allows ions to migrate to the battery's anode and cathode and react, producing a flow of electricity.

electrons: mythical beasties that purportedly occupy the region surrounding the atomic nucleus. Their whereabouts are unknown and presumed to be unknowable, but their group behavior is predictable, allowing us to enjoy the Electronic Age. Large numbers of them flowing together comprise electricity. See also **current.**

EMR (Electro Magnetic Radiation): Whenever electrons flow, a magnetic field is created, whether around the nerves in our bodies or around the high-voltage powerlines that crisscross our land. Evidence suggests that long-term exposure to strong electromagnetic fields is not good for living things. I keep the big dirty radiators, such as electric blankets and inverters, at a distance from where I live, work, and sleep. EMR can easily be measured with a meter designed for that purpose.

end-use losses: inefficiencies in the household or place of business that are paid for by the customer, as distinct from transmission and distribution losses that are absorbed by the utility.

energy audit: an accounting of the flow of energy over time, for the purpose of seeking and prioritizing economies.

envelope: a membrane that encloses something; when speaking of a building's envelope, the outer skin of the building.

environmentally friendly work: work that may be carried out without damaging the biosphere in any way.

equalizing: a periodic, systematic overcharging of batteries to ensure that every cell comes up to full charge and is pulling its weight.

equipment cost: the total sum expended to purchase the hardware. This term is usually used together with "operating cost" or the cost to operate the equipment. A car might cost $10,000 (equipment cost) but over its useful life might consume as much in fuel and half as much in maintenance and repairs (operating costs). In this case, equipment costs represents only 40% of total cost of ownership.

externalities: cost considerations other than those normally accounted as cost. A typical externality is the health costs incurred by respiratory patients in communities downwind from and under the plume of a large, dirty coal-burning electricity generation facility.

externalities adder: an arbitrary and generally insufficient charge to get the externalities back into the accounted costs. In Vermont, it appears that 30% must be added to the cost of electricity just to cover the health costs of dirty generation.

face time: a term used to describe the time that many employees, and particularly bosses, need to have "face to face" with their co-workers. Work-at-homers must often plan for activities at the office to get enough face time with their associates and supervisors.

false economy: something that looks cheaper than it really is. False economies result from buying things that almost fit because they are on sale. Cheap windows, for example, may cost 40% less installed, but over their lifetime leak energy worth ten times the saving over better windows that cut that leakage by two thirds.

ferroconcrete: a construction technique using iron-reinforced concrete, usually to build water tanks and boats.

first-growth: trees that have never been harvested. An "ancient forest" of old-growth redwoods may have been undisturbed for three millennia.

fission: splitting the nuclei of atoms.

flat discount rate: accounting for the cost of energy as if the small user and the very large user contribute equally to the cost of generation and transmission infrastructure.

float: referring to a battery: When not much energy is being taken out of storage, and the charging source keeps the batteries "topped off" at full charge, the charge controller "floats" the batteries.

flush toilet: the kind we're used to, where you depress a lever and 1.6 or more gallons of potable water flushes the unmentionables away.

forebay: the pool above a hydroelectric generation plant from which it draws water. The water travels through a headrace where it enters a penstock and flows down to the turbines, then debouches into the tailrace and the afterbay.

Frankenstein knife switches: big mechanical switches with one or more conductive blades connected to an insulated handle. When switched "on," the blades fit between spring-loaded receivers to make contact.

fuse: a carefully designed metal conductor that melts (fuses) if too much current flows through it. Fuses are usually encased in glass or ceramic so if they "blow" enthusiastically, they will not splatter bits of molten metal.

Gaia: the name for the large planetary organism that includes all life on Earth, including us. The Gaian Hypothesis suggests that all life forms are interdependent and linked together so that whatever is done to the smallest finally affects everything and everyone.

generator: a device that converts mechanical, rotational energy such as might be produced by a water wheel, turbine, windmill, bicycle crank, or (worst case) internal combustion engine, and converts it into electricity by spinning a coil in a magnet field, or vice versa.

gen-set: shorthand for a motor-generator pair, usually a small gasoline motor and house-current generator.

geothermal: correctly used, this refers to energy derived from steam trapped below the ground, often in volcanically active areas such as Yellowstone and California's Geysers. Steam wells are drilled and capped, and the steam is made to drive turbines as it escapes. This term is also sometimes used incorrectly to describe the inefficient practice of pumping heat into uninsulated ground during the summer, then retrieving it in the winter.

gnomon: the indicator on a sundial.

golf-cart batteries: 6-volt batteries designed to withstand the deep-cycling of a golf cart, discharged for 18 holes, then recharged, repeatedly. Such batteries are fairly good for storage in a solar-powered home energy system.

gotcha!: the unanticipated consequence.

gradient: a steady slope, or, in electricity, a charged field that changes more or less linearly with distance, typically a very small distance across the thin dimension of a semiconductor.

gravity flow: the pressure provided in a water system, when the supply is sufficiently above the point of use (the showerhead, for example), so that no energy needs to be expended to set the water flowing. For each 2.1 feet of elevation, 1 pound per square inch (psi) is generated, and we Americans like a 20 psi shower, so (with a little for pipe friction) a 50-foot tower gives us the gravity we need.

greywater: all kinds of wastewater except from the toilet—kitchen sink (if you don't cut or wash meat, fish, or other putrescibles), shower water, lavatory water, laundry water are all shades of grey. If you think about it, you'll understand why, in a zero-defects sanitary regime, all waste water is treated as black. When necessity (in the form of too many people and not enough water) intervenes, greywater can be handled safely, and great gardens result.

grid, the: the network of electrical transmission and distribution lines that link 99+% of the homes in the U.S. and Canada.

grid-excited inverter: an inverter that exports locally harvested electricity into the power grid and only operates when the grid is energized.

half-life: a chemical term that describes how certain active chemicals lose half of their activity over a characteristic period of time. Radioactive materials "decay" in this way, some with half-lives measured in the nanoseconds (billionths of a second) and others much longer. Plutonium's half-life is just over 22,000 years.

head: in a hydroelectric system, the elevation of the intake over the generator.

headrace: in a hydroelectric system, the channel connecting the forebay to the entrance of the penstock.

heritage fuels: part of the greater endowment of heritage materials that make the planet earth such a rich and varied place. Heritage fuels, also called fossil fuels—coal, petroleum, and natural gas—represent a surplus of carbon on the planet that took Gaia 300 million years to get safely stored underground. Humans have dug up and burned about half of the petroleum, a third of the natural gas, and a quarter of the coal in about 100 years.

high-tech glass: the best fenestration that money can buy, usually two (or even three) layers of glass, or two layers with a plastic inner film, sealed in a frame with the interior spaces filled with an inert gas such as argon. A single-pane window has an R-value of 1.0, while the best high-tech glass has an R-value exceeding 8.0 and therefore loses an eighth as much energy.

hole: a metaphor used in the Alice-in-Wonderland world of subatomic physics to describe the way electrons wander around in a semiconductive or conductive material. If the electrons wander from negative toward positive, then the holes meander from positive toward negative—or else the law of conservation is violated.

HUD: the federal Department of Housing and Urban Development, the agency that oversees the quality of American housing.

HVAC: the architects' shorthand term for Heating, Ventilation, and Air Conditioning.

hydraulic ram: a simple device that uses the "water hammer" effect of a large vibrating column of water to pump a small amount of water uphill surprisingly large distances.

hydrometer: literally, "water measurer," used to measure the amount of water in a denser liquid, such as battery electrolyte (or beer).

hydronic floor: a floor mass, typically of concrete or tamped earth, with tubes through which heated fluid is pumped. The massive floor takes up the heat and gently radiates it upward into the room. This gentle form of heating is the most efficient method for maintaining comfortable space.

hypocaust: the ancient Roman term for a floor, usually made of slabs of marble or similar stone, constructed with the flue from a wood fire *under* the floor. Heat from the smoke heated the floor. Ruins of hypocaust-heated buildings can be found in France and England.

incandescent bulb: any lightbulb that makes its light by heating a filament to incandescense. Such bulbs create more heat than light, and consume about 80% of their energy making heat. The incandescent lightbulb is one of the most inefficient devices we use. Use compact fluorescent bulbs instead!

infiltration: a cold war problem: cold comes creeping in through the glass, around the windows, through the walls, out of the outlets, wherever a crack or opening or lack of insulation allows cold air to enter the heated home. Temporary measure may help—caulk, door snakes, storm windows—but only good design and craftsmanship when the home is being built really solve the problem for good.

infra-red scanner: a modified video camera tuned to "see" in the infra-red spectrum so that it can make an image of the cold places where infiltration is taking place (from inside the house) or (from outside the house) hotspots where heat is escaping.

infrastructure: a fancy word for the artifacts that are necessary to support civilization: streets and highways, reservoirs, water and sewage treatment facilities, water mains and sewers, powerlines of all sizes and their associated substations, switching and generation facilities, all the communications towers and stations, railroads, canals, and so forth.

insolation: the amount of sunshine on a spot, usually "an annual average expressed as n hours of insolation," meaning that the place gets the equivalent of n hours of unclouded sunshine per day averaged over a year.

insulation: not to be confused with insolation; any material used to keep energy from getting away. The insulation on wire protects the wire from mechanical damage, and keeps the electricity from escaping to neighboring conductive objects like our hands. The insulation in roofs, walls, and floors keeps warmth in during winter and out during summer. Both kinds of insulation share a single characteristic: They are very resistant to the flow of energy.

integral: multiple functions are married into a single unit.

integrated roofing: the weather surface of the roof also serves as the carrier for PV material, so the roof sheds water in bad weather and harvests electricity on sunny days.

inverter: a device that converts DC electricity, as produced by PVs and stored in batteries, into AC house current, the kind used by most familiar household devices.

Jacobs windmill: one of the early mass-produced windmills, much favored by remote farmers prior to rural electrification.

kilowatt: one thousand watts; ten 100-watt lightbulbs burn a kilowatt of electricity.

Laughlin-type soil: a very poor soil almost completely devoid of nutrients.

lead-acid batteries: the commonest and most cost-effective form of storage batteries, found in vehicles, uninterruptible power supplies, and renewably powered home systems.

lineal foot: the unit of sale for objects usually delivered in long pieces, like pipe or 2x4.

load center: once called the fuse box, this is where the energy is distributed via the bus bar and circuit breakers to various circuits.

load: in electricity, any device that consumes electricity, so an electric water heater is a big load, and a clock or a night light is a little load.

low-e windows: one form of high-tech glass, these are notable for their low emissivity, which means they incorporate a coating or plastic film that reduces the passage of radiant energy.

low-voltage: in electrical terms, less than house current, typically 12 or 24 volts.

lumens: an exact measure of quantity of light. A 60-watt incandescent lightbulb and an 18-watt compact fluorescent lightbulb each produce about 1,000 lumens.

manometer: an instrument for measuring the pressure in pressurized systems such as ammonia-based refrigerators.

micro-hydro: household scale hydroelectric generation, typically a two-inch "penstock" delivering water with a head of a hundred feet or less to a turbine the size of half a grapefruit connected to a modified automobile alternator; such a rig can easily keep a household in electrons day and night.

micro-PV: small self-contained PV-energized devices such as calculators, walkway lights, street numbers, or emergency roadside call-boxes.

molecular assembly: bio-engineering techniques that assemble a product one molecule at a time, often using low energy under environmentally safe conditions.

monetized costs: putting a dollar value on costs that have in the past gone unvalued—for example, what is the dollar loss when a cultural site is inundated by a hydroelectric reservoir?

multicrystalline: many crystals of silicon in a semi-chaotic state, typical of medium-grade, medium-efficiency photovoltaic material; contrasts with high-grade, single-crystal, high-efficiency PV cells sliced from a single crystalline boule of purified silicon, and with low-efficiency thin-film amorphous modules.

mutagenic: known to cause mutation.

n-type silicon: negatively doped silicon.

Neo-Luddite: the disciples of Ned Ludd broke machinery and blocked access to factories at the start of the Industrial Revolution; since then, anyone who opposed industrialization has been called a Luddite or neo-Luddite.

net metering: an increasingly popular energy billing arrangement put forward state by state in which a small power producer, for example a renewably energized home, can feed surplus power directly through the electric meter and into the grid, making the meter spin backwards, effectively using the grid as a 100% efficient storage device. In 1993, the first edition of this book called for net-metering, and now almost half of U.S. electricity consumers have that opportunity; before that, two meters were required, and the utility paid only the "avoided cost," a fraction of the value for electricity generated by consumers.

nickel-iron batteries: see **Edison cells**.

nominal voltage: the actual voltage of low-voltage energy systems varies widely during operation, but they usually hover above a basic, or theoretical voltage. A nominal 12-volt system would be in trouble at 11.8 volts, and fully charged at 13.8 volts; the energy sources—PV, wind-spinner, micro-hydro, AC-powered charger, or gen-set—used to charge the batteries in such a system might produce 17 volts. Equipment meant for these systems must tolerate such fluctuations without grumbling.

off-peak: times when the demand for electricity

from a system is not at its greatest, or peak. Peak demand is typically at 2:00 PM on a sunny week day, revealing bad architecture and overuse of air conditioning.

off-peak hours: times defined by the utility as having the greatest system capacity.

oikos: the Greek word for home, and the base of words beginning eco-.

operating cost: the cost of operating a system or appliance over its useful life. See **equipment cost.**

over-voltage: a circumstance in which the voltage is too high—sun shining on a PV array on a cold day with snow on the ground might produce more voltage than a charge controller can handle—so the over-voltage protection cuts the connection.

overcurrent protection: a mechanism or circuit that protects a system from too much current—the washing machine running at the same time a circular saw starts up will most likely trip the overcurrent protection of a small inverter.

p-type silicon: positively doped silicon.

pathogens: materials that cause illness.

payback: the time it takes for the total of equipment and operating costs of a more efficient unit to fall below the total for a conventional unit, when comparing two systems, usually a conventional low-price, high-maintenance system and a more expensive-to-buy but cheaper-to-operate system.

pelton wheel: a turbine designed by an engineer to be driven by energetic streams of fluid issuing from one or more nozzles. The cups of the turbine are designed so that the water splatters clear of the stream from the nozzle after transferring a maximal amount of energy to the spinning wheel.

penstock: the pipe leading from the forebay and headrace to the generator in a hydroelectric power plant.

performance-based rebate: the idea that a profes-sional's compensation might include an ongoing bonus for designing systems that save operating money. Such a rational feedback loop rewards professional excellence and punishes incompetence, and so is (of course) hotly opposed by professional organizations.

permaculture: landscape design and plantings intended to last more or less permanently. The best way is to restore whole natural living systems using flora native to the habitat.

phantom loads: small "convenience" appliances such as calculators, lighted-dial phones, clocks, and so forth; individually they use tiny amounts of power but in a fully accessorized household, their total load is considerable.

photon: another mystical beastie, this time a "packet" of light, but something no one has ever isolated, described, or measured individually in any way. Since the "wave theory" of light is unable to explain the way light travels across the vast vacuum of space, scientists needed a "particle theory." It has proved a useful model, as it also explains the way the photoelectric effect works. Physicists find it very convenient to be able to switch back and forth between models when explaining the way light works. Some day they may find a unifying theory. Meanwhile, light is a wave, or a photon, depending . . .

photovoltaic module: a silicon lattice connecting photovotaic cells, which generate an electrical charge when exposed to sunlight. Those of us who know them well call them PVs.

piezo-electric ignition: a mechanism for lighting a stove burner or oven using materials that give off a spark when struck by a small spring-loaded hammer.

pioneer species: a plant or animal that tolerates harsh conditions and can grow on disturbed soil. Fireweed is a classic example: after a forest fire, it grows up first, and shelters the later seedlings that will reconstitute the forest.

PMS: Perpetual Midnight Syndrome, or the flashing 12:00 one sees on VCRs and other electronic clocks that lose their sense of time when the power fails, and whose owners don't bother to reset them. Evidence of a power thief in the act.

potential: in electricity, the difference in voltage between two points in a circuit. The difference in potential between the poles of a golf-cart battery may be as much as 6.8 volts. The battery has the potential to deliver energy at something like that voltage. One never knows how deep the potential goes until the energy becomes kinetic.

power center: an all-in-one control and load center integrating several balance-of-system functions in one big box. A typical power center contains one or more charge controllers, metering, and over-current protection (in the form of circuit breakers) for branch circuits.

power cubes: power thieves that are always on and always using energy even when the equipment they power is off. Many phantom loads are fitted with power cubes. A typical power cube converts household current (nominally 117 volts AC) into low-voltage DC at an efficiency of something like 50%, with most of the waste being given off as heat.

power-conditioning: the process of "cleaning up" the power from a dirty source, such as a gen-set or the grid, so that the spikes and brown-outs will not damage sensitive equipment.

precession: the way a top's axle wobbles before it stops spinning. The same phenomenon explains the 46-degree "wobble" that governs the earth's seasons.

public power: where the electricity is delivered by a publicly owned entity.

putrescibles: substances such as meat and fish that rot and attract vermin before they compost.

PV(s): shorthand for photovoltaic module(s).

PV array: a group of PVs wired to work together.

PV system: a stand-alone electrical system in which the majority of the energy is harvested by photovoltaic modules.

PV-direct: a system in which the load is directly powered by a PV array without any storage or back-up. A PV-direct pump is a good example.

R-value: a measure of a material's resistance to thermal transfer.

radiation: a means by which a quality is carried across a medium without interacting with the medium; light reaches us from the sun across the vacuum of space.

radon: a radioactive (and therefore carcinogenic and mutagenic) gas.

rare earth: uncommon metals used for doping semiconductors and fabricating specialized alloys, including magnets. Rare earths include cerium (Ce), dyprosium (Dy), gandolinium (Gd), lanthanum (La), praseodymium (Pr), neodymium (Nd), samarium (Sm), terbium (Tb), and ytterbium (Yb).

rate: when buying energy, the cost per unit; in electricity, the cost per kilowatt-hour. Often utilities mask the true cost by separating out a metering cost, transmission costs, and other obfuscations, but the *true* cost is the total dollars paid divided by the total number of kilowatt-hours consumed.

rate structures: tariffs or other tables that determine how much different classes of consumers pay for different amounts of electricity. A typical rate structure will have the aluminum plant consuming gigawatt-hours paying an order of magnitude less for a kilowatt-hour than a single-family household.

rectify: literally, "turn it right side up"—and so, rectifying AC electricity (which varies from +170 volts to -170 volts 60 times each second) means changing it so it no longer alternates, but always yields a positive potential. A typical power cube steps the voltage down to 12 volts AC and then rectifies and conditions it to12 volts DC.

resistive loads: electrical devices that do their work with heating elements, such as electric stoves, hot plates, heaters, and hair-fryers. In most cases, these are inappropriate uses of electricity in an independent home.

root mean square: the mathematical calculation determining that AC electricity, which fluctuates periodically from +170 volts to -170 volts, is really 117-volt house current. An appliance doesn't "see" the rapid fluctuations, but sees works as though it were receiving a steady flow of current.

Rural Electrification Administration: the federal agency responsible for stringing the wires to remote farmsteads during the 1930s "recovery."

safety disconnect: a switch that allows sources of power such as PVs and batteries to be disconnected from household circuits and control components so that they may be worked on safely.

sanitarian: an expert in maintaining public health, particularly with respect to the proper disposal of blackwater.

self-sufficient: in the extreme case, a system that is wholly reliant on its own abilities to gather its necessities within its own boundaries. In practice, an independent home may be self-sufficient in one or more specific areas, electricity, heating fuel, or possibly building materials. Complete self-sufficiency is a major challenge.

semi-conductor: a material that is electrically conductive under some circumstances, and resistant under others. Silicon, the basis of so many of our conveniences from solid-state electronic devices to PVs, is a semi-conductor.

septic tank: the typical rural destination for blackwater and most greywater. A septic tank is a cement or plastic tank holding a thousand gallons more or less for each bathroom, fitted with baffles so that the aerobic and anaerobic decay process takes place and reduces the solid material to sludge, which sinks to the bottom of the tank, the floating material is kept inside the tank, and the effluent, which is fairly sterile, flows out into a leach field where soil bacteria takes care of any pathogens before the water

returns to the water table. In some areas, such as the Tahoe Basin, where rural development is reaching groundwater saturation, septic tanks are being phased out, and alternative systems are being evaluated.

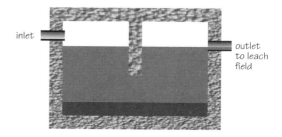

service-entrance panel: the electrical apparatus between the service drop and the household system. Usually the electric meter is located here.

setpoints: in an electrical system, the parameters that govern how a component performs. Setpoints in a charge controller determine the high voltage at which the controller stops sending energy to the batteries, and the low voltage at which it starts charging them again.

silicon: a semi-conductor.

silicon lattice: the orderly arrangement of silicon atoms in crystalline silicon.

single-glazed windows: windows covered with a single pane of glass.

sinusoidally: varying as the sine wave, sinuously.

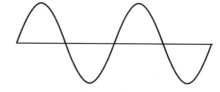

skid-mounted PV system: a home-energy system mounted on a portable platform that fits onto a trailer or in the bed of a pickup truck, so that the system provider can assemble or service the system in the shop, then drive to the job site, skid the system onto the ground, plug it in, and away you go!

solar assist: using the sun's energy to help a process along. Solar water heaters provide only a portion of the heat desired, so they give domestic hot water systems a solar assist.

solar oven: an insulated box designed to capture and hold the sun's heat and use it for cooking. A home-made solar oven can easily maintain oven temperatures in excess of 400°F.

solar panels: any flat surface meant to gather solar energy, usually by converting sunshine into hot water. The term is also used (inexactly) to describe PV modules.

solar window: the opening between trees, buildings, and other obstructions through which the sun can shine on a given spot.

southern exposure: the side of a house with the best exposure to the sun.

specific gravity: the measure of the density of a liquid, with water defined as 1.0.

specific heat: the measure of a material's ability to retain heat. Wood has very poor specific heat, while masonry, concrete, and rock are good.

step-down: to reduce the voltage potential by directing the electricity through a transformer. A power cube typically contains a step-down transformer that reduces the house current to 12 volts. On the utility scale, large step-down transformers reduce high voltage from kilovolt transmission lines to lower voltage distribution lines, where other smaller step-down transformers—the cans hanging on power poles—reduce distribution voltage to house current.

stranded costs: a term used by utility planners in reference to costs associated with bad investments in uneconomic or prematurely decommissioned nuclear or other power plants. In many states, utilities were able to persuade lawmakers that taxpayers should share these stranded costs (thereby saving their stockholders a bundle).

strategic: a militaristic euphemism for resources that might have value during an armed conflict and that we Americans are supposed to be willing to send our young to war to protect our global right to plunder.

sub-transmission: of utilities, a step down from long-distance transmission. Long-distance high-voltage transmission lines carry electricity from a generation facility to the outskirts of the city, where it is stepped down and carried by sub-transmission lines to sub-stations, where it is again stepped down to distribution voltages and along the streets.

sun: a unit equivalent to the brightness of the sun, used to measure the effectiveness of a concentrator, a mirrored or lensed array focusing light on a PV cell. A typical concentrator array will focus the light of ten suns on a cell.

swing season: the time between abundant sunshine and abundant rainfall or wind, generally spring or autumn.

switch gear: the equipment used by utilities to change the configuration of the transmission and distribution lines.

switched outlets: wall outlets wired so that they (and anything plugged into them) can be turned off with a wall switch.

tailrace: the chaotic water just outside the hydroelectric generator, before the water calms down and enters the afterbay.

teratogens: materials that are known to cause birth defects.

thermocline: a sharp change in temperature between surface and deep water in the ocean; a particularly marked thermocline off the west coast of Hawaii has been used experimentally to generate electricity.

thermosiphon: plumbing that takes advantage of the fact that cold water is heavier than hot water (see **convection**), and will induce water to circulate as long as the heat differential is maintained. If a solar panel is placed below a storage tank, a thermosiphon will initiate itself, with colder water in the storage tank making its way

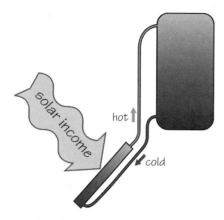

into the panel, there to be heated and replaced by cold water. Gradually, this process heats all the water in the storage tank.

Title 24 standards: guidelines promulgated by HUD to encourage builders hoping for federal government financing to implement minimal standards of energy efficiency.

tracker: a device that pivots on one or two axes to keep a PV array pointed squarely at the sun. Trackers can be passively aimed by captive CFCs or actively driven by actuator motors under the control of photocells. A one-axis tracker follows the daily motion of the sun; a two-axis tracker also follows the sun's seasonal motion. A tracker can improve PV yield by about 30%, depending on season and latitude, and usually costs about 30% of the value of the modules it tracks, so in many cases it is more cost-effective to buy 30% more modules and firmly affix them to a well-oriented surface.

true cost: the whole cost, including exploration, exploitation, habitat restoration, transportation, refinement, disposal of impurities, packaging, distribution, advertising and sales, purchase price, installation, operation and maintenance, decommissioning, and recycling costs. The sticker price often neglects to mention some of the costs that should be included in the purchase price, and our decisions to buy seldom takes these costs or the differential cost to operate into account. When we do consider them, we usually buy more wisely. Critics of "true cost accounting" argue that some of these costs are difficult to monetize accurately, and therefore should be ignored.

turbine: a wheel with vanes or cups on it designed to transfer energy from an energetic stream of gas or liquid to rotational energy at its axle.

turbine generator: a turbine with its axle connected to a generator so that the stream of gas or liquid spins the generator and induces a flow of electrons.

under-voltage: a condition where the potential is inadequate to power equipment, and may damage it. "Brown-outs" are forms of under-voltage caused by excessive demands on a utility's resources. In a home or utility system, controls are installed to turn the power off when it is sufficiently under-voltage.

unmonetized externalities: costs that can easily be ignored, such as the known carcinogenic effects of diesel exhaust, because they are not directly borne by the immediate user, but by the tolerant and largely unaware society at large.

utility interconnect: describes the way a stand-alone home energy system can use the grid for storage by selling surplus electricity and buying it when the domestic supply is inadequate. Other terms for this are grid-interactive, grid-intertie, and grid-connected.

utility-scale: large and complex electricity generation and distribution schemes serving a large community, as a contrast to simpler home-scale or neighborhood-scale power systems.

valence band: the outer "shell" or region of reactive electrons in an atom. Electrically conductive elements have many electrons in their valence bands, while insulators have very few; semiconductors, as you might expect, fall in between.

variable selectivity window: a window with a special film on it that changes reflectivity depending on electrical signals. Think of it as venetian blinds on a molecular level.

vigas: larger straight round rafters on top of which the roofs of adobe homes are constructed.

watt: a measure of electrical power, calculated as the mathematical product of voltages times amperage. A 5-watt, 12-volt halogen light—a typical low-voltage walkway light—consumes about four-tenths of an amp.

weather-stripping: usually metal-reinforced felt gaskets or seals that reduce infiltration around windows and doors.

whole cost: see **true cost.**

whole-house fan: a powerful fan that is capable of changing all the air in a house several times an hour. Whole-house fans are used instead of air conditioning in climates where it is warm by day and cooler by night. During the evening hours, the whole-house fan circulates cooler air throughout the house, charging its thermal masses with coolness. During the day, the house is closed up tight, and the thermal masses "radiate coolth" and keep the indoor temperatures bearable.

windmill: a wind-driven turbine used to perform a mechanical task such as pumping water. Windmills usually spin more slowly than windspinners.

windspinner: a propeller connected to a generator; also known as a windcharger or a wind generator.

yurt: a round portable building made of felt.

BIBLIOGRAPHY

The Alternative Building Sourcebook*: Traditional, Natural and Sustainable Building Products and Services.* Steve Chappell, ed., 1998 (ISBN 1-889269-01-8). Innumerable resources for building appropriate shelters.

The Apple Grower*: A Guide for the Organic Orchardist.* Michael Phillips, 1998 (ISBN 1-890132-04-7). Johnny Appleseed meets himself on the next rung of the evolutionary ladder. Once again, by building a strategy based on close analysis of the basic elements, we find that common sense and self-sufficiency provide a golden path.

Architect's Sketchbook of Underground Buildings*: Drawings and Photographs.* Malcolm Wells, 1990 (ISBN 0-962187-81-X). A good argument for leaving Nature on top, and building underneath.

Bamboo Rediscovered. Victor Cusack, 1998 (ISBN 0-959588-98-1). If you live with bamboo, or where native trees yield no usable timber, then bamboo is a perfect building material.

Beyond the Limits*: Confronting Global Collapse, Envisioning a Sustainable Future.* Donella H. Meadows, Dennis L. Meadows, and Jørgen Randers, 1992 (ISBN 0-930031-62-8). "Essential reading for everybody who is concerned with the central issue of our times: how to achieve a transition to a sustainable global future," according to Gro Harlem Brundtland, former Prime Minister of Norway and present chairman of the World Commission of Environment and Development. The authors lay out the mathematical and conceptual rules by which the world game will be played whether we like it or not.

The Bread Builders*: Hearth Loaves and Masonry Ovens.* Daniel Wing and Alan Scott, 1999 (ISBN 1-890132-05-5). As a once-and-future baker, I love this book. Like my book, it starts by examining the smallest bits of its task, and builds a conscious, well-understood loaf. A good loaf, it turns out, can only be built with the best ingredients and a knowing baker (of course), but a proper oven is also required. By the time you finish this book, you will want to build one.

Building a Win-Win World: Life Beyond Global Economic Warfare. Hazel Henderson, 1997 (ISBN 1-576750-27-2). Hazel leads a faction who believe that our accepted present-day global economy is predicated on some wrong and damaging assumptions. If you're seeking a well-presented explanation of how to get beyond all that, try this book.

Build it With Bales. S. O. MacDonald and Matts Myhrman, 1998 (ISBN 0-964282-11-9). A more basic and technical work on the subject than the classic *The Straw Bale House.*

Cob Builders Handbook. Becky Bee, 1998 (ISBN 0-965908-20-8). Building with cob is like grown-up mud pies. Concepts such as "straight" and "corner" are elective, and the form is perfect for beginning builders.

Community Design Primer. Randolph T. Hester, Jr., 1990 (ISBN 0-934203-06-7). A concise book about creating community thoughtfully— what a concept!

Complete Book of Cordwood Masonry Housebuilding: The Earthwood Method. Rob Roy, 1992 (ISBN 0-806985-90-9). Rob Roy, who also wrote about building *Mortgage-Free*, shows off his favorite building strategy. In the woods, building with cordwood is an excellent, and seldom-considered, technique.

Consumer Guide to Home Energy Savings (6th ed.). Alex Wilson and John Morrill, 1991 (ISBN 0-918249-31-7). One more of those little books that might give someone a good start on a better energy path.

Consumer Guide to Solar Energy: Easy and Inexpensive Applications for Solar Energy. Scott Sklar and Kenneth G. Sheinkopf, 1995 (ISBN 1-566250-50-1). Covers some of the same territory as *Consumer Guide to Home Energy Savings*, but from a different angle.

The Contrary Farmer. Gene Logsdon, 1995 (ISBN 0-930031-74-1). Gene Logsdon manages to inform me while making me laugh (and, sometimes, flinch.) This book and its partner (below) helped me understand the common sense aspects of agriculture. Here, Gene tells how he gardens by his wits instead of following convention.

The Contrary Farmer's Invitation to Gardening. Gene Logsdon, 1997 (ISBN 0-930031-96-2). Gene shows us all how to grow our food, and makes me wonder why I didn't start doing so much earlier.

Cooking with the Sun: How to Build and Use Solar Cookers. Beth and Dan Halacy, 1992 (ISBN 0-962906-92-1). A delightful culmination to the cycle. The sun nurtures us and the food we eat; why not use it to cook with, too?

The Earth-Sheltered House. Malcolm Wells, 1998 (ISBN 1-890132-19-5). Hunkering down beneath earthen berms or building underground makes sense in places where the winds roar and winter is endless. Malcolm is a master of making caves seem airy and homey, and this book is an excellent sample of his work.

Eco-Renovation: The Ecological Home Improvement Guide. Edward Harland, 1999 (ISBN 0-930031-66-0). If you already have a house, and want to make it work better, this book shows you how. Originally written for the British market, you will have to do a little translating, but the principles and ideas are sound and appropriate.

End of the Road: From World Car Crisis to Sustainable Transportation. Wolfgang Zuckermann, 1992 (ISBN 0-930031-51-2). Cars are our most glaring offense against nature. Worldwatch president Lester Brown says "No amount of minor readjustment will return [our transportation system] to working order; the system needs an overhaul. . . . With *End of the Road,* we now have a clear, comprehensive, and eminently readable manual for the overhaul."

Energy-Efficient Houses. Fine Homebuilding, 1993 (ISBN 1-561580-59-7). *Fine Homebuilding Magazine's* "Great Houses" series is fun to look at, although the builders of these manses suffer from the Edifice Complex. You can certainly see that living energy efficiently doesn't prevent us from living graciously.

Finding and Buying Your Place in the Country. Les and Carol Scher, 1996 (ISBN 0-793117-85-2). After a busy working life in property law and real estate, and several revisions, this book is encyclopedic—and readable! If you have questions or misgivings about buying country land, get this book and study it; you'll save a bundle.

The Flower Farmer: An Organic Grower's Guide to Raising and Selling Cut Flowers. Lynn Byczynski, 1997 (ISBN 0-930031-94-6). Another way to make homesteading pay for itself.

Four-Season Harvest: How to Harvest Fresh Organic Vegetables from Your Home Garden All Year Long. Eliot Coleman, 1999 (ISBN 0-930031-57-1). Important ideas about solving one of the main self-sufficiency problems in the garden: what to eat in the winter. Writing from Maine, Eliot knows whereof he speaks.

The Fuel Savers: A Kit of Solar Ideas for Your Home, Apartment, or Business. Bruce Anderson, ed., 1991 (ISBN 0-962906-90-5). A good starting place for improving home energy efficiency.

Gaviotas: A Village to Reinvent the World. Alan Weisman, 1998 (ISBN 0-930031-95-4). Alan tells the story of a small community in one of the world's most troubled countries that has found a solution. This is an exciting book, full of hope and pragmatism.

The Good Life: Helen and Scott Nearing's Sixty Years of Self-Sufficient Living. Helen and Scott Nearing, 1990 (ISBN 0-805209-70-0). This inspirational work glorifies self-sufficiency from the ground up. All who live in independent homes owe a debt to these trailblazers.

The Good Woodcutter's Guide: Chain Saws, Portable Sawmills, and Woodlots. Dave Johnson, 1998 (ISBN 1-890132-15-2). Managing the woodlands and woodlots does not come naturally to city-bred tree-huggers. Dave bases his book on years of experience and thoughtful inquiry.

Hammer. Nail. Wood.: The Compulsion to Build. Thomas Glynn, 1998 (ISBN 1-890132-06-3). One highly personal and thoroughly enjoyable account of the building of a house.

Hearts Open Wide: Midwives and Births. Susan Mazur and Pam Wellish, 1987 (ISBN 0-914728-54-7). Yeats advised, "make them fill the cradles right" and that's what this book tells us about: Out among the independent homes, there is a new way to come into the world, and this is how it feels.

Heaven's Flame: A Guide to Solar Cookers. Joseph Radabaugh, 1998 (ISBN 0-962958-82-4). As important an idea as the clothesline, but seldom applied. This book can help you get started.

Home Ecology: Simple and Practical Ways to Green Your Home. Karen Christensen, 1990 (ISBN 1-555910-62-9). Also other current titles of this genre, which is alive and well at your bookstore. Almost every such book gives me new ideas. Books like this are good gifts for those who are lagging behind on the conservation frontier.

Homeschooling for Excellence. David Colfax and Micki Colfax, 1988 (ISBN 0-446389-86-2). David and Micki did it—raised, homeschooled, and launched several excellent students—and then wrote about it. Wise words for anyone considering educating children at home.

Homing Instinct: Using Your Lifestyle to Design and Build Your Home. John Connell, 1999 (ISBN 0-446516-07-4). This book helps tease out those crucial but easily missed ideas from the corners of your mind—and the minds of your family—and into the design for your perfect home. Don't try to build a home without this book!

Hope, Human and Wild: True Stories of Living Lightly on the Earth. Bill McKibben, 1997 (ISBN 0-316560-64-2). This hopeful sequel to his chilling book *The End of Nature* is summed up by David Ehrenfeld, founding editor of *Conservation Biology*: "Those who care about both the wild Earth and its civilized people will greet this book with joy. It shows that good things can come not only from planning, but unexpectedly—even arising out of devastation and evil. McKibben lets us understand that in this time of gathering environmental and cultural tragedy, hope is not only necessary, it is also reasonable." Confirming my suspicions, Rush Limbaugh endorses the author thusly: "He is an environmentalist wacko." My Dad gave me this book, saying "I need you to read this," and I say the same to you. Thank you, Dad.

How Buildings Learn: What Happens After They're Built. Stewart Brand, 1994 (ISBN 0-140139-96-6). The important part of a structure's life begins when it is finished. That's when the builders and architects go home, and this book's story starts. If I had read this book before I started building, I wouldn't have so much rebuilding and maintenance to do now.

How to Build an Underground House. Malcolm Wells, 1991 (ISBN 0-962187-83-6). Malcolm turns the way we think about home on its ear, and that is a very good thing. His book is a primer for thinking about homes in a new light.

Independent Builder: Designing and Building a House Your Own Way. Sam Clark, 1996 (ISBN 0-930031-85-7). This book carries on the tradition, while updating the ideas and adding a New England perspective.

Introduction to Permaculture. Bill Mollison and Reny Mia Slay, 1997 (ISBN 0-908228-08-2). Sustainability is being able to keep on doing what we keep on doing, and so permanence and low maintenance are desirable. This book explains how to make this happen in the garden.

The Man Who Planted Trees. Jean Giono, 1987 (ISBN 0-930031-06-7). A heartening fable about healing the earth one acorn at a time.

Material World: A Global Family Portrait. Peter Menzel and Charles C. Mann; photographs by Sandra Eisert, 1994 (ISBN 0-871564-37-8). For a striking visual refresher course on the diversity and disequilibrium of wealth in the world, take a good look at this book. Somewhere between Texas and Tibet there must be a sustainable medium we can all enjoy!

Mortgage-Free! Radical Strategies for Home Ownership. Rob Roy, illustrations by Malcolm Wells, 1998 (ISBN 0-930031-98-9). In this book, Rob explains exactly why independent homesteaders don't want to invite the bank to join in the land acquisition partnership, and then he shows how to get land and create a homestead without debt. This is an important, practical, and liberating book.

The Natural House Book: Creating a Healthy, Harmonious, and Ecologically Sound Home Environment. David Pearson, 1998 (ISBN 0-671666-35-5). This book is illustrated by the brilliant Malcolm Wells, and delivers exactly what the title says. This book has lots of good ideas in it, and we never have too many of those.

The New Organic Grower: A Master's Manual of Tools and Techniques for the Home and Market Gardener. Eliot Coleman, 1995 (ISBN 0-930031-75-X). The best book on organic gardening I've found.

The New Solar Electric Home: The Photovoltaics How-To Handbook. Joel Davidson, 1987 (ISBN 0-937948-09-8). Another excellent book about the nuts and bolts of solar electricity.

The Owner-Built Home. Ken Kern, 1992 (ISBN 0-686312-20-1). Ken started the sensible-building revolution, and encouraged me to become a builder. Kern's commonsense approach to things like air circulation and doors made me realize that we had better do things because they make sense, rather than because we've always done them "that way."

The Passive Solar House. James Kachadorian, 1997 (ISBN 0-930031-97-0). A regional passive solar book—the ideas are best for northern tier states, especially New England—that shows how to use the sun as your furnace. This book has a strongly individualistic approach, reflecting the style of the author, and contains many good ideas that could easily be adapted for other styles and regions.

Passport to Gardening: A Sourcebook for the 21st-Century Gardener. Katherine LaLiberté and Ben Watson, 1999 (ISBN 1-890132-00-4). A wonderfully encyclopedic book that will help any gardener improve the yield and beauty of the garden. This book has a northeastern flavor, although it has plenty of information for any gardener.

A Pattern Language: Towns, Buildings, Construction. Christopher Alexander, Sara Ishikawa, and Murray Silverstein, 1977 (ISBN 0-195019-19-9). The housing industry set out to reinvent homebuilding and discarded centuries of building wisdom as a first step. This book attempts to organize what we used to know about the built environment. This is a fascinating book for those of us who want to create a fitting and sensitive home place, village, or city.

Permaculture: A Designers' Manual. Bill Mollison, 1997 (ISBN 0-908228-01-5). This is a more theoretical work than *Introduction to Permaculture,* explaining the "why" as well as the "how." This is an excellent book for those of us who want to get serious about planting for the ages. Mollison is from Down Under, but his ideas work extremely well in the northern hemisphere.

Pocket Ref. Thomas J. Glover, ed., 1989 (ISBN 1-885071-00-0). I use this book constantly—How many square feet in an acre? What's that relationship between watts and volts? What does it mean when my computer beeps twice? Irreplaceable.

A Primer on Sustainable Building. Dianna Lopez Barnett and William D. Browning, 1995 (ISBN 1-881071-05-7). Lays out simply and concisely the rationale for building sustainably. Abundant examples of successes. Inspirational. Even enemies of sensible building will learn something from this book.

The Rammed Earth House. David Easton, 1996 (ISBN 0-930031-79-2). The sculptural and thermal qualities of the walls of a rammed earth house make for extraordinarily comfortable shelter. David's homes are spectacular as well as being, well, earthy.

Rays of Hope: The Transition to a Post-Petroleum World. Denis Hayes, 1977 (ISBN 0-393064-22-0). We seem to be having a lot of trouble envisioning a future without fossil fuels, and Denis has given it a lot of thought, much of which is distilled into the book.

The Real Goods Solar Living Sourcebook. John Schaeffer, Doug Pratt, and the Real Good Staff, eds., 1999 (ISBN 0-930031-82-2). Two revisions after I left Real Goods, this book still has a lot of me in it. Therefore (of course) lots of good advice. It is also a shameless catalog and a great, but generally over-priced, compendium of possibilities and solar schlock, so prepare for the hard sell!

Retreats: Handmade Hideaways to Refresh the Spirit. G. Lawson Drinkard and Lawson Drinkard, 1997 (ISBN 0-879057-98-X). This book shows that building is fun, and you don't necessarily have to live there!

Retrofitting for Energy Conservation. William H. Clark II, 1997 (ISBN 0-070119-20-1). This book focuses specifically on changes that improve the energy performance of existing houses.

The Sauna. Rob Roy, 1996 (ISBN 0-930031-87-3). Another book by Rob Roy, applying his cordwood technique to a special purpose. The author's personable, narrative style makes this a fun book to read even if you never want to build a cordwood sauna.

Saving the Planet: How to Shape an Environmentally Sustainable Global Economy. Lester R. Brown, Flavin Postel Brown, and Sandra Postel, 1991 (ISBN 0-393308-23-5). The authors ask, "How can we create a vibrant world economy that does not destroy the ecosystem on which it is based?" In this important book they answer this question meaningfully and well.

Scott Nearing: The Making of a Homesteader. John A. Saltmarsh, 1991 (ISBN 1-890132-21-7). A biography of the great teacher, who from his homestead, with his partner Helen at his side, gave us a shelf full of iconoclastic and influential books about sustainable living as an antidote to consumerism and environmental destruction.

Sharing the Harvest: A Guide to Community-Supported Agriculture. Elizabeth Henderson with Robyn Van En, 1999 (ISBN 1-890132-23-3). Carrying organic gardening to the next step: growing enough to share with your neighbors.

Shelter. Shelter Publications, 1996 (ISBN 0-898153-64-6). More ideas about home from around the world. Includes important information like how to load a camel and the exact number and dimensions of the chords in a dome. An epic poem to making your own place.

A Shelter Sketchbook: Timeless Building Solutions. John S. Taylor, 1983, 1997 (ISBN 1-890132-02-0). I find a great deal of inspiration in the examples of brilliantly adaptive architectural solutions found all over the planet and drawn by Taylor in this wonderful little book. Good ideas! You will surely find several you can adapt into your home.

The Slate Roof Bible. Joseph Jenkins, 1997 (ISBN 0-964425-80-7). If you live where there is slate, then you can have an indigenous roof. This book focuses on commercial slate, but the serious do-it-yourselfer could easily interpolate for found materials.

The Smart Kitchen: How to Design a Comfortable, Safe, Energy-Efficient, and Environment-Friendly Workspace. David Goldbeck, 1989, 1994 (ISBN 0-960613-87-0). Kitchens are the common workspace shared by all homes—and often the least well-designed and most dominated by appliances and their corporate agenda. David writes sensibly about making an efficient, appropriate, comfortable, and productive cooking space.

The Solar Electric House: Energy for the Environmentally Responsive, Energy-Independent Home. Steven J. Strong and William G. Scheller, 1994 (ISBN 0-963738-32-1). An accomplished solar architect's perspective on the electromechanical reality you'll need to master.

The Solar Home: How to Design and Build a House You Heat With the Sun. Mark Freeman, 1994 (ISBN 0-811724-46-8). Another passive solar regional book.

Solar Gardening: Growing Vegetables Year-Round the American Intensive Way. Leandre Poisson and Gretchen Vogel Poisson, 1994 (ISBN 0-930031-69-5). Another approach to season extension, the great conundrum of northern-tier self-sufficiency.

Straight-Ahead Organic: A Step-by-Step Guide to Growing Great Vegetables in a less than Perfect World. Shepherd Ogden, 1999 (ISBN 1-890132-20-9). Based on two generations of organic gardening know-how. I like Shep's book for its attitude against "control-crazy garden practices." A wonderfully useful accumulation of wisdom.

The Straw Bale House. Athena Swentzell Steen, Bill Steen, David Bainbridge, and David Eisenberg, 1994 (ISBN 0-930031-71-7). Straw bale building has become an intriguing example of an old building material rediscovered out of necessity. In California, for example, rice-growers can no longer burn the stalks, but we can build with them. Many good questions should be asked about straw bale—across how many mountain passes can it be trucked and still be sustainable? Where will the necessary thermal mass come from? But there is no question that straw is a wonderful building material, as can readily be seen in this book. Stories, explanations, and lovely pictures.

Sunset Western Garden Book: 40th Anniversary Edition. The editors of *Sunset Magazine*, 1995 (ISBN 0-376038-51-9). If you live in the west, this wondrous work is indispensable. Especially strong on the west's diverse climatic and growing zones.

Timeless Way of Building. Christopher Alexander, 1979 (ISBN 0-195024-02-8). This book is about the craft of building, and adds another dimension to Alexander's work in *A Pattern Language*.

Whole Foods Companion: A Guide for Adventurous Cooks, Curious Shoppers, and Lovers of Natural Foods. Dianne Onstad, 1996 (ISBN 0-930031-83-0). This book is an encyclopedia of good food sense, and it should grace every kitchen.

Who Owns the Sun? People, Politics, and the Struggle for a Solar Economy. Daniel M. Berman and John T. O'Connor, 1996 (ISBN 0-930031-86-5). A striking explanation of how little power you have over your own energy life if you live on the grid.

Wind Energy Basics: A Guide to Small and Micro Wind Systems. Paul Gipe, 1999 (ISBN 1-890132-07-1). Less technical and more accessible than Paul's earlier *Wind Power for Home and Business*, this book tells most of us all we need to know about choosing a safe and reliable wind system.

Wind Power for Home and Business: Renewable Energy for the 1990s and Beyond. Paul Gipe, 1993 (ISBN 0-930031-64-4). The authoritative resource for America's most abundant renewable energy source.

INDEX

CHELSEA GREEN

Sustainable living has many facets. Chelsea Green's celebration of the sustainable arts has led us to publish trend-setting books about organic gardening, solar electricity and renewable energy, innovative building techniques, regenerative forestry, local and bioregional democracy, and whole foods. The company's published works, while intensely practical, are also entertaining and inspirational, demonstrating that an ecological approach to life is consistent with producing beautiful, eloquent, and useful books, videos, and audio cassettes.

For more information about Chelsea Green, or to request a free catalog, call toll-free (800) 639-4099, or write to us at P.O. Box 428, White River Junction, Vermont 05001. Visit our Web site at www.chelseagreen.com.

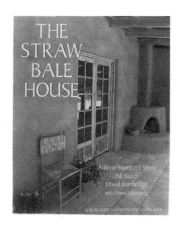

The Straw Bale House
Independent Builder:
  Designing & Building a
  House Your Own Way
The Rammed Earth House
The Passive Solar House
The Earth Sheltered House
The Sauna
Wind Power for Home & Business
Wind Energy Basics
The Solar Living Sourcebook
A Shelter Sketchbook
Mortgage-Free!
Hammer. Nail. Wood.

Four-Season Harvest
The Apple Grower
The Bread Builder
Keeping Food Fresh
Simple Food for the Good Life
The Flower Farmer
Passport to Gardening
The New Organic Grower
Solar Gardening
Straight-Ahead Organic
Good Spirits
The Contrary Farmer
The Contrary Farmer's Invitation
  to Gardening
Whole Foods Companion

Believing Cassandra
Gaviotas: A Village to Reinvent
  the World
The Man Who Planted Trees
Who Owns the Sun?
Global Spin: The Corporate
  Assault on Environmentalism
Seeing Nature
Hemp Horizons
Genetic Engineering, Food and
  Our Environment
Scott Nearing: The Making of a
  Homesteader
Loving and Leaving the Good Life
Wise Words for the Good Life